P9-DNF-999

IMPROVING SAFETY IN THE CHEMICAL LABORATORY: A PRACTICAL GUIDE

IMPROVING SAFETY IN THE CHEMICAL LABORATORY: A PRACTICAL GUIDE

Edited by

Jay A. Young
Chemical Safety and Health Consultant
Silver Spring, Maryland

A WILEY-INTERSCIENCE PUBLICATION
JOHN WILEY & SONS
NEW YORK · CHICHESTER · BRISBANE · TORONTO · SINGAPORE

Copyright © 1987 by John Wiley & Sons, Inc.

All rights reserved. Published simultaneously in Canada.

Reproduction or translation of any part of this work
beyond that permitted by Section 107 or 108 of the
1976 United States Copyright Act without the permission
of the copyright owner is unlawful. Requests for
permission or further information should be addressed to
the Permissions Department, John Wiley & Sons, Inc.

Library of Congress Cataloging in Publication Data:

Improving safety in the chemical laboratory: a practical guide /
  [edited by] Jay A. Young.
    p. cm.
    "A Wiley-Interscience publication."
    Includes bibliographies and index.
    ISBN 0-471-84693-7
    1. Chemical laboratories—Safety measures. I. Young, Jay A.
QD51.I48 1987   87-20587
542'.0289—dc19   CIP

Printed in the United States of America

10 9 8 7 6 5 4

Contributors

Margaret-Ann Armour
University of Alberta
Edmonton, Alberta, Canada

Janet Baum
Payette Associates
Boston, Massachusetts

Ernest I. Becker
University of Massachusetts
Boston, Massachusetts

Leslie Brethcrick
Woodhayes, West Road
Bridport, Dorset, England

Jack J. Bulloff
New York State Legislative Commission on Science and Technology
Albany, New York

Louis DiBerardinis
DiBerardinis Associates, Inc.
Wellesley, Massachusetts

Rudolph Gerlach
Muskingum College
New Concord, Ohio

Stephen K. Hall
Medical College of Ohio
Toledo, Ohio

W. K. Kingsley
J. T. Baker Chemical Co.
Phillipsburg, New Jersey

Andrew T. Prokopetz
National Institute of Environmental
   Health Sciences
Research Triangle Park, North
   Carolina

Michael J. Reale
Lonza, Inc.
Fair Lawn, New Jersey

Patricia A. Redden
St. Peter's College
Jersey City, New Jersey

Flo Ryer
Formerly, U.S. Department of Labor,
Occupational Safety and Health
   Administration
Washington, DC

G. Thomas Saunders
Geneva Research
Durham, North Carolina

Stephen T. Springer
Richardson-Vicks
Shelton, Connecticut

Robert E. Varnerin
Boston University
Boston, Massachusetts

Douglas B. Walters
National Institute of Environmental
   Health Sciences
Research Triangle Park, North
   Carolina

J. R. Williamson
Eastern Michigan University
Ypsilanti, Michigan

Jay A. Young
Chemical Safety and Health Consultant
Silver Spring, Maryland

Preface

To do it better today than yesterday—to improve safety in the chemical laboratory—is the theme of this book. Underlying this theme, and underlying the content that makes this theme practical, are the first and second laws of safety, their corollaries, and a conclusion.

Thus, the first law of safety states: To occur, an accident requires that at the same location at least two mistakes occur simultaneously, or almost simultaneously.

Corollary to the first law: If the required mistakes are not confluent in time and space, one has either a "close call" or a "nonevent."

A close call may or may not be noticed. Close calls involving toxic overexposure, for example, are rarely dramatic and may be missed entirely; a close call that is a small fire instead of a large explosion usually is noticed, of course.

A nonevent is either a single mistake by itself, hence no accident, or the two or more mistakes are sufficiently separated in time and/or space as to not even be a close call.

The second law of safety: All accidents prophesy; like other coming events they cast their shadows before them.

Corollary to the second law: The prophesy that an accident will happen appears in the form of a close call or nonevent.

The conclusion is obvious and triplefold:

1. By eliminating the causes of close calls and nonevents, accidents will be eliminated.
2. If we learn how to identify that a close call or nonevent has in fact happened, we can then know that there was a cause, or causes, which can be identified.
3. Once identified, a cause can be eliminated.

Stated differently, it is the intent of this book to help you learn how to discern that close calls or nonevents have happened, and to help you first identify and then eliminate their causes.

This book consists of five chapters. The chapters are practical; by following their recommendations, causes of accidents can be identified in advance and eliminated. Some of the recommendations are indeed formidable. For those who would like a bit of help, the rest of the book provides a start in the four appendices.

Appendix 1 deals with the hazards that chemicals and the handling of chemicals present in the laboratory and the precautions that are appropriate to these hazards: flammable, reactive, corrosive, toxic, and physical.

Appendix 2 discusses particular precautions that, through enacted laws and the related regulations derived from those laws, society has determined we must incorporate in our laboratory work.

Appendix 3 facilitates. It treats the essentials of space design—what kind of buildings, and what kind of facilities in those buildings, we ought to have to fulfill more efficiently our responsibilities to minimize laboratory hazards.

Appendix 4 further facilitates, describing some of the resources from which still more help can be obtained.

Appendix 5 is not a part of this book. It is instead locally prepared—*your* laboratory safety operations plan. This brief outline of one such plan may be useful:

   I Statement of management commitment to safety
        Authority and responsibilities
            Reserved to top management
            Delegated to subordinates
            Procedures for fulfilling responsibilities

(To be reviewed at regular intervals and if necessary revised. Designed to be perceived by laboratory employees as a commitment that is present in management's decisions and actions, as well as in the written statement itself.)

  II Competence in the practice of safety by laboratory employees
        Criteria for competence desired in new employees
        Employee training
      (For technical staff, secretarial staff, janitor staff, other support staff.)
        Competence of instruction
        Evaluation of effectiveness of instruction
        Revisions to fit state of the art
 III Standard operating procedures, written format
        Review and acceptance requirements
        Revision policy
  IV Laboratory maintenance practices
        For laboratory apparatus and equipment
        For built-in fixtures
        Safety for maintenance contractors' employees
  IV Safety audits and inspections
        Frequency, by whom conducted
        To whom reported, actions to be taken
        Follow-up procedures to insure actions effective

                                                    JAY A. YOUNG

Silver Spring, Maryland
November 1987

Contents

CHAPTER 1

Organization for Safety in Laboratories

Ernest I. Becker

University of Massachusetts, Boston, Massachusetts

1.1 INTRODUCTION

The functions of a safety program in an institution are to reduce the possibility of accidents so as to protect people and property. The benefits of a safety program are to reduce lost work time resulting from such accidents, to have more effective and predictable production schedules, to reduce the cost of insurance and litigation, and to create and maintain a motivated work force.

Who is responsible for safety? "Everyone" is the simplistic answer. Management must create a positive safety-conscious atmosphere throughout the institution. Laboratory workers are responsible for handling and processing of chemicals safely. Intermediate management has the responsibility to see that safety regulations are followed by laboratorians and that management is made aware of deficiencies in safe construction, safety equipment, or rules for safe operation.

1.2 MANAGEMENT RESPONSIBILITIES

A sound safety program begins with support at the highest management level in the institution. The president or chief executive officer of a company, or the president or chancellor of a university, must issue a statement of purpose for the safety program and its chain of command.

In academic institutions the organization may be different because of greatly differing needs by various study groups. Thus it is clear that those studies entailing extensive laboratory work should have a direct line of communication by the respective departments, whereas most humanities, social science, and management study areas may need to be recognized only by one representative per large study unit. Thus the natural sciences, engineering, geology, art, certain aspects of social sciences, public health, and environmental studies should be individually recognized in the chain of command. Other areas, such as english, foreign languages, history, and economics may need to be recognized only by groups or schools; for example, school of social sciences, school of foreign language studies, school of management. One precaution; that is, the needs for specific laboratory safety instruction in humanities and in social sciences, which use chemicals and carry out chemical operations, may vary from one institution to another depending on whether laboratory experimentation is carried out extensively in a given unit. Thus an art department encompassing painting, sculpture, and/or photography may, indeed, qualify for its own safety representative. Maintenance and housekeeping (janitorial) departments should always have safety representation, since their use of and handling of hazardous chemicals is extensive.

In institutions with laboratories the chain of command might be the vice president in charge of safety as the president's deputy followed by the research director, then section heads, group leaders, and the laboratory workers.

The need for safety representation in units without major investment in laboratory experimentation does not reduce to zero. There are still safety and health considerations concerning all residents of the institution such as access and egress to buildings, ventilation, and fire hazards, to name a few issues that need general consideration.

Also reporting to the vice president for safety is the safety office, headed by a safety engineer, who is degreed in some aspect of health and safety, such as safety engineering or public health. In academic institutions this is generally called the campus Safety Office. Its more specific functions are discussed later.

1.2.1   Vice President for Safety

The vice president for safety has the responsibility for laying out broad policy matters for each of the units responsible to him/her. Some of these policy matters may be

1. Forming an institutional safety committee.
2. Defining functions that the layer of administration below the institution safety committee should follow:
   (a) Selecting divisional and departmental safety committees.
   (b) Adopting safety manuals for the various laboratories.
   (c) Consultation with the safety officer on the availability of safe facilities, equipment, and procedures for the following:
      i.   Carrying out laboratory experimentation.
      ii. Waste disposal.
      iii. Seeing that regulations for handling of hazardous and/or toxic chemicals are adopted.
   (d) Directing that copies of all purchase orders for chemicals be sent to the safety office for review.

(e) Recommending that accident records be collected, analyzed, and recorded in a computer readable manner so that annual reports on trends in safety may be prepared.

(f) Ensuring that plans for new construction of laboratory buildings and for renovation of existing laboratories or laboratory buildings be reviewed by the safety office.

(g) Requiring that periodic safety reports be prepared at departmental, section head, or research unit level.

(h) Requiring that periodic safety surveys are carried out by the institution safety committee.

(i) Preparing an institution-wide annual safety report for the president of the institution.

(j) Recommending use of external consultants on safety issues, which may be outside the competencies of the safety committee and the safety office.

(k) Requiring that the safety office communicate the requirements of federal, state, and local laws and regulations to all levels of personnel responsible for safety.

3. Requiring that all recommendations approved by the safety committees, which are not covered by (2), be sent to all affected units for review before institution-wide adoption.

1.2.2 Department Heads and Heads of Other Units

Department heads and heads of special research units in the institution will at least

1. Select a safety committee to report to the head of the unit that they represent. Thus department heads and heads of special research units would select the members of the safety committee reporting to the Director of Research. In academic institutions, heads of natural sciences departments, including geology, would select the safety committee. Dean's would have the responsibility of convening heads of departments responsible to them for safety discussions.

2. See that regulations and decisions of the institution safety committee are studied and brought to the next lower level responsible for safety.

3. See that needs relative to safety are budgeted and passed upward to the next higher level of management.

4. See that recommendations from lower operating levels are discussed and forwarded to higher levels of management.

5. See that laboratory workers

   a. Are informed as to safe laboratory practices.

   b. Are informed on proper waste disposal.

   c. Record safety data and procedures in notebooks where appropriate.

d. Record all accidents, whether to personnel or to equipment or facilities, in their notebooks and on separate accident report sheets to their supervisors.

e. Are advised on both the potential hazards or risks to be expected in reactions to be performed and on the precautions to be followed.

6. Require periodic safety inspections and audits with written reports and recommendations.

7. See that laboratorians are instructed in the above, including the elements of toxicology where appropriate.

1.2.3   Laboratory Workers

The individual laboratory workers must be informed as to the hazardous nature of each compound that they handle in their work. In the event compounds to be made are unknown or such information is not available for the compounds to be used, the laboratory workers should treat such substances as if they were severely hazardous. They are also responsible for instructing technicians and other assistants on safe procedures for handling hazardous substances and carrying out hazardous reactions.

1.2.4   The Safety Officer

The safety officer, head of the institution's safety office, properly reports to the vice president for safety. The safety officer's functions might be

1. To serve as a source of information to those individuals who request it.

2. To keep the institution posted on local, state, and federal laws and regulations pertaining to safe operations and waste disposal.

3. To see that the institution complies with the matters raised in (2).

4. To receive all accident reports and analyze them in accord with the requests of the institutional safety committee.

5. To maintain in the safety office or in the institutional library a set of references and journals on safety matters.

6. To write, or assist departments in writing, safety manuals pertinent to their departmental needs.

7. To meet with the institutional safety committee as an information resource.

8. To prepare safety budgets for their office and assist departments in preparing their safety budgets.

9. To review all new construction plans and plans for modification or renovation for proper safe design.

10. To give seminars on safety issues to departments, the vice president for safety, and others.
11. To keep a file of the material safety data sheet (MSDS's) for materials used in the institution.

1.3 HAZARD AND PRECAUTIONARY INFORMATION

Omitted from this discussion is the impact of the right-to-know and related regulations on the laboratory management. Employees have the right to be informed about the hazards and precautions that are known about any chemical substance with which they are required to work. Large institutions may have such data, but smaller institutions may not. It is, then, required that supervisors do a literature search to provide such information. When such a search is beyond the capacity of the institution to carry it out, then an option may be to use outside consultants to carry out such searches.

1.4 THE BOARD OF DIRECTORS

In either an industrial or academic institution, the board of directors should have a safety committee (corporate safety committee) whose function it is to oversee overall efficiency of the safety program, to monitor its progress, and to see that ineffectiveness is corrected.

CHAPTER 2

Precautionary Labels and Material Safety Data Sheets

Michael J. Reale

Lonza, Inc., Fair Lawn, New Jersey

and

Jay A. Young

Chemical Safety and Health Consultant Silver Spring, Maryland

2.1 INTRODUCTION

In the preamble(1) to the Hazard Communication Standard, the Occupational Safety and Health Administration (OSHA) discusses the importance of including laboratories in the scope of the standard. The testimony gathered and comments received demonstrate that laboratory employees are exposed to hazardous chemicals, and that they are at risk because of such exposures. In many respects laboratories can present an environment that is as hazardous as that found in the manufacturing sector.

The fact that laboratories are typically under the supervision of highly qualified and experienced scientists does not by itself at all ensure that a laboratory is being managed safely. Thus OSHA stated, the fact that laboratory supervisors are trained in conducting chemical research or other laboratory operations does not mean that they are adequately trained in, or concerned with, the hazards of the substances they are working with. As Dr. Daniel Teitelbaum testified (1):

> In my own experience as a university professor, having worked in research laboratories and run research projects, I can say to you that many research scientists are less than informed about the toxicity of the materials with which they work in spite of their doctoral degrees.

7

2.2   HAZARD COMMUNICATION: PURPOSE

Providing information about chemicals so that hazards are recognized and their effects eliminated or controlled is the *raison d'etre* for labels and material safety data sheets (MSDSs) wherever the chemicals are used and handled, in the laboratory or elsewhere. Thus in the laboratory, labels and MSDSs are tools that a supervisor utilizes to be effective in improving the overall safety in the laboratory, thereby protecting the health and welfare of both laboratory workers and the environment.

Although labels and MSDSs share these common objectives, they differ in form, in approach, and, in some respects, target audience. For these reasons, each will be treated separately.

2.3   THE MATERIAL SAFETY DATA SHEET (MSDS)

An MSDS is a technical document, which usually consists of two or more pages of printed information comprising a compendium of relevant chemical, physical, and toxicological data, procedures for safe handling, storage and use, and implementation of emergency practices. Material safety data sheets have traditionally been prepared with the health care and safety professional in mind. Thus industrial hygienists and occupational health specialists use the information and recommendations in formulating safe work practices and in developing protocols for employee exposure monitoring. Medical personnel draw upon the health information in preventing, as well as treating, overexposure. Engineers can determine proper materials of construction and placement of equipment by making reference to the data provided.

2.3.1   The MSDS and Employee Training

An MSDS, however, can also serve as the nucleus around which the laboratory supervisor can build safety training programs. The Occupational Safety and Health Act (OSHA) Hazard Communication Standard (2,(h)) requires employers (including laboratories) to

1. Maintain copies of MSDSs for all hazardous chemicals in the workplace.
2. Train employees about the methods that may be used to detect the presence of a chemical in the workplace, and to alert them when a release has occurred.
3. Teach employees about the chemical and physical hazards of chemicals in the workplace.
4. Instruct employees in measures to protect themselves from these haz-

ards, such as work practices, personal protective equipment, and emergency procedures.

5. Explain MSDSs to employees and inform them how to use the relevant hazard information.

Academic and other nonmanufacturing laboratories are clearly subject to the Hazard Communication Standard. Consequently, these five elements are obviously useful to directors of all laboratories in their training of the personnel they supervise, whether they are employees directly, employees of a subcontractor, research associates, or students.

2.3.2 MSDS Format and Content

Until issuance of the OSHA standard, the preparation of MSDSs was generally voluntary, and the format and content varied from manufacturer to manufacturer. The format of an MSDS is still optional (for a suggested format, see Fig. 2.1), but the standard mandates (2,(g)) that MSDSs contain at least the following information:

1. Identity of the chemical. This may be a chemical name, code name or number, trade name, brand name, or generic name. Whatever is selected as the identity must permit cross references to be made among the MSDS, the product label, and a required list of hazardous chemicals found in the workplace.

2. The chemical name or names. If the MSDS covers a single substance, then both its chemical and its common name (if any) must be shown.

If the MSDS is for a mixture of chemicals, the requirement for naming each component depends on the percentage present, the degree of hazard each presents, and on whether the hazard of the mixture, *as a mixture* has or has not been determined. Further, under certain specified conditions related to the protection of trade secrets, the name(s) of the components of a mixture may be withheld except under certain emergency conditions.

3. Physical and chemical characteristics. Examples include vapor density, vapor pressure at one or more temperatures, melting point, flash point, flammable limits, fire point, odor, specific heat, heat of combustion, coefficient of cubical expansion, electrical resistivity, molecular or formula weight.

4. Description of physical hazards. These classes of chemicals are among those that present physical hazards that must be described: organic peroxides, oxidizers, pyrophorics, flammables, combustibles, explosives, water reactivity, and those that possess the potential for fire, explosive or reactive hazard.

5. Description of health hazards, including both acute and chronic ef-

fects. Examples of such chemicals by class include: carcinogens, reproductive toxins, sensitizers, irritants, corrosives, hepatotoxins, nephrotoxins, neurotoxins, agents that act on the hematopoietic system, central nervous system, agents that act on other organ systems, other toxic agents, and agents that damage the eyes, skin, lung, or other mucous membranes.

The description of health hazards must also indicate those signs and symptoms that are indicative of an exposure to the hazardous chemical and must also specify any medical conditions that are generally recognized as being aggravated by exposure to the hazardous chemical.

6. The primary route(s) of entry of the chemical into the body. Thus, gases and the vapors of volatile liquids and solids enter primarily by inhalation—as may also a solid or liquid likely to be used or handled so as to generate a dust or mist. Corrosive liquids and solids generally affect an individual through skin contact. Ingestion would rarely be a primary route of entry but should be considered for any substance that is highly toxic by ingestion.

Note that only primary routes of entry are required to be stated. It is incorrect to conclude that harmful effects will occur only via the stated route(s). All health hazard information should be assessed when determining safe work practices.

7. The recommended exposure limit. Usually, this will be the American Conference of Governmental Industrial Hygienists threshold limit value (ACGIH TLV®) and/or the OSHA PEL. Clearly, this information is particularly useful in training laboratory workers and in planning for engineering controls, monitoring, and the use of personal protective equipment.

8. Whether the chemical has carcinogenic potential. If named as such in OSHA regulations or in certain other lists, or if a valid study has shown such potential, then it must be so identified. Certain exceptions may apply (3,p.4).

9. Safe handling and use procedures. These include appropriate hygenic practices, protective measures to be followed during maintenance and repair of contaminated equipment, and procedures for cleanup of spills and leaks.

10. Control measures. This includes information on one or more means, such as engineering controls, work practices, or personal protective equipment employed to reduce or eliminate exposure to the hazardous chemical. Not all of several mentioned measures need be employed by every user of that chemical.

11. First aid and other emergency procedures. The instructions should cover all reasonably foreseeable employee exposures, not necessarily restricted to exposure by primary route of entry, discussed previously. Generally, first aid instructions are described for eye and skin contact, inhalation, and ingestion. Note that if these instructions call for the use of

antidotes or of unusual or specialized apparatus that is not routinely supplied as part of the laboratory first aid equipment, that chemical should not be used until the deficiency is remedied. Similarly, if the first aid and/or emergency instructions call for any nonroutine treatment or equipment to be used by emergency personnel, the chemical should not be used until those agencies have indicated that they are so prepared.

This section of the MSDS also provides information on the correct procedures and substances to be used in fire fighting and in spill/leak cleanup that might involve the chemical. It is crucial to ensure in advance that the appropriate quantities of the recommended fire extinguishing media are present and in place in the immediate laboratory area and that personnel are informed of and have practiced their use.

The importance of advance planning based on this section of the MSDS cannot be overemphasized. Local personnel should be well trained, emergency personnel should be fully informed, and their equipment known to be ready for use if needed.

12. The date of preparation of the MSDS or of the last change to it.

13. The name, address and telephone number of the chemical manufacturer, importer, or other responsible party from whom additional information can be obtained during normal working hours. As part of advance planning before the use of a hazardous chemical, since errors do occur, the telephone number should be verified.

Figure 2.1 is an example of MSDS format. This particular example was prepared by OSHA as a model; it can be used to comply with the standard. It is not mandatory that this model be used; it is required that all 13 elements listed here be present (12 elements, item 8 may be omitted, if there is no reason to identify the chemical as a carcinogen as specified in the standard). If no relevant information is known for a given element, the space allocated to that element cannot be left blank; the entry "None found," or "Not established," or appropriate equivalent must appear.

2.4 LABELS

Chemical label means any written, printed or graphic material affixed to or accompanying containers of a chemical, . . . whose purpose is to communicate in language or other symbols, in color or form, specific information and emotional/perceptual stimuli designed to affect the perceptions and behaviors of human beings who are prospective users of the chemical (4,p.6).

Lirtzman (4,p.7) lists eight common uses for a label: To identify the product and its manufacturer; to provide hazard warning; to disclose remedial measures; to alert certain users to specific personal information; to identify the function of a product; to assist in the marketing and promotion of a product; to provide directions for use; to educate the user.

Material Safety Data Sheet May be used to comply with OSHA's Hazard Communication Standard, 29 CFR 1910.1200. Standard must be consulted for specific requirements.	U.S. Department of Labor Occupational Safety and Health Administration (Non-Mandatory Form) Form Approved OMB No. 1218-0072
IDENTITY *(As Used on Label and List)*	*Note: Blank spaces are not permitted. If any item is not applicable, or no information is available, the space must be marked to indicate that.*

Section I

Manufacturer's Name	Emergency Telephone Number
Address *(Number, Street, City, State, and ZIP Code)*	Telephone Number for Information
	Date Prepared
	Signature of Preparer *(optional)*

Section II — Hazardous Ingredients/Identity Information

Hazardous Components (Specific Chemical Identity; Common Name(s))	OSHA PEL	ACGIH TLV	Other Limits Recommended	% *(optional)*

Section III — Physical/Chemical Characteristics

Boiling Point		Specific Gravity (H$_2$O = 1)	
Vapor Pressure (mm Hg.)		Melting Point	
Vapor Density (AIR = 1)		Evaporation Rate (Butyl Acetate = 1)	
Solubility in Water			
Appearance and Odor			

Section IV — Fire and Explosion Hazard Data

Flash Point (Method Used)	Flammable Limits	LEL	UEL
Extinguishing Media			
Special Fire Fighting Procedures			
Unusual Fire and Explosion Hazards			

(Reproduce locally)   OSHA 174, Sept. 1985

FIGURE 2.1a.   An example of a material safety data sheet, first page.

The purpose of a precautionary label for a hazardous chemical is less broad and more sharply delineated: To prevent injury.

As discussed further in Section 2.7.5, the American National Standard for Hazardous Industrial Chemicals—Precautionary Labeling (5), also known as ANSI Z129.1-1982, the content of a precautionary label:

Section V — Reactivity Data

Stability	Unstable		Conditions to Avoid
	Stable		

Incompatibility (*Materials to Avoid*)

Hazardous Decomposition or Byproducts

Hazardous Polymerization	May Occur		Conditions to Avoid
	Will Not Occur		

Section VI — Health Hazard Data

Route(s) of Entry:	Inhalation?	Skin?	Ingestion?

Health Hazards (*Acute and Chronic*)

Carcinogenicity:	NTP?	IARC Monographs?	OSHA Regulated?

Signs and Symptoms of Exposure

Medical Conditions
Generally Aggravated by Exposure

Emergency and First Aid Procedures

Section VII — Precautions for Safe Handling and Use

Steps to Be Taken in Case Material Is Released or Spilled

Waste Disposal Method

Precautions to Be Taken in Handling and Storing

Other Precautions

Section VIII — Control Measures

Respiratory Protection (*Specify Type*)

Ventilation	Local Exhaust		Special
	Mechanical (*General*)		Other
Protective Gloves		Eye Protection	

Other Protective Clothing or Equipment

Work/Hygienic Practices

✩ U S G P O 1986–491–529/45775

FIGURE 2.1b. An example of a material safety data sheet, second page.

1. Is based on the foreseeable use, handling, and storage of the chemical as these pertain to the hazardous characteristics of the chemical itself,
2. Consists of, as and if appropriate,
   a. The identity of the chemical or hazardous component.
   b. A *signal word,* Danger, Warning, or Caution.

    c. Statement(s) of hazard(s).

    d. Precautionary measure(s).

    e. Instructions in case of contact or other exposure.

    f. Specification of antidote(s).

    g. Instructions to physicians.

    h. Instructions in case of fire, spill, leak, or other emergency.

    i. Instructions for container handling and storage.

2.5 PRECAUTIONARY LABELS AND MATERIAL SAFETY DATA SHEETS: COMPARED

Differences between a precautionary label and an MSDS are apparent. As pointed out previously, the target audience of an MSDS is the health care and safety professional; the label is aimed at the user of the product who may or may not be technically trained. The MSDS is a technical document prepared to present information; the precautionary label generally presents instruction in a nontechnical fashion and uses simpler language than an MSDS. The purpose of an MSDS is to inform; the purpose of a precautionary label is to teach the user.

An MSDS generally comprises two or more printed pages. The MSDS provides many categories of information; its content can be described as detailed, even encyclopedic.

On the other hand, a label is constrained in size because it normally must be affixed to a container of limited dimensions. Since a label must serve many users, the amount of text that can be devoted to specific precautionary language is limited. If an attempt is made to include more information by reducing the type size, the label can become unreadable. Furthermore, incorporating a lengthy text on a label can lead to information overload and this too is counter productive.

Another factor that distinguishes a precautionary label from an MSDS was alluded to previously; that is, precautionary language on a label must often be read and understood by workers who have limited reading ability. Compared to an MSDS, the precautionary label must use plain words and yet be precise, concise, and clear in describing the precautionary information that applies to a chemical.

2.6 PRECAUTIONARY LABELS

Laboratory directors should recognize the limitations of a precautionary label. A competently constructed precautionary label reflects the result of professional judgment in sifting all of the information known about a chemical. Thus, only those certain, selected hazards, data, and precautionary measures reasonably judged to be needed by and therefore known to a worker comprise the content of a precautionary label. Clearly, a precautionary label is a limited source of information; it must be supplemented with training based on the MSDS or equivalent.

Training is especially important in the case of hazardous research chemicals for which there often is little or no hazard information on the label. The label writer may have assessed potential hazards by analogy with known similar materials (and may or may not have so stated on the label); such a label should indicate that the material is experimental and the toxicological and other hazardous properties have not been investigated fully. In such cases the laboratory director should train employees and initiate the procedures and protocols as described in Chapter 9.

2.7   PRECAUTIONARY LABELING SYSTEMS

What follows is a synopsis of selected precautionary labeling systems. For a more detailed discussion, see O'Connor and Lirtzman (4). Some labeling systems are voluntary, while others are mandated by government statute or regulation. All derive directly from the original voluntary system, the *LAPIC Guide,* that originated from the work of the Labeling and Precautionary Information Committee of the Manufacturing Chemists Association—now known as the Chemical Manufacturers Association. The LAPIC *Guide to Precautionary Labeling of Hazardous Chemicals* is out of print.

2.7.1   The OSHA Warning Label

The OSHA Hazard Communication Standard (2,(f)) mandates the use of a warning label, that is; this mandated label need convey only warnings, not necessarily other information such as information related to precautions that eliminate or control the hazards. In detail, the mandated label consists of these three required parts:

1. The identity of the chemical. The identifying word(s) or set of coded symbols that constitute the *identity* must be the same as those shown as the identity in the MSDS.
2. Appropriate hazard warnings. These are defined as any words, pictures, symbols, or combination of these that convey the hazards of the chemical(s). The warnings must provide specific information regarding the hazard, that is, *causes eye burns* or *flammable.* Phrases such as *Danger* are not acceptable by themselves, since they only provide an indication of the degree of severity of the hazard; such signal words do not specify the hazard involved.

Not every hazard will appear on the label. The OSHA states, ". . . determination of the hazards to be highlighted on the label will involve

some assessment of the weight of evidence regarding each hazard reported on the data sheet (3,p.10)."

3. Name and address of the chemical manufacturer, importer, or other
responsible party.

The OSHA standard specifically requires that labels on incoming containers of hazardous chemicals not be removed or defaced (2,(a)(3)(i)).
Laboratory supervisors should establish appropriate policies. Although
other OSHA label provisions(8) do not apply to laboratories, it is desirable
that all containers of chemicals in the laboratory are labeled with the chemical identity and at least the appropriate statements of hazard. An example
from the OSHA rule making record is instructive. A Mr. Frank Baird
worked in a research laboratory. During his employment in that laboratory,
he was unknowingly exposed to high levels of mercury vapor. He contracted
mercury poisoning from that laboratory work. His description demonstrates
the need for informative labels on all containers of hazardous chemicals in
the laboratory (1):

> The mercury I worked with came in unmarked glass bottles. There were no
> warning labels, no markings saying poison. It didn't even have a label saying
> mercury. The Greeks knew that mercury was poisonous, the Romans knew
> that mercury was poisonous, even my employers knew that mercury was poison-
> ous, but they may not have known just how poisonous its vapors were or how
> badly my exposure exceeded the toxic limits set by the U.S. recommended
> standard in 1942.

The laboratory supervisor should also be familiar with certain substance-
specific labeling requirements, for > 20 chemicals, that OSHA has issued in
its Occupational Safety and Health Standards (2, Subpart Z).

2.7.2   The DOT Shipping Label

The U. S. Department of Transportation (DOT) labels comprise another
mandatory labeling system that is well known and widely recognized. This
system primarily uses diamond shaped labels of different colors affixed to
packages of hazardous materials for the purpose of communicating the hazard of the chemical to workers in the shipping and transportation industry.
The DOT labels use color, shape, graphic art, and words to describe a
chemical hazard.

Laboratory supervisors should be aware of inherent shortcomings in the
DOT labeling system compared to precautionary labels. Thus, the DOT
system recognizes only acute hazards that can be presented by a material
during transportation. The symbols are not designed to provide information
on hazards associated with the storage, use, or handling of the material in
question. The DOT labels do not differentiate varying degrees of severity of

hazard, nor route of entry. These labels do not describe the precautions to be employed that will control the hazards presented by the material when it is used.

The DOT label is required to be affixed to the outer packaging. Hence, if a shipment consists of an inner container inside an outer box, the DOT labeling is lost whenever the outer box is removed and discarded.

2.7.3 The Hazardous Material Identification System

The hazardous material identification system (HMIS) was developed by the National Paint and Coatings Association. It is a voluntary system particularly applicable to the precautionary labeling of paints, varnishes, and related coating materials (6). Although it is sometimes used for other kinds of materials, that application is somewhat limited by the nature of the system. The HMIS is a comprehensive system that uses a combination of colors, numbers, letters, and symbols to communicate hazard information on the following:

1. Chemical identity
2. Nature and degree of hazard
3. Recommended personal protective equipment
4. Chronic health effects.

Chemical identity is established by chemical or common name, code name or number, or any other means that clearly identifies the material. It is easily cross referenced to an MSDS.

The HMIS provides numerical ratings, 0 through 4, least to most serious, in colors, blue, red, and yellow, respectively, for health, fire, and reactivity hazards.

Chronic health hazards are noted by using an asterisk after the health hazard rating (blue colored, above) or by a written warning.

Recommendations for personal protective equipment are printed on a white background portion of the label using letters to represent certain specific kinds of equipment. Thus, an A signifies safety glasses; a C signifies safety glasses, gloves, and synthetic apron; an X signifies situations requiring special handling, which is further described in words elsewhere. Typically, the personal protection letter designation is left blank on the label and is to be filled in by the purchaser, for example, the laboratory supervisor.

The HMIS has limitations. It recognizes broad categories of hazards but not specific kinds of hazards, the distinction between a corrosive and a sensitizer, for example, or the use of an asterisk but with no description of the specific chronic effect. As with all other label systems, users must be trained. The HMIS cannot, nor is it intended to carry the whole burden of total hazard communication.

2.7.4   Other Labeling Systems

Other mandatory labeling systems that can be encountered in a chemical laboratory come under the Federal Insecticide, Fungicide, and Rodenticide Act (FIFRA), the Toxic Substances Control Act (TSCA), the Resource Conservation and Recovery Act (RCRA), and the regulations of the Consumer Product Safety Commission (CPSC). Consult O'Connor and Lirtzman (4) for details.

A voluntary labeling system familiar to most laboratory workers has been informally adopted by most of the major suppliers of laboratory chemicals. The labels in this system consist of text with added graphic art. However, the symbols used by different suppliers are not all the same. Many of these labels also show a colored area intended to indicate a recommended manner of storage. And again, the color code of each supplier is different. Some of the information about hazards and precautionary measures that is conveyed by the added graphic art or the color of the colored area contradicts the printed text on the label or in the corresponding MSDS. Until all suppliers use the same symbols, the same color code, and all of which is consistent with the label and MSDS texts, users should make their own careful judgments as to the reliability of the information furnished.

2.7.5   The American National Standard Institute (ANSI) System

By far, the most common precautionary labeling system with which laboratories will come in contact is the voluntary American National Standard for Hazardous Industrial Chemicals—Precautionary Labeling (5). The current edition, ANSI Z129.1-1982, consists of four sections and two appendices.

Section 1 identifies the scope of the standard, the preparation of precautionary labeling for hazardous chemicals used under industrial occupational conditions. By hazardous chemical is meant (5):

> A chemical (or mixture of chemicals) that is either toxic or highly toxic; an irritant; corrosive; a strong oxidizer; a strong sensitizer; combustible; either flammable or extremely flammable; dangerously reactive; pyrophoric or pressure generating; or that otherwise may cause substantial acute or chronic personal injury or illness during or as a result of any customary or reasonably foreseeable handling or use . . .

The ANSI standard specifically is silent on when or where a label should be attached to a container.

Definitions for the terms, toxic, highly toxic, and so on, listed previously, and for other words and phrases are found in Section 2. In general these definitions correspond to those in the OSHA Hazard Communication Standard.

Section 3 of the standard provides general requirements for precautionary labels. This section discusses each of the parts of a label that are listed as the content of an ANSI label in Section 2.4 of this chapter. Section 4 is the source of the phrases seen on many labels: Thus, familiar statements of hazard such as *Harmful if swallowed, Causes eye irritation,* and *Extremely flammable* are statements of hazard from Section 4. *Keep container closed, Do not get in eyes,* and *Wash thoroughly after handling* are examples of precautionary measures from this section.

Appendix A shows examples of labels for a variety of differently hazardous chemicals using the words and phrases from Section 4. Appendix B deals with the labeling of serious chronic hazards. Neither appendix is a part of the standard.

A precautionary label that meets the ANSI standard is composed of the following parts:

1. The name of the hazardous chemical, or if a mixture the name(s) of the hazardous component(s).

2. A single *signal word* to indicate the overall degree of hazard; in decreasing degree, Danger, Warning, or Caution. *Poison* is not a signal word; its use is reserved for highly toxic chemicals and appears on labels with the signal word, Danger, and a depiction of a skull and crossbones.

3. One or more statements of hazard, as appropriate. If the statements of hazard given in Section 4 of the standard do not fit, suitable statements are to be developed following the style and tone of the examples in Section 4.

4. One or more precautionary measures. Usually, there will be at least one precautionary measure for each statement of hazard. However, for some hazardous chemicals a single precautionary measure suffices to protect against more than one kind of hazard, and occasionally a given hazardous property properly requires more than one precautionary measure. When suitable precautionary measures cannot be found in Section 4, precautionary measures should be developed, as indicated previously for statements of hazard.

5. Remedial measures. These include first aid instructions in case of contact or other exposure, such as inhalation; specification of antidotes, if an antidote may be administered by a lay person; special instructions to physicians, particularly for antidotes that should only be administered under a physician's care; fire fighting methods, procedures for cleanup of spills and leaks, where these are not obvious; and instructions for container handling and storage.

2.7.6   The ANSI System; Limitations

The ANSI standard is silent on several matters that seem at first sight to be critically important. Thus, as stated previously, it is outside the scope of the standard to specify when and where a label should be placed on a container of a hazardous chemical. The standard does not specify type size, or color, or need for a contrasting background. It is silent regarding the physical dimensions of a label. There is nothing in the standard, according to its detractors, on the obvious need to insure that the words and phrases are capable of being understood by a user.

Some critics aver that the signal word, Danger, should always be printed in red—forgetting that many people are color blind. Some critics would apply readability tests, designed by competent testing psychologists for quite different purposes, to the content of a label. Clearly, when and where to put a label on a container, how big the label should be, whether the type size and color of the printing contrasts with the background, whether or not a label text is understandable to the ordinary person—all of these, and related matters—are properly outside the scope of the standard since they all depend on common sense.

On the other hand, as with a label that conforms to any other standard, there are limitations. For example, the obvious need for brevity and clarity that is demanded by the standard tends to produce a label that is something like the now no longer extant telegraphic 10-word message. And, like those telegrams, the content of such a label is not always thoughtfully pondered; its intended meaning may not be fully comprehended in the kind of cursory examination one gives to a 10-word telegram. It is as a consequence of this that the myth that "people do not read labels" is believed to be true. Labels are indeed read by users, but they are also indeed sometimes read too quickly.

The solution, of course, is training during the course of which the hazards and precautions are thoroughly treated. Then, the labeled information serves as a reminder—its proper function, rather than as the single source of information—a function that a label never could nor was ever intended to fulfill.

Perhaps because of oversight, the 1982 standard does not require that either the identity of the manufacturer, or an informational telephone number, be placed on the label. The standard makes no reference to the need for training except for an optional precautionary measure that makes reference to an MSDS. Target organs are required to be named in some statements of hazard, for example, skin, eyes, but other equally appropriate target organs or organ systems, for example, liver, kidney, central nervous system, need not be mentioned according to the standard. The standard is silent on the question of whether other languages than English should be used on a precautionary label. The text of the standard itself is singularly obscure in its brief instructions on how to construct a precautionary label for a hazardous chemical.

2.8 EVALUATION OF LABELS AND MATERIAL SAFETY DATA SHEETS

More than a few precautionary labels and MSDSs are deficient and cannot be relied on despite the requirements of governmental regulations. Several tests, none of them definitive, can be applied to indicate that a label or MSDS might be questionable. These include:

Internal Inconsistencies. For example, the chemical is correctly described as flammable, but either the label or MSDS, or both, do not identify likely sources of ignition, or there is no mention of vapor density compared to air and the significance of that comparable density. Or, a chemical is correctly described as poisonous, but there is no information on first aid treatment in case of ingestion or inhalation or other probable route of entry. Another common example of internal inconsistency is a lack of parallel between stated hazardous properties and suitable precautionary measures. For example, a label may recommend separation from combustible materials without identifying that the hazardous chemical is a strong oxidizer.

Unqualified General Statements. A label recommends using with adequate ventilation but the MSDS does not clearly indicate how to tell if the ventilation is adequate, that is, such that the breathing air concentration is well below the TLV® or other appropriate limit. Or the label recommends that contact with skin be avoided but the MSDS does not specify how this is to be accomplished, or it recommends gloves but does not specify the glove material.

Confusing Format. The text of a precautionary label should be organized, for example, stated in the orderly fashion described as the one through five parts of an ANSI label in Section 2.7.5; each part should be clearly separate from the other parts, especially the precautionary measures should not be interspersed among statements of hazard (this is a common error).

Technical Errors. The information is simply wrong. Thus, some labels and MSDSs recommend soapy water or vinegar instead of water to wash a corrosive from the skin or ipecac or salt solution instead of water to induce vomiting.

Competitive Comparison. When competitive labels and MSDSs from different suppliers are compared, it is often at least possible to determine that one supplier is less reliable than another.

The only certain method for evaluating the quality of a label or MSDS is to first know all of the hazards a chemical presents and to know all of the precautionary measures and remedial measures that are appropriate to those hazards. Then the label or MSDS can indeed be evaluated. When this laborous process is not practical, use the tests just described.

REFERENCES

1. *Federal Register,* Vol. 48 (No. 228), November 25, 1983, p. 53288.
2. *Code of Federal Regulations,* Title 29, Chapter XVII, Part 1910.1200.
3. *OSHA Instruction CPL 2-2.38A,* May 16, 1986, Appendix A.
4. C. J. O'Connor and S. I. Lirtzman, Eds., in *Handbook of Chemical Industry Labeling,* Noyes, Park Ridge, NJ, 1984.
5. ANSI Z 129.1-1982, *American National Standard for Hazardous Industrial Chemicals—Precautionary Labeling,* American National Standards Institute, New York, 1983.
6. *Hazardous Materials Identification System Revised Implementation Manual,* National Paint and Coatings Association, Washington, DC, 1985).

CHAPTER 3

Doing It Right

Margaret-Ann Armour

University of Alberta, Edmonton, Alberta, Canada

3.1 INTRODUCTION

There may be 101 ways of performing a manipulation in the laboratory or of responding to an emergency, but there is only one right way of doing it. What are the most effective methods of teaching this right and safe way? The systems described in this chapter have been developed from programs shown to work by the companies, universities, and colleges who use them. The most effective programs, however, are those developed in-house. Therefore, the ideas for safety drills, training courses, and seminars presented here are intended as a base to be adapted and built on. It is hoped they will stimulate continuing improvement of existing programs and development of new ones. Whatever system is adopted should encourage professional behavior and willing cooperation with safety procedures.

3.2 SIMULATED DRILLS

To respond appropriately to an emergency in a chemical laboratory requires prior experience in a simulated situation. Few of us can show self-control, quick thinking, and correct response without such practice.

Several drills are described in Section 3.2.3.1 to 3.2.3.4. For all of them, generalized instructions are given, which need to be adapted to meet each specific laboratory situation. The most basic of the drills, evacuation of the building, is presented at some length. Suggestions for other drills are less detailed to encourage these to be developed to suit circumstances in the laboratory. Drills require considerable preplanning and training of supervisory personnel, and take time to perform. However, they simulate real situations, they involve everyone in the laboratory, and they provide a test of the effectiveness of safety training programs. Therefore, their practice is strongly recommended for a variety of eventualities.

Drills may involve employees, all of whom have been informed and trained in the procedure, or they may include undergraduate students who are more dependent on being told where to go and what to do. In all cases, a laboratory supervisor or other designated person must be prepared to take control and issue directions calmly and authoritatively. This is the person who should be ready to react to the unexpected and to make on-the-spot decisions. One of the things drills help to show is whether the designated person is the right one for the job.

Only evacuation drills need affect everyone in the building at the same time. The other drills are effectively practiced by a small group, such as all the occupants of one laboratory. It may be desirable to arrange that many such small groups perform the same drill at the same time, for example, fire drill. In other cases, it is more practical to have one group at a time carry out the drill, for example, rescue from a smoke-filled laboratory, where there are limited numbers of self-contained breathing apparatus.

The drills described subsequently can be held either as announced or surprise events. There is value in both methods and if a drill is being held for the first time, announcing it beforehand gives everyone the opportunity to think about how they should react. After the first time, however, drills should be held at regular intervals and should be a surprise. Certainly, unannounced sounding of the fire alarm is essential to test reaction to an unexpected event.

3.2.1 Practical Hints for Organizing Drills

Both announced and surprise drills must be planned carefully in order to obtain maximum benefit from them and to minimize negative reaction of the participants. Ways of obtaining maximum benefit will be addressed in the discussion of each drill and its evaluation. Negative reactions can be minimized if surprise drills are timed so that, as far as possible, experiments are not ruined, important meetings are not disrupted, and it is not always the same lecturer's class that is interrupted. After the drill is over, a note or word of appreciation from the organizer to the laboratory supervisors helps to maintain cooperation.

If the building alarm is connected to an on-site or off-site fire department or to a central security control, appropriate warning must be given of a planned drill during which the alarm will be activated.

In large organizations, experts such as the safety officer and fire protection officer can help with the planning and evaluation of drills within a division or department. Small companies and institutions can often call on these people for advice and assistance, or they may wish to employ consultants to aid in setting up a program.

3.2.2   Frequency of Drills

Drills are held so that reaction to an emergency is appropriate and fast. Therefore, they must be held often enough that the procedure is familiar and readily called to mind. There are several factors that should be considered when deciding how often to schedule drills.

Some drills are basic and need to be held more often than others. Evacuation of the building is necessary in a number of emergencies, therefore drills should be practiced at least every 4 months. If there is a more frequent turnover of personnel in the laboratory, the frequency of the drill should be increased. When there is an influx of new workers, for example, if painting or maintenance is being done, or if summer students are arriving, a drill should be scheduled.

Fire drill, as the phrase is being used here, involves testing the reaction of laboratory occupants to a supposed fire and may or may not include evacuation. Such a drill, which would usually involve a small group such as all of the occupants of one laboratory should be performed not less than yearly with each group, or when there are a number of new laboratory occupants.

The frequency of drills such as removing an occupant from a smoke-filled laboratory, treatment of a chemical spill, or other specialized drills related to particular situations, will depend on the applicability of these drills to the laboratory. Thus, if highly toxic gases are in constant use, the drill for reacting to a leak of gas will be practiced regularly.

3.2.3   Types of Drill

Suggestions for performing drills are discussed in subsequent paragraphs.

3.2.3.1   Evacuation of the Building.   This drill is basic to all emergencies which require that a building be cleared of occupants, such as fire, leak of toxic gas, or threat of explosion. To comply with most legal requirements, evacuation routes are clearly posted within buildings, all laboratory managers have a plan of evacuation and will have tested this plan. Before arranging an evacuation drill, the laboratory manager should ensure that the laboratory supervisors, or others designated as in charge of individual laboratories in the event of emergency, are familiar with the sound produced by the fire alarm and trained in the appropriate procedures. When the alarm sounds, the person responsible calmly but authoritatively takes charge of the situation, directing the occupants on how to leave the building and where to assemble outside. Everyone stops what he or she is doing, gas taps are closed, and other energy-producing or energy-consuming devices are switched off. After the last person has left the laboratory, the supervisor quickly checks to make sure that there is no one remaining hidden behind benches or equipment, and leaves the room, closing the door. Once outside, at the assembly point, the supervisor accounts for everyone in the group.

The instruction is given to remain outside until a prearranged signal is received allowing return to the building.

Although on some occasions it may be desirable to announce that an evacuation drill will take place at a certain time, or during a certain week, it is essential also to hold this drill when as few people as possible are aware that it is a drill. There can be a very different reaction to an expected drill than to an alarm that has to be assumed to indicate a genuine emergency. An evacuation drill is best held at irregular intervals to avoid any tendency to react slowly or even ignore the alarm as "just another evacuation drill." In fact, it has to be impressed on everyone that whenever the alarm sounds, evacuation *must* take place.

The timing of the evacuation drill should allow as many circumstances as possible to be covered, for example, when some personnel are out of the laboratory, at a central storeroom, in an instrument room, in a solvent storageroom, or even in a washroom, and when visitors are present in the building.

3.2.3.2   Fire Drill.   This drill allows a rehearsal of what to do should fire break out in a specific area. It is best practiced with a small group, for example, the occupants of one laboratory. The supervisor would announce that a fire had occurred in a certain section of the laboratory and, as the drill proceeds, would describe the progress of the fire fighting effort. The drill can be directed by the supervisor or by the person in whose area the supposed fire had occurred. One person is told to go and activate the fire alarm and/or to make the appropriate telephone call and one or more people are directed to grab and, if desired, to use the fire extinguishers while the rest of the laboratory occupants leave the area. This drill can be the precursor to evacuation drill, section 3.2.3.1, or the supervisor can announce that the fire was small and has been put out using the fire extinguishers.

3.2.3.3   Rescue from a Smoke- or Vapor-Filled Laboratory.   Fire is the most common emergency in a laboratory and is likely to fill the room with toxic gases. Further, spills can occur in almost any laboratory. Therefore, this drill provides important practice in the use of self-contained breathing apparatus and tests knowledge of the correct procedure for removing someone trapped in a smoke- or vapor-filled laboratory.

Like a fire drill, it is best practiced with a small group. The laboratory supervisor announces that the laboratory has filled with toxic fumes and directs everyone to leave. The "victim" collapses, and the rescue drill begins. Someone is sent for the self-contained breathing apparatus and someone else to phone for an ambulance and medical help. The rescuer is aided into the breathing apparatus, enters the laboratory and drags the victim out into fresh air. Artificial respiration can be included in this drill. It is useful to assume that a second person is needed to complete the rescue. A rope is found, or several belts are tied together and attached to the second rescuer with the other end being held by someone outside the laboratory, so that he/she can be pulled to safety should the need arise. The second rescuer remains in the laboratory for only as long as breath can be held.

3.2.3.4 Specialized Drills for Specific Laboratory Situations. Supervisors have a responsibility to recognize potential emergency situations in their laboratory that require drills other than those described in Sections 3.2.3.1 to 3.2.3.3. Some supervisors may wish to conduct first aid drills, especially if the laboratory is remote from professional medical help. Others may prefer to allow workers to attend first aid courses run by professionals such as the local Red Cross chapter or local ambulance association.

Chemical spills can present hazards; use "spilled water" to allow practice in having someone guard the spill, in finding spill absorbent, and in disposing of the residue.

3.2.4   Evaluation of Drills

What means can be used to evaluate a drill? For most drills some objective evaluation is possible. The time taken to completely evacuate the building can be measured and questions asked such as is the time reasonable, and is it decreasing as the drill is repeated? Similarly, the rescue of a victim from a smoke-filled laboratory can be timed. Subjective evaluation is also necessary and valuable. A form titled with the name and date of the drill, to be completed by each laboratory supervisor should include questions such as the following:

How many people were in the laboratory?

Was there any panic or confusion?

Did people know how to react correctly?

How could reaction to the emergency be made more efficient?

What improvements could be made in the procedure?

The form should always have space for and solicit other comments by the supervisor. After reading the forms, the drill organizer should discuss the responses either at a meeting or individually with the supervisors and with as many as possible of the others involved in the drill. Not only does this provide helpful information, it also encourages laboratory supervisors to take a responsible interest in the drills. From watching the drills, from reading the responses on the forms, and from verbal comments, the laboratory manager or administrator can judge whether the procedures for the particular emergency are practical and appropriate, whether the training programs for these procedures are effective, and whether the laboratory supervisors can control the situation.

3.3 TRAINING PROGRAMS

Training in safe practices in the chemical laboratory is an integral part of becoming a professional chemist. Ideally, this training is continuous, from the example and teaching of the laboratory supervisor, through printed information posted on bulletin boards, in safety manuals and texts, to formal safety programs. Although this chapter describes the formal programs, the importance of the less formal component, especially the example of the laboratory supervisor, cannot be overemphasized.

Chemical safety programs for employees in industrial and government laboratories have somewhat different criteria from those in teaching institutions.

To comply with the Hazard Communication Standard and right-to-know regulations (4), see also Chapter 12, certain employers must provide laboratory workers with education about the chemicals to which they are exposed and training in their safe use. Although this requires laboratory managers to reevaluate and develop their safety programs, it also means that employees can be expected to take advantage of these programs and to comply with the safety guidelines. In the industrial or government research and development laboratory employing a large proportion of graduates, a dictatorial approach to safety training is likely to be met with strong resistance. This defeats the aim of the training, which is willing cooperation with safety procedures. Therefore, the program must be conducted by respected safety staff whom the graduates regard as experts in their field and it should be designed to persuade them to comply with the guidelines using rational argument.

In academe, pressures to provide quality safety training to undergraduates and graduate students are increasing. Prospective employers are urging teaching institutions to include safety as part of the curriculum of chemistry graduates and university administrators are aware of the legal liabilities of the chemistry laboratory. Also, with increasing knowledge of relationships between the hazards of a chemical and its structure, the study of chemical safety is becoming an acceptable academic discipline. In academe, there is the opportunity for systematic training in the principles of laboratory safety, in addition to teaching safe practices for specific manipulations. However, safety training must compete for time in the crowded undergraduate curriculum and the busy graduate's schedule and forever tight resources. Therefore, curriculum committees need to be persuaded of the necessity and value of including courses, and the whole academic community needs to be committed to the safety program. Often, one member of a department, willing and able to give the time, first to developing what is seen to be a high-quality program, and then to selling it to colleagues, can succeed. One of the major requirements of such success is perseverance.

The programs for industrial or government employees and those for teaching institutions will be discussed separately.

3.3.1   Goals

Safety training programs in government, industrial, and teaching laboratories have one purpose, that is, the reduction of accidents and near accidents. To be successful, the programs must achieve several goals. One of the most obvious, and probably the easiest to reach, is the provision of information. Before knowledge is put to effective use, however, an awareness of the need to operate safely is required, as is a positive attitude to safety. The analogy is often drawn to driving a car. One can know how to drive and yet not be a safe driver. Thus, there are two goals for the training program on chemical laboratory safety: to raise the consciousness of the participants to the need for safety, and to encourage their conscious enthusiasm for safe practices. That is, laboratory personnel should analyze for hazards both in their projects and in their way of operating.

3.3.2   Programs for Industrial or Government Employees

Two essential components of a safety training program for laboratory employees are the initial orientation and regular, ongoing sharing with the laboratory supervisor and safety personnel of the company. A valuable addition to this regular program is a safety seminar with visiting speaker.

3.3.2.1   Initial Training for New Employees.   First impressions are long remembered. Initial safety training makes clear to the new employee that the company is committed to making the laboratory a safer place to work. For most employees, this will be a morale booster during a stressful time. In fact, it has been suggested that the concern of an employer with safety should be raised before the first day of work, even as early as the interview of a prospective employee.

The level of safety awareness of the new employee will depend on background and experience. The recruit may be transferring from another company or may be straight from high school, technical college, or university when the level will be affected by the training received in these institutions. Sensitivity of supervisory personnel to this difference is important. During the first days of a new job, an employee will be introduced to supervisor and colleagues, will receive a job description, and will be familiarized with the workplace. New workers also need to have the basic safety rules of the laboratory and specific safety information relating to their particular responsibilities clearly explained by the safety officer or their direct supervisor. Such orientation would include details of emergency procedures with provisions for hands-on experience where practicable and an introduction to the safety organization within the company. Where a number of new employees join the organization at the same time, the information may be presented in a series of lectures, demonstrations, or discussions.

All employees should be provided with a safety manual, a key document. The contents of this manual should include the following:

Company safety policy and organization
Emergency procedures and accident reporting
Protective clothing and equipment
Safety rules for the laboratory
Appropriate specific information on chemical and equipment hazards.

If a safety manual is not available, provide a written handout containing information on at least emergency procedures and basic laboratory safety rules. Prepare and distribute a safety manual as soon as possible.

Some employees will be using materials or equipment for which special training is required before they start work. Training provided in-house is likely to be the most effective since it can be designed to meet specific needs. However, it requires up-to-date knowledge and good communications skills on the part of the instructor. In large companies, this training may be provided by the safety officer or other member of the safety organization. Small companies may prefer to send the employee to take a commercially developed course (11,12), or use an off-the-shelf training program (13,14).

Depending on the sophistication of the employee and his/her job responsibilities, the initial training may include an introduction to general safety information, again, either provided in-house by the safety department, or by sending the employee to take a commercially available course. The basic subjects covered would include

Emergency responses
Safety and health
Accident prevention
Reviewing a project for possible hazards
Finding information

3.3.2.2 Regular Program. Whatever ongoing safety training program is developed by a company, it should be systematic and regular and have input from employees.

Many large companies organize their employees on a divisional basis and regular weekly meetings are held between small groups of employees and their direct supervisor. These meetings provide the ideal setting for the presentation and discussion of safety information. Therefore, safety should be a permanent item on the agenda so that the meetings become an integral part of the safety training program. It is at this time that the supervisor can present information about any new chemicals being handled, distribute and discuss manufacturer's safety data sheets, show videotapes about an aspect of safety, examine current safety practices, and so on. It is the opportunity for employees to ask questions and to make suggestions for improvement. Practicial and beneficial suggestions must be seen to be given consideration by management and implemented where possible. If they are not acted upon, full discussion of the reasons should be presented at a subsequent meeting. Some companies have a different employee each month present a short paper on some aspect of safe practice related to the employee's work. When appropriate, staff from the company's safety office or guests who are experts in some area of chemical safety may be present.

At these meetings, accidents or near accidents can be examined for the circumstances that led to the problem. Here is the chance for near accidents to be discussed and documented, thus allowing factors in common to such incidents over a period of time to be recognized and rectified.

Essential features of this type of program are the small size of the group so that there can be an informal sharing of knowledge, ideas and concerns, and the ability of the supervisor to direct the thoughts and actions of the group towards improving existing situations and bringing good suggestions to the attention of management. It is recognized that many topics as well as safety must be discussed at the meetings and that the time available is limited. When held on a regular basis, short, well thought out, concise presentations with time allowed for discussion, can be highly effective training.

Small organizations may not have the regular meetings or the personnel or resources to mount a sustained formal safety program. However, they too must have communication between their workers and the laboratory supervisor, who may even be the company president. In any case, safety should form a regular topic both of discussion and of written memos.

Employees of large and small organizations should be encouraged to take advantage of safety training programs available to them at local teaching institutions or presented by commercial firms and should be allowed both time off to attend and have their expenses paid.

A valuable addition to the regular program is a 1-day or shorter safety seminar, which is conducted by a visiting speaker on some aspect of safety. Such an event underlines the commitment of management to making the laboratory a safer place to work and to ensuring that all employees are informed. A safety seminar for academe is described in detail in Section 3.3.5; it is adaptable to the industrial or government laboratory.

3.3.3 Programs for Undergraduate Students

Postsecondary institutions practice different types of safety training of undergraduate students. Some have developed courses on chemical laboratory safety; others teach safety as part of the chemistry laboratory courses; still others hold regularly scheduled safety seminars. In some cases, safety training programs combine two or more of these approaches. Although there are strengths to each approach, this author believes that well-taught safety training in the laboratory is highly effective. The ideal is a combination of this training with a safety course and/or safety seminar.

3.3.3.1 Safety Training in the Laboratory.

This approach to the safety training of undergraduates will be described in some detail since it can be modified to fit other training programs. Note that its strength lies in the relevance of the presented material to the activity taking place and in the fact that it is on-going. A potential weakness, however, is that some aspects of safety can be missed. Finally, the quality of the instruction depends not only on the skill of the instructor but also on the instructor's commitment to safety, whether the instructor be a teaching assistant in academe or an industrial laboratory supervisor.

First year students, especially those from rural schools, may have little practical laboratory experience and be nervous about performing experiments. Therefore, introductory safety information should be presented in such a way that the students recognize that by doing things the right way, they can work at the bench safely and confidently.

When students arrive at their first laboratory class, they should be introduced to the safety equipment present in the laboratory with, if practical, a demonstration of how it works. The basic safety rules relating to the course should be written on the blackboard and explained. The explanation should appeal to the intellectual curiosity of the student. For example, eye protection must be worn while in the laboratory since eye injuries can be very serious and the damage is usually nonreversible. Cigarette smoking not only may ignite flammable liquids, but contact of vapors with the hot cigarette end can result in the production of carcinogenic chemicals. The rules should also be printed in the laboratory manual or distributed as a handout. These rules should be few in number, briefly but clearly worded, and strictly enforced. (For such rules see, e.g., Ref. (15).) The procedure for evacuation of the laboratory should be described, including the location of the fire alarm and an indication of the sound it makes.

As the term proceeds, appropriate safety information is provided. Thus, before beginning each laboratory experiment, information about the physiological properties of the chemicals to be used and instructions for handling them are presented. The right (and safe) way to perform the manipulations and to use the equipment required in the upcoming experiment are discussed. The students should be involved in the discussion, and it may be useful to have them take turns to find information in the literature about the chemicals and to present their findings to the class. Before using flammable solvents, the fire triangle can be introduced; also students should be reminded of the location and use of fire extinguishers. When an experiment requires the use of the fume hood, ways of testing the air flow rate and what that rate should be are appropriate topics. As the students become more experienced, their suggestions for improving safety in the laboratory can be solicited and discussed. Also, they should be given more responsibility for recognizing potential hazards in a procedure and taking safety measures. Many attractive posters containing safety information are available from chemical manufacturers. These can usefully be displayed in the laboratory.

Students believe that important topics will appear on examination papers. If this topic is to be recognized as an integral part of the curriculum, laboratory quizzes and end of semester tests, both practical and written, must contain safety-related questions.

3.3.3.2 Safety Training through Courses.

Formal courses on chemical laboratory safety for undergraduates have been developed in several postsecondary institutions. Students enrolled in these programs should be taught by an instructor who is both knowledgeable about and committed to safety in the laboratory. There is a danger that the course will be seen as another academic exercise. Where the course is linked to training in the regular laboratory class described in section 3.3.3.1, this danger is minimized. In many institutions, the course is an elective and has as a prerequisite a course in organic chemistry. In some, especially technical colleges, the course is required for graduation with a diploma in chemical technology. Such courses have to be recognized by the administration as an integral part of the curriculum for undergraduates majoring in chemistry. This is becoming easier with the concern over liability for laboratory mishaps and the development of intellectually challenging courses on laboratory safety. However, the production of a quality course requires much time and effort of a committed faculty member. Published details of courses being offered are very helpful (16–19,24,26,29).

One of the problems of teaching safety in the laboratory is lack of time. Students are eager to begin the experiment and often require most of the time available to complete it. Also, whereas instruction in the laboratory addresses mainly the practical aspects of safety, there is opportunity in a course to relate the hazardous and toxic properties of chemicals to their molecular structure, the flammability of solvents to their flash point and ignition range, and to provide training related to general rather than to particular situations. Such topics as toxicology, industrial hygiene, and occupational health can be addressed. There is also an opportunity during courses to assign projects to the students such as the following:

1. Preparing material safety data sheets for a list of chemicals
2. Reviewing a laboratory with which they are familiar for its safety equipment, and making suggestions for improvement
3. Preparing a purchase order for safety equipment for a new laboratory, knowing what the equipment would be used for and the budget that was available.
4. Analyzing a published experiment for hazards.
5. Identifying direct and indirect causes of a real or hypothetical accident and then suggesting how it could have been avoided.

These exercises help prepare the student to be responsible in future employment.

3.3.4 Program for Laboratory Teaching Assistants

In any teaching institution that has laboratory classes, the laboratory supervisors have both moral and legal responsibilities towards the safety of their students. They also have considerable influence on the attitude these students have toward chemical safety. Thus, the training of laboratory teaching assistants, who in large institutions are frequently graduate students, is very important, both from the point of view of preparing them for their own careers, and in the encouragement of safe practices in future generations of laboratory workers.

3.3.4.1 Initial Training Session. Before beginning to teach in the laboratory, teaching assistants need safety training to encourage them to be knowledgeable, responsible and, hopefully, enthusiastic about safety. Although this training should be mandatory for incoming graduate students, it also should be required annually for all teaching assistants, so that memories are refreshed on emergency procedures before classes start. The more senior the professor who issues the directive to attend, the better. Several sessions lasting 1 or 1 ½ hours each are preferable to one long session.

In the sessions, the teaching assistants are familiarized with the safety equipment in the laboratory and are given hands-on training in its use. Printed handouts detailing emergency procedures are distributed and discussed. An evacuation drill should be scheduled very early in the term so that teaching assistants have the chance to put into practice what they have learned. The legal liability of the teaching assistants in the laboratory is explained. An introduction to laboratory management is needed so that teaching assistants know how to control the students under their supervision, and how to deal with behavior that is unacceptable in a safe laboratory.

After this basic and in-house material has been covered, the aim should be to present relevant safety information in as interesting a manner as possible. In many instances, teaching assistants are around for 3 or 4 yr or more. The experience of senior teaching assistants can be utilized by involving them in the training session, for example, to take part in demonstrations or to make brief presentations on selected topics of which they have special interest or knowledge.

Also, the staff supervisors responsible for the various laboratory courses are useful presenters and resource people for the seminar. In fact, focusing in successive years on safety aspects related to one of general introductory, organic, inorganic, and analytical chemistry provides variety of material in a 4-yr cycle. In addition to this variety, creativity is also required in the format. Many audiovisual materials are available, one of the most appropriate being the safety module of Project Teach, which includes a videocassette and "What's wrong here?" slides (30). Use of slides such as these requires audience participation, something to be encouraged as much as possible.

The acting out of emergency situations that result in *injury* takes time and effort to plan but it is very worthwhile, especially if, after the performance, the audience is asked what they would have done in the same circumstances.

One of the most important aspects of these initial safety training sessions for teaching assistants is the visibility and participation of all faculty responsible for laboratory courses.

3.3.4.2 Regular Safety Training. Initial safety training is essential, but once-a-year exposure is not enough. Teaching assistants need to become used to considering the safety aspects of every experiment. The regular weekly or biweekly meetings of teaching assistants with the faculty supervisor provide an excellent opportunity for discussion of safety-related problems in the just completed undergraduate experiment and of safety aspects of the upcoming experiment. For this to happen, time for such discussion during the meeting must be allotted on a regular basis, so that the teaching assistants expect to be asked about problems and are prepared to discuss them. By documenting minor problems or near accidents reported at these meetings, recurring difficulties become obvious and remedial action can be taken before the experiment is repeated. Teaching assistants become more aware of potentially hazardous situations and ready to report them with suggestions for improvement. Not only does this encourage graduate students to train undergraduates to "do it right," but also sensitizes them to recognize potential hazards in their own research and to take appropriate action.

When safety aspects of the upcoming experiment are examined, appropriate emergency procedure information should be repeated. For example, the first time a concentrated acid is being used, first aid for a spill on the skin should be reviewed and the use of the safety shower and eyewash fountain discussed.

Providing written information to teaching assistants about the properties of the chemicals to be used in the next experiment and about the safe use of any equipment is more effective than only oral presentation. It is likely that at least some of this information will be in the laboratory manual. Discussing it at the meeting emphasizes its importance and fulfills the spirit of the *right-to-know* and *hazard communication* requirements. Teaching assistants then are encouraged to share the information with the students in the class during their prelaboratory lectures.

The safety training resulting from these meetings is directly related to the curriculum of the course with which the teaching assistants are associated. To fill in the gaps and provide the broader picture of chemical safety, some departments run a weekly chemical safety seminar that first year graduate students must attend. Not only does this provide comprehensive safety training, it also teaches the fledgling researchers to anticipate hazards in their own bench work.

3.3.5 Safety Seminar

Although basic training in safety occurs most effectively in small meetings with supervisors who have a strong commitment to safety, or in formal well-designed courses, an infrequent high profile event is a valuable part of a safety training program. Such an event publicizes the commitment of the department administration to safe practices within the department. Holding such an event on a regular once or twice yearly basis means that it becomes an expected part of the department's activities. It provides a valuable opportunity to involve people from the academic community, the government, local industry, and science teachers and school administrative staff from surrounding jurisdictions. Therefore, notice about the seminar can be circulated as widely as possible and the invitation to attend should come from a top departmental executive.

With a potentially large audience of varying backgrounds, the safety seminar must be carefully planned and professionally run. A successful format includes a theme speaker who is well known as an authority on some aspect of chemical safety. It can be helpful to have as the theme speaker a person who is highly respected for their research in a field of chemistry other than safety. This not only assures an audience of colleagues in the field, but also means that the safety talk will be heeded. The safety seminar is an excellent place for interaction between the teaching institution and local industry and government, and representatives from these spheres are usually happy to take part.

Wide publicity for the safety seminar will result in an audience with broad interests. Therefore, basic generalized information is an important part of the agenda chosen from such topics as local occupational health and chemical hazard regulations; legal responsibilities and liabilities; personal protection; emergency procedures; chemical hazard information; procedures for surplus and waste disposal; first aid. One topic area can be chosen for discussion in greater detail, the topic perhaps being dependent on the area of expertize of the theme speaker. As with the teaching assistant initial training sessions, Section 3.3.4.1, the program has to be sufficiently interesting and attractive as well as being informative, to hold the attention of the audience. Many excellent audiovisual materials are available (31–40); panel discussions and presentations which involve the audience are all needed. Demonstrations of safety equipment and of first aid treatment are useful. Time for questions is best allowed after each section and at the end of the seminar. Although the topic is deadly serious, a humorous approach in some seminars can provide variety (41).

A good safety seminar can have many valuable spin offs. The cooperation on safety matters between industrial and government experts and members of academe is fostered with exchange visits being arranged. Requests for safety seminars tailored to meet the needs of nonacademic staff, high school chemistry teachers, and groups who work with chemicals in other departments often follow.

Thus, the seminar meets at least one of the goals of a safety training program in that it raises awareness of the need to do things the safe way.

3.3.6   Evaluation of Training Programs

Holding safety training programs is necessary to meet legal requirements. They also result in a sense of "doing something" by the laboratory supervisor. But does the program work? This question can be answered by discovering whether laboratory accidents and near misses have been reduced since the program was implemented. This is easy to do if statistics have been carefully kept over an extended period. If such has not been done, then the enquiry will help encourage it to happen.

All participants in safety training courses or seminars should be requested to complete evaluation forms so that ideas for improvement of the program can be shared.

The ability of students to answer safety-related questions in quizzes and end-of-term tests is another measure of the success of the program. On a less objective level, there is the perceived awareness of safe practices of laboratory workers in the organization and the number of questions or comments directed to the laboratory supervisor or safety officer.

3.4   CONCLUSION

It is much easier to write about ideal safety training programs than it is to put them into practice and then to keep the momentum going. Successful programs have several points in common: they are ongoing, systematic, and relevant, with practice drills and hands-on experiences; they include both written directions and strong verbal communication among all levels of safety staff, supervisors, and employees; they are strengthened by audiovisual aids and creative presentations.

Programs having these elements and organized by knowledgeable and enthusiastic personnel cannot fail to improve safety in the laboratory.

3.5   SELECTED BIBLIOGRAPHY

3.5.1   Drills

1. *Emergency Response in the Workplace,* OSHA Publications, Room N-4101, Frances Perkins Building, Third Street and Constitution Avenue, NW, Washington, DC 20210.

2. N. V. Steere "Fire, Emergency and Rescue Procedures" in *Handbook of Laboratory Safety*, 2nd ed., N.V. Steere, Ed., CRC Press, Boca Raton, FL, 1971, p. 15.

3. D. J. Van Horn, "Education and Training", in *Chemical Emergencies in Laboratories—Planning and Response*, Vol. 6, NIH Research Safety Symposium, 1982.

3.5.2 Training Programs—Industrial or Government

4. "Informing Workers of Chemical Hazards," and "Hazard Communication: Worker Right-to-Know," ACS Office of Federal Regulatory Programs, 1155 Sixteenth Street, NW, Washington, DC 20036.

5. E. Thompson, "Managing People," in *Health and Safety in the Chemical Laboratory*, Royal Society of Chemistry, London, 1984, pp. 139–151.

6. New Employees, in *BDH Safety News*, Vol. 2, No. 1, Autumn 1982.

7. C. J. O'Connor, J. A. Young, and L. W. Berlein, Eds., *How to Plan an Effective Employee Hazard Communication Program and Safety Data Sheet and Label Program for "Right-to-Know" Regulations*, Labelmaster, Chicago, 1984.

8. *Informing Workers of Chemical Hazards: The OSHA Hazard Communication Standard*, American Chemical Society, Washington, DC, 1985.

9. D. J. Van Horn, "New Professional Employee Safety Orientations," in *Chemical Emergencies in Laboratories—Planning and Response*, Vol. 6, NIH Research Safety Symposium, 1982.

10. Working Safely in a Laboratory: An Orientation for the New Employee, Safety Manual, Research and Development, Smith Kline and French Laboratories, SmithKline Corporation, reprinted in *J. Chem. Educ.*, **53**, A159 (1976).

3.5.3 Courses

3.5.3.1 Courses—Live

11. ACS Short Course in Laboratory Safety and Health, ACS Education Division, 1155 Sixteenth St. N.W., Washington, DC 20036.

12. J. T. Baker Co. short courses, "Right-to-Know/Hazard Communication," "Hazardous Chemical Safety," "Safety for Supervisors," "Management of Hazardous Chemical Wastes," and "Hazardous Chemical Spill Response," J. T. Baker Chemical Co., Phillipsburg, NJ 08865.

3.5.3.2 Courses—Taped

13. J. Kaufman, "Laboratory Safety and Health," American Chemical Society, ACS Education Division, Washington, DC 20036.

14. "Chemical Hazard Training for Students," American Lung Association, 766 Ellicott St., Buffalo, NY 14203.

3.5.4   Training Programs—Academe

15. *Safety in Academic Chemistry Laboratories,* 4th ed., American Chemical Society, Washington, DC, 1985.
16. L. J. Nicholls, *J. Chem. Educ.,* **59** A301 (1982) (An Undergraduate Chemical Laboratory Safety Course).
17. K. A. Simpson, *J. Chem. Educ.,* in press (Safety Course for Chemical Technologists).
18. R. Bayer, "Chemical Safety and Health Courses—Colleges," in *Proceedings of the Symposium on Educational Approaches to Safety and Health in Chemistry,* J.J. Fitzgerald, Ed., in University of Connecticut, Storrs, CT, 1984.
19. L. J. Nicholls, "Chemical Safety and Health Courses—Universities", in *Proceedings of the Symposium on Educational Approaches to Safety and Health in Chemistry,* J.J. Fitzgerald, Ed., University of Connecticut, Storrs, CT, 1984.
20. D. W. Brooks, "Lab Instruction of High-Volume General Chemistry Courses", in *Proceedings of the Symposium on Educational Approaches to Safety and Health in Chemistry,* J.J. Fitzgerald, Ed., University of Connecticut, Storrs, CT, 1984.
21. C. F. Wilcox, Jr., "Lab Instruction of High-Volume Organic Chemistry Courses", in *Proceedings of the Symposium on Educational Approaches to Safety and Health in Chemistry,* J. J. Fitzgerald, Ed., University of Connecticut, Storrs, CT, 1984.
22. M. M. Renfrew and J. M. Shreeve, *Acc. Chem. Res.,* **17,** 201 (1984) (Toward a Safer Laboratory).
23. K. M. Reese, *Health and Safety Guidelines for Chemistry Teachers,* American Chemical Society, Washington, DC, 1979.
24. M. C. Nagel, *J. Chem. Educ.,* **59,** 791 (1982) (An Innovative Course in Lab Safety).
25. R. E. Uhorchak, *J. Chem. Educ.,* **60,** A41 (1983) (An Operational Safety and Health Program).
26. W. H. Corkern and L. L. Munchausen, *J. Chem. Educ.,* **60,** A296 (1983) (Safety in the Chemistry Laboratory: A Specific Program).
27. N. T. Freeman and J. Whitehead, *Introduction to Safety in the Chemical Laboratory,* Academic, New York, 1982.
28. A. A. Fuscaldo, B. J. Erlick, and B. Hindman, *Laboratory Safety—Theory and Practice,* Academic, New York, 1980.
29. G. G. Lowry, *J. Chem. Educ.,* **55,** A235 (1978) (A University Level Course in Laboratory Safety).
30. Project Teach, Module IX, "Safety," Department of Chemistry, University of Nebraska, Lincoln, NE 68588-0304, 1980.

3.5.5   Audiovisuals*

31. *28 Grams of Prevention,* Fisher Scientific Co., 711 Forbes Ave., Pittsburg, PA 15219.

*See also Chapter 17.

32. *In the Movies It Doesn't Hurt,* Central Office of Information, Hercules Rd. and Westminster Bridge Rd., London, England.

33. *Chemistry Lab Safety Procedure,* Instructional Development Services, Mission College, 3000 Mission College Blvd., Santa Clara, CA 95054.

34. *Safety—Isn't It Worth It?,* Fisher Scientific Co., 711 Forbes Ave., Pittsburgh, PA 15219.

35. *Eye and Safety Protection in the Chemical Lab,* National Society for the Prevention of Blindness, Ed. Feil Productions, 4614 Prospect Ave., Cleveland, OH 44103.

36. *Using Fire Extinguishers the Right Way;* NFPA, Quincy, MA 02269.

37. *If Only You Knew;* Chemical Industries Association Ltd., London.

38. *Compressed Gases Under Your Control,* Matheson Gas Products, Secaucus, NJ 07094.

39. *Laboratory Hazards and Safety Procedures,* U.S. Geological Survey, U.S. Department of the Interior, Denver, CO 25046, 1979.

40. *Science: Live to Tell About It,* National Science Film Library, Ottawa, Canada.

41. *The Saga of Safety Sue and Her Dealings with Danger Drew,* by D. M. Nichols and J. T. Roberts, in *Chemical Safety Supplement,* Smith Kline and French Laboratories, SmithKline Corporation, Philadelphia, PA 19101.

CHAPTER 4

The 95 Percent Solution

Jay A. Young

Chemical Safety and Health Consultant
Silver Spring, Maryland

4.1 INTRODUCTION

All accidents are caused; none are fortuitous. When the cause is known in advance, the accident is forseeable and can be prevented. In principle, were we wise enough, all causes could be known in advance. This chapter lists identified causes of laboratory accidents; some foreseen and prevented, some not. Clearly, the list is not complete—nor could it be; readers should add other items; with these additions, perhaps about 95% of all the causes of chemistry laboratory accidents would be included in the list, at least indirectly. Use the list to help eliminate the causes of accidents in your laboratory; use it also as a prompt, to stimulate identification of other causes not mentioned here. Suggestions for incorporation into the next edition of this book are invited.

4.2 THE LIST OF CAUSES OF LABORATORY ACCIDENTS

The predominant cause of laboratory accidents probably is the laboratory worker who knowingly takes a chance. Except for item 4.2.9, Lack of Personal Responsibility, there is no order of importance in List 4.2; what might well be critical in one laboratory is of minor importance in another.

4.2.1 Lack of Protective Equipment

- Eye protection not continuously worn. Eye protection did not meet American National Standards Institute (ANSI) or equivalent standards. Wrong type of eye protection, for example, safety glasses instead of goggles.

- Face protection not worn, or did not meet accepted standards. Face protection inadequate—half-shield leaving mouth, chin, neck exposed, shield too narrow to protect ears and ear canal.
- Gloves not worn. Improper gloves worn, short gauntlet does not protect bare arm, glove material not tested for rate of permeation by substances being handled, gloves not checked for pinholes, interior of gloves contaminated by prior use. Foreign substance enters gloved enclosure through loose gauntlet end—muff not used over gauntlet end and underlying sleeve to close the open end.
- Protective clothing inappropriate for the hazard—fabric not fire resistant, or not resistant to corrosive material being handled. Protective clothing not commensurate with the risk—lab apron worn instead of coveralls; shoes or boots unsuited for materials being handled.
- Unsuitable personal dress—loose sleeves, long tie, unrestrained long hair, hanging jewelry, full cut blouse, open toe shoes, wristwatch with porous or absorbent strap.
- Safety shield placed on bench so as to protect the individual but no shielding at sides and rear to protect persons nearby.
- Equipment intended for emergencies used routinely—respirators donned in lieu of using hood or in lieu of designing work so as to prevent release of toxin. In lieu of a work plan that would prevent a fire, fire extinguisher removed from assigned location so as to be handy just in case—and hence not available to other laboratory workers.

4.2.2   Lack of Communication

- Material safety data sheets ignored, precautionary labels not read. Material safety data sheets and/or labels silent on or minimize degree of hazard or appropriate precautions. Double labeling, new label for different substance affixed over original label, and new label falls off revealing original (and incorrect) label.
- Laboratory personnel do not know what to do when alarm bell sounds, nor how to distinguish from sound of other signals.
- Personnel occasionally work alone in laboratory, with nearest person in building too far away to hear a call for help; or colleague in nearby laboratory departs without informing other person who is now alone in the building.
- Laboratory worker poorly informed on what to do if overexposed.
- Personnel do not routinely inform co-workers of their plans to carry out hazardous procedure.
- Local fire department unaware of current hazard conditions in laboratory.

- Nearby hospitals uninformed of specifics for treatment of exposures to chemicals currently in use. Local physician not prepared in advance to treat foreseeable injuries resulting from exposures to chemicals in current use.
- Results, conclusions of safety audits, or inspections not available to the general public.
- Hot surfaces not labeled or otherwise identified as hot.
- No one in the laboratory has been recently updated in first aid, CPR, and so on.
- National Fire Protection Agency (NFPA) "704 Diamond" that is painted on outside of door to the laboratory no longer applies to existing fire hazards now in the laboratory.

4.2.3   Lack of Proper Ventilation

- Hoods in poor condition, inadequate draft or no draft, no recent measurement of hood performance, qualitative or quantitative. Hood draft excessive, turbulence drives fumes into laboratory space. Hood exhaust improperly located and fumes are recirculated back into laboratory or other occupied space.
- Improper hood use, equipment setups, containers of reagents, general clutter in hood precludes effective use of hood to capture noxious fumes. Hood used for long term storage of chemicals and also used as a fume hood. Perchloric acid used in hood not designed for such use. Source of fumes placed too close to hood face and fumes that are generated flow into laboratory instead of out, through hood ducting.
- Glove boxes leak when used, contaminating laboratory spaces. Glovebox exhaust ducting allows contaminants to enter laboratory spaces.
- Flexible ducts (elephant trunks) do not capture noxious fumes as intended.
- Infrequent or nonexistent sampling of breathing air for presence or absence of toxic substances. Records from past sampling in poor condition, not useable for retrospective studies of personnel exposures.

4.2.4   Personal Hygiene Problems

- Some personnel regularly apply cosmetics while in the laboratory. Eating is permitted in the laboratory. Food is stored in laboratory refrigerator. Eyewash fountain is used as a source of drinking water. Coffee, tea, or soft drinks are consumed within the laboratory.
- Pipetting by mouth is allowed. Beakers, flasks, watch glasses, petri

dishes, and so on, are used in place of cups, saucers, and bowls for food and drink.

- Personnel do not habitually wash thoroughly whenever they leave the laboratory.

4.2.5 Electrical Hazards

- Electrical wiring violates National Electric Code, or, is not known to be in compliance. Some sockets support "octopus" array of plugged-in wires; frayed wires are not removed from service; some equipment wires are immersed in puddle of water or other spilled liquid. Or, equipment wires pass near a source of intense heat, or open flame, or near oxidizing reagent likely to be spilled. Wire sockets and plugs corroded and become hot in use.
- Ground fault circuit interruption protection is not available for all circuits that serve the laboratory.
- Inadequate provisions to minimize buildup of static electrical charge.

4.2.6 Storage Problems

- Flammable liquid storage not in accordance with NFPA Manual 30. Too large a volume stored in nonsafety can, or other acceptable container; container is corroded; container is improperly labeled; no provision for storage in approved flammable liquid storage cabinet.
- Compressed gas cylinders not secured against toppling over. Protective caps not in place on cylinders stored awaiting use or empty and awaiting return to supplier.
- Some bottles of chemicals have been on laboratory shelves so long untouched that no one knows who ordered that chemical, nor the purpose for which it was purchased or synthesized. Other reagents are stored, unused, in the warehouse or stockroom because the unit cost was less if a greater than needed quantity was purchased. Still others are stored against the day long in the future when someone might need it for a reason now not foreseeable.
- Incompatible chemicals are not segregated in storage locations.
- A non-"explosion-proof" refrigerator is used to store liquids that produce vapors that could explode.
- Storage shelves are not fitted with an edging lip. Reactive chemicals, in beakers, bottles, and flasks, are placed near the edges of a laboratory bench.
- Boxes are stored one on top of the other instead of on shelves. Laboratory aisles, aisles in storage rooms, blocked by overflowing stored

equipment or chemicals. Stored materials block access to safety shower, to eyewash, to designated exit door, to fire extinguisher.

4.2.7 Inadequate Emergency Procedures and Equipment

- Telephone number emergency list not up to date or inaccessible to person using telephone. Emergency telephone not available after normal working hours.
- Fire extinguishers not regularly maintained, records not available, personnel untrained in proper use, wrong type of extinguisher provided in accessible location so that it will be used instead of proper type, more distantly located. Laboratory personnel unacquainted with which type of extinguisher to use on different kinds of fires.
- No plans exist in advance for shutting off utilities in case of emergency.
- Plans exist for fire, explosion, toxic exposure emergencies, but no plans for other foreseeable events such as corrosive spill, riot, or hurricane.
- Emergency evacuation plans do not include provision for specified meeting place of evacuees, away from building, nor for responsible head count.
- No single person designated as sole authority to approve reentry.
- Safety showers and eyewash fountains not recently tested; records of infrequent prior testing not available. Portable eyewash fountains that provide only a few minutes of flowing water relied on instead of copiously flowing fixtures that provide at least 15 min of water flow. Fire extinguishers not checked regularly for full charge, records of such checking in poor condition.
- Laboratory personnel unaware of need for 15-min minimum eyewash in the event of exposure; also do not know to hold upper and lower eyelids away from eyeball and to roll eyes while washing affected eye.
- Fire doors are routinely blocked open, or blocked closed. Exit doors are securely locked.
- Emergency drills rarely carried out; results of drill are not constructively evaluated and applied to changing drill procedures.
- No one knows how to use the equipment for treating emergency spills.

4.2.8 Management Responsibilities Inadequate

- Housekeeping is poor, refuse disposal puts janitorial staff at risk. Hazard analysis of new procedures is casual or nonexistent; when it is done, records are not kept. Analysis of accidents sometimes seems to protect persons at fault instead of preventing future accidents.
- Correctly or incorrectly, safety inspections and safety audits are perceived to be more vindictive than constructive.

- Chemical hazards are well controlled but physical hazards—unguarded moving belts and pulleys, unattended heat sources, untaped Dewar flasks, generation of static charges, and clogged drains, are not controlled.
- Safe practice policies are in place on paper but they are not enforced for casual visitors, for necessary visitors such as plumbers, painters, sales persons, nor for other department personnel such as janitors or office staff; occasionally laboratory workers take liberties.
- There is no safety and health operations manual, or chemical hygiene and safety plan, or equivalent, or the existing document is out of date.

4.2.9 Lack of Personal Responsibility

- A laboratory worker says "Even though I know better, I'll do it—just this once!"

CHAPTER 5

Safety Inspections, Safety Audits

Jay A. Young

Chemical Safety and Health Consultant
Silver Spring, Maryland

5.1 INTRODUCTION

Every well-managed laboratory conducts safety inspections on an organized, planned basis. Other laboratories also conduct safety inspections; typically, after a dramatic close call with no serious injuries nor marked damage to property, or after a serious accident. The purpose in either case is the same: To prevent future accidents. This requires that a safety inspection be effective; but every laboratory is unique, only a safety inspection designed to fit the needs of the individual laboratory is likely to effectively prevent accidents. Guidelines for such particulars cannot be found in any book but must instead be developed by the persons involved in their own laboratory environment. In this context the purpose of a series of safety audits is to generate competence in laboratory personnel to design and conduct safety inspections that are uniquely suited to that laboratory.

5.2 ACCIDENT PREVENTION

Prevention of accidents is a two-step process, indentification of hazard followed by control or elimination of the identified hazard. It is estimated(1) that 10% of unidentified hazards cause a near miss or, at most, a minor injury that requires simple first aid treatment, 1% cause injuries for which responsible supervisors require that a written record be made, 0.1% are disabling for periods of a day to several weeks, and 0.003% cause an accident that is either permanently disabling or fatal. Thus, for each near miss or simple first aid accident, 10 not yet identified and controlled hazards are present in the work environment. For each recordable accident, there are 100 unknown or uncontrolled hazards existing, and so on. Although it is obvious that these conclusions, drawn from a generalized statistical summary, do not apply literally to a particular laboratory environment, it is also obvious that even a near miss demonstrates the existence of unknown or uncontrolled hazards which, because they are unknown or uncontrolled, require safety inspections or safety audits, or both, to be eliminated.

Since it is clear that an accident that is about to happen may be fatal or a near miss or somewhere in between, it is wiser to identify and control or eliminate hazards with regular safety inspections and audits in advance of the event rather than after the fact. The problem is what to look for.

5.3   ACCIDENT CLASSIFICATION

Safety audits reveal what to look for in a specific laboratory workplace; these generalizations can serve as a guide in developing safety audits and in carrying out safety inspections. Accidents can be classified in seven ways:

1. *"Struck by" Exposures*. An object in motion strikes a person. Laboratory examples include an object dislodged from a high shelf, a flying shard of glass from an exploding or imploding vessel, horseplay involving thrown objects.
2. *"Strike Against" Exposures*. A moving person collides with a fixed object. Laboratory examples are rare.
3. *Fall Exposures*. A person falls from a higher to a lower level or trips and falls at the same level. In the laboratory, spills of liquids on floors with consequent slipping and falling are the most common examples.
4. *"Caught on" Exposures*. A person or their clothing is caught on a stationary or moving object. Thus, in a laboratory, a loose sleeve, or long hair or a dangling tie can be caught in equipment with unguarded, rotating parts. One common example is entanglement in the unguarded belt and pulley of a vacuum pump; another is a long necklace looped around the post of a ring stand, toppling it when the leaning necklace wearer straightens up.
5. *"Caught between" Exposures*. Two moving objects or one moving and one stationary object pinch, crush, or otherwise catch a person between the objects. Laboratory incidents are rare.
6. *Overexertion Exposures*. Self-explanatory; excessive strain causing a sprained back while lifting or moving, say, a heavy instrument is an example from the laboratory.
7. *Chemical   Exposures*. Also   self-explanatory.   Laboratory   examples include:

   (a) Toxic gases, vapors, dusts, mists in the air.

   (b) Irritating or corrosive gases, liquids, solids in contact with mucous membranes or skin.

   (c) Temperature extremes, intense cold from cryogenic materials, intense heat from a runaway reaction or fire, or from very hot liquids or solids.

(d) Sources of radiation such as lasers, radioactive specimens, and UV sources.

(e) An enclosed space deficient in oxygen.

5.4   HAZARD CATEGORIES

Continuing with generalizations, hazards are often classified into two categories, those due to unsafe acts by persons and those due to unsafe conditions in the workplace environment. Of the two, hazards arising from unsafe acts are more prevalent than hazards related to unsafe conditions. These laboratory examples are illustrative:

Unsafe Acts

1. Violations of safety rules.
2. Operating equipment without proper training, or without authority.
3. Altering safety devices so as to make them inoperative.
4. Using equipment that is in a defective condition.
5. Servicing or altering electrically energized equipment.
6. Using unsuitable protective equipment or clothing, or not using such at all.
7. Taking shortcuts.
8. Horseplay.
9. Failure to warn or to protect co-workers while adequately protecting one's self.
10. Poor record keeping.

Unsafe Conditions

1. Inoperative emergency equipment (fire extinguishers, safety showers, eyewash fountains).
2. Unsatisfactory training in the use of emergency equipment.
3. Poor housekeeping.
4. Narrow clearances in passageways, spaces between laboratory benches or between bench and wall, exit doorways, area in which emergency equipment is located.
5. Improperly designed storage areas, inadequate shelving.
6. Insufficient illumination.
7. Crowded lab bench surfaces.
8. Improper electrical wiring.
9. Inoperative warning systems.

10. Mechanical equipment operating marginally due to inadequate maintenance.

5.5 POINTED QUESTIONS

Yet another general way to develop safety audits and to carry out safety inspections is to prepare detailed written answers to a short set of pointed questions, such as:

1. What problem, what area, should be inspected or audited?
2. Why should this problem or area receive priority?
3. Identify the critical factors involved in the problem or area.
4. Identify other factors that are involved; clarify why they are not critical.
5. Who will develop the audit or carry out the inspection? Justify the selection of the person or persons assigned to the task.

5.6 SAFETY INSPECTIONS

Although desirable, it is not necessary to incorporate both safety inspections and safety audits in a laboratory safety program. When safety audits are not used, a safety inspection is typically a walk through by one or more members of "the safety committee" who pause here and there, duly noting violations of published safety rules and marking off a check list derived from someone's ideas of hazards. Clearly, there are better ways to conduct a safety inspection.

A safety inspector should at least be minimally conversant with good safety practices both as these are discussed in reference works and as they are carried out in other laboratories. It is helpful if, in addition to noting infractions, the inspecting team also commends observed exemplary safe acts and conditions.

Doubtless, a single safety inspection can range over a broad variety of concerns. It is more effective to limit the scope of a single inspection to one or at most a few matters selected from topics generated by thoughtful reflection upon one or a few of the accident classifications, hazard categories, answers to the questions above, or recommended by a safety audit. A subsequent inspection might review the same matters as the preceding inspection, or it might look to a different, but focused, area.

Every safety inspection should have a built-in follow-up requirement to insure that the recommendations of that inspection are implemented. Always, a competent safety inspection also addresses the follow-up actions mandated by the most recent prior inspection and a given safety inspection may well include an evaluation of the adequacy of follow-up mandates of several prior inspections.

The question is often asked whether a safety inspection should be announced in advance or conducted in a fashion to surprise the inspectees. Were it the purpose to embarrass laboratory workers, surprise inspections would be the rule. Since this is far from the purpose, and given the well-known characteristics of human nature, surprise inspections either should never be used or used only when it is clear that a surprise inspection will serve to improve safety practices better than other administrative techniques. Stated differently, the purpose of a safety inspection is not solely to improve safety, its purpose is to improve safety in such a manner that eventually good safety is practiced whether or not an announced inspection is imminent.

Another commonly asked question relates to the use of a check-off list. Proponents affirm that without a list, important matters may be overlooked. Opponents aver that the use of a list can tend to be mechanical, with the various items checked off in a desultory, offhand manner such that the result is equivalent to overlooking the matter. Hence, if the totality to be inspected is so large that a list is needed by the inspectors as a reminder, then by all means either use a check-off list or focus the content of the inspection to a more manageable number of details that can be remembered without a list. Or, if an inspector will likely use a check-off list in a desultory way, either retrain the inspector or select a more competent individual as the inspector.

Yet another question deals with the frequency of laboratory safety inspections. Because each laboratory is different, and further is different at different times, no fixed inspection frequency can be recommended. On the other hand, once the question is asked the prudent answer probably would be, "More frequently than the current frequency." The cost of safety inspections is a proper budget item; if there is no current budgeted support for safety inspections, it is likely that the current frequency of inspections could prudently be increased by a factor of five, or more. If currently budgeted and there is some well-founded concern that the allowed amount is unrealistically small, then probably it would be prudent to carefully consider increasing the frequency by a factor of two or three.

Safety inspections can of course be conducted by one person. Many laboratories are large enough to make such inspections a burden for the individual. Further, the combined wisdom of a small group of inspectors is greater than the sum of the contributions from all of the individuals. The same guidelines apply to selecting the members of a safety inspection committee that apply to determining the constitution of any committee; these include considerations of individual competence and interest, the priority demands of other duties, personal ideosyncrasies, opportunity for training new and inexperienced laboratory workers, and, not least, internal political considerations. Usually, ancillary staff and supporting personnel are not part of a safety inspection committee; when they are participants, the results are often surprisingly constructive. Thus, consider the janitor, the purchasing agent, the stenographer, the sales manager, a vice president, as candidates for membership.

It is sometimes useful to divide a safety inspection committee into two unequally sized groups, the larger group conducts the inspection and the smaller group is charged with enforcing the follow-up mandates. It is also useful occasionally to include outside persons as part or as the whole composition of a safety inspection committee. Of course, committee membership should be rotated at timely intervals.

5.7 SAFETY AUDITS

Safety audits serve the same purpose, accident prevention, as safety inspections but in other respects a safety audit does not resemble a safety inspection. A safety audit is a planned, organized process in which documented answers to safety questions are collected, to the end that all personnel are inculcated with positive safety attitudes and habits. Stated differently, safety audits provide a basis for action by stimulating the initiation, maintenance, review, and assessment of records that describe and define current safety policies, practices, and attitudes.

The results from a safety audit can be used to plan and guide a safety inspection. For example, if the scope of a safety audit included concerns about warning bells and whistles, the documents and records produced by the audit might serve as information to be used in planning and conducting a safety inspection focused on laboratory workers' recognition of the sound of the fire alarm bell and their ability to distinguish that sound from the sound of the toxic gas release alarm as judged by their actions when one of the alarms is sounded in a drill.

The results from a safety audit can serve other purposes; for example, to evaluate administrative policy, to find ways to improve the training of new employees, to determine criteria to be used in the design of new laboratory facilities, or to identify better ways to control laboratory hazards.

5.7.1 The Safety Audit Committee

A safety audit is conducted by a committee. From the specification of the scope of the audit, the committee identifies the documents and records that should be reviewed, the personnel to be interviewed, and the questions to be asked. After the document and record review and the interviews, the committee prepares a report with recommendations. The scope of an audit can be broad—hazard control, accident management, for example, or more narrowly focused—fire extinguisher maintenance, use of fume hoods, training in handling corrosives. Obviously, a safety audit is useless if the recommendations are not implemented; hence, the scope of a safety audit might well include, in addition to the matter of concern, a review of the effectiveness of implementation of the immediately preceding safety audit. When this is not feasible, the scope of some safety audits might well be restricted to an evaluation of the effectiveness of implementation of several prior audits.

A safety audit report should be brief, usually < 500 words, supported when appropriate by perhaps a voluminous appendix. Tendencies toward numerous recommendations ordinarily should be suppressed; usually, a few concise recommendations are more likely to be implemented than several less precisely stated aspirations. Good safety after all is necessarily a continuously developing process; some matters are often best left to future safety audits. However, at least annually a safety audit should be directed toward the identification of gaps and omissions in the combined scopes of, say, the prior years' safety audits.

The safety audit committee must have the authority to interview persons whose duties fall within the scope and as well to acquire records that fall within the scope of the audit. As with safety inspections, at least one committee member should be conversant with good safety practices and some of the other members may usefully be drawn from nonchemical staff and support backgrounds. Since safety audits can tend to become subjective, even self-congratulatory, occasionally one or more members of the audit committee should be from the outside, not an employee of the organization.

5.7.2 Safety Audit Scope

This list suggests the variety and range of scope of safety audits:

- Is the safety budget adequate?
- To whom should the chief safety officer report?
- Who is responsible for preventing serious injury?
- To what persons are copies of the organization's safety policy distributed and what actions do they take upon receipt?
- Who formulates the safety policy of the organization and upon what principles is the policy based?
- What use is made of accident reports?
- What evidence is there showing that the procedures used to identify and control hazards are effective?
- Is it possible to identify an unknown hazard?
- How should a safety audit be conducted?
- What steps are now taken to protect nontechnical staff and visitors (janitors, plumbers, salespersons, office personnel) from laboratory hazards?
- Should the current list of safety rules be revised?

- How is emergency equipment (fire extinguishers, safety showers, eyewash fountains, alarm bells) maintained? Are all components presently in operating condition?
- Should the cost of inventory control be part of the safety budget?
- What criteria are used to determine the physical location on the shelves in the chemical storage area of each chemical that is stored there?
- What information is supplied in advance of an incident to the physicians and hospital emergency personnel concerning treatment for possible overexposures in this laboratory? Are those facilities now prepared to efficiently handle an overexposure?
- Does the current safety training program produce desirable and permanent changes in the behavior of both laboratory workers and janitorial staff?
- What changes were instituted as a result of the most recent visit of local fire fighting officials?
- In the event of a serious accident, are the procedures now in place sufficient to minimize the likelihood of panic on the part of persons living or employed nearby? What would we tell the inquiring reporter or TV interviewer? Who would respond to their questions?
- What criteria are used to recognize an employee's distinguished contribution to good safety?
- What is the purpose of a safety meeting?
- Should a member of the corporate board (or equivalent) participate as a member of a safety audit committee?

These questions are intended to stimulate locally generated specifics that will define the scope of a safety audit. Thus, the question on recognition of employee contributions to safety does not in any way imply that there should or should not be such recognition. Instead that question is in the list as an example of a safety audit topic that might be overlooked if we think of safety audits as applicable only to unsafe acts and unsafe conditions. Although indeed in one sense these two comprise the whole of safety; unimaginatively interpreted such a restrictive view misses the whole point—people are at the heart of safety. Further, it may not be evident to those who have never participated as a member of a safety audit committee, but reliance on written records and responses in interviews is at the heart of the power of an audit. Thus, consider the previous question as to whether the safety showers are now in operating condition. The obvious way to find out is to leave the safety audit committee room and check those showers; the revealing way to improve safety is to rely on whatever information on the topic may or may not be made available to the audit committee members in that room and their questions about the nature and extent of that information.

Finally, use the questions in the previous list to stimulate your own creative questions that will be useful in your unique laboratory situation. Thus, perhaps not the question on the adequacy of the safety budget but rather, should there be a safety budget? Or, not what changes did the local fire chief recommend when he last visited 2 years ago after that big fire, but should the chief have been invited to that informal conference last week when we discussed which type of fire extinguisher to purchase in connection with the new research project?

REFERENCE

1. *Guidelines for a Chemical Plant Safety Program and Audit, SG-21,* Chemical Manufacturers Association (Manufacturing Chemists Association), Washington DC, 1978.

APPENDIX 1

CONTROLLING THE HAZARDS

CHAPTER 6

Flammability, Combustibility

Rudolph Gerlach

Muskingum College, New Concord, Ohio

6.1 INTRODUCTION

Fire codes are prepared by the National Fire Protection Association (NFPA), the federal Department of Transportation (DOT) has established standards, the Occupational Safety and Health Administration (OSHA) has adopted some codes in its regulations, and most local communities have established their own local codes and standards.

As is often the case with chemicals, other hazards, some obvious and some insidious, are present when dealing with flammable materials. Thus, closed containers of flammable chemicals can develop, as a result of boiling, an internal pressure that can cause a boiling liquid expanding vapor explosion (BLEVE). Containers of corrosives can also explode, spreading hot corrosives. Fires involving compressed gas cylinders are particularly hazardous because the cylinders can explode or rupture, resulting in the release of large quantities of mechanical energy. The released gases may also be corrosive or toxic, for example, chlorine, phosphine, or flammable, for example, hydrogen. The rupture of a compressed gas cylinder containing such gases could well introduce a hazard equal to or greater than the original fire.

6.2 DEFINITIONS AND TERMINOLOGY

Unfortunately, there exists confusion over some of the terms common to a discussion of flammability. This confusion is at least partially the consequence of many different organizations establishing their own terms. In this section we shall examine some of these terms. Since a large percentage of fires are fueled by liquids, it is important to fire prevention and fire fighting to recognize it is the vapors of these liquids that actually burn. The distinction between the terms vapor and gas is often blurred. Commonly, a gas is defined as a substance that can exist only in the gaseous state at standard temperature and pressure, 0°C and 760-Torr pressure. Vapors are generated by substances that can exist as liquids under these conditions. It is important from a safety point to note that, with few exceptions, vapors are more dense than air. Therefore, at room temperature most vapors tend to settle to the lowest possible elevation. Some common gases, for example, hydrogen, methane, and ammonia, are less dense than air and thus tend to rise when released from their container.

The U.S. Department of Transportation has defined what constitutes *burnable* or oxidizable materials. These are published in the *Code of Federal Regulations* (CFR)(1). Table 6.1 summarizes these definitions and gives the particular sections of the CFR where a more complete discussion can be found. Note that liquids are divided into two classes, combustible and flammable; combustible being the less hazardous and flammable the more hazardous, having the lower flash point (defined later). Since there exist tens of thousands of materials that fit into these two classes, it stands to reason to say that there also exists a wide range of flash points within each class. The NFPA has subdivided each class, combustible and flammable, into more definitive classes so that safer storage of these materials can be achieved(2). Table 6.2 summarizes the NFPA classification scheme.

As noted previously, the flash point is a prime factor considered when the fire hazard of a liquid is being determined. Flash point, sometimes referred to as flash temperature, is the lowest temperature under controlled (laboratory) conditions to which a liquid must be heated so that enough vaporization will occur to create an ignitable fuel-air mixture at the liquid's surface when an ignition source is introduced. The fire need only be of momentary duration.

When using the literature values for flash points one needs to remember

Table 6.1. Classes of Ignitable Materials

Hazard Class	Definition[a]
Combustible liquid	Any liquid having a flash point above 100°F (37.8°C) and below 200°F (93.3°C) (Sections 173.115 (d) and 173.115 (b))
Flammable liquid	Any liquid having a flash point below 100°F (37.8°C) (Sections 173.115 (a) and 173.115 (d))
Flammable gas	A compressed gas that satisfies the criteria for flame projection, lower flammability limit, and flammability range (Section 173.300 (b))
Flammable solid	A nonexplosive material that is capable of producing fire as a result of friction, heat retained from production or which, if ignited, produces a serious transportation hazard (Section 173.150)

[a]Section numbers refer to the U.S. Department of Transportation's *Code of Federal Regulations* (see Ref. 1).

Table 6.2. NFPA Classification of Flammable and Combustible Liquids for Storage Purposes

Classification	Term	Flash Point and Boiling Point if Applicable
Class I	Flammable liquid	Below 100°F (37.8°C)
Class IC	Flammable liquid	At or above 73°F (22.8°C) Boiling point below 100°F
Class IB	Flammable liquid	Below 73°F (22.8°C). Boiling point at or above 100°F (37.8°C)
Class IA	Flammable liquid	Below 73°F (22.8°C) Boiling point below 100°F (37.8°C)
Class II	Combustible liquid	At or above 100°F (37.8°C)
Class III	Combustible liquid	At or above 140°F (60°C)
Class IIIB	Combustible liquid	At or above 200°F (93.4°C).

that the flash points can be measured by more than one method. The American Society for Testing and Materials (ASTM) has established standard criteria for the measurement of flash points using these methods. Table 6.3 lists these methods of determining flash points. Table 6.4 gives closed-cup flash points for some common chemicals.

The Underwriters Laboratories (UL) have described the use and application of the Tag closed cup and Pensky–Martens closed cup and a special apparatus with a method for determining flash points $< 16°F$ $(-9°C)$(3). The apparatus is described as follows:

Table 6.3. Standard Methods for Determination of Flash Points

Flash Point Test Name	Description[a]	ASTM Test Method
Tag closed cup	For most liquids with a low viscosity, e.g., $9.5 \times 10^{-6} m^2/s$ at 25°C and with a flash point $< 93°C$	D56
Cleveland open cup	For petroleum products (except fuel oils) provided the flash point is $< 70°C$	D92
Pensky–Martens closed cup	For fuel oils and related liquids	D93
Tag open cup	For liquids with flash points between -17.8 and 168°C	D1310
Setaflash closed cup	For paints, varnishes, and similar products, if they have a flash point between 0 and 110°C	D3278

[a]For details, see the appropriate ASTM test methods.

Table 6.4. Flash Points, Lower and Upper Flammable Limits, and NFPA Ratings of Some Commonly Used Chemicals

Chemical	Closed-Cup Flash Point (%F)	Flammability Range (% Volume in Air)		NFPA Fire Rating
		Lower	Upper	
Acetone (dimethyl ketone)	− 4	2.6	13	
Acetylene		2.5	100	4
Acrolein (acrylic aldehyde)	− 15	2.8	31	3
Acrylonitrile, (vinyl cynanide)	32	3.0	17	3
Allyl chloride	− 25	2.9	11.1	3
Arsine (hydrogen arsenide)		4.0	74	
Butadiene (divinyl)		2.0	12	4
Butyl acrylate	118	1.5	9.9	2
Diborane (boroethane)		0.8	88	4
1,1-Dichloroethane	22	5.6	11.4	3
Diethyl ether (ethyl ether)	− 45	1.9	36	4
Ethane		3.0	12.5	4
Ethylene		2.7	36	4
Heptane	25	1.0	6.7	3
Hydrazine	100	4.7	100	3
Hydrogen		4.0	75	4
Hydrogen sulfide (heptic gas)		4.0	44	4
Methanol (wood alcohol)	53.6	6.0	37	3
Methyl ethyl ketone (butanone)	44.6	1.8	11	3
Methoxyethanol	102	2.3	24.5	2
Methyl hydrazine	17	2.5	92	3
Naphthalene (white tar)	175	0.9	5.9	
Octane	56	1.0	6.5	
Pentane	− 40	1.5	7.8	4
Styrene, monomer (phenylethylene)	90	1.1	6.1	3
Tetrahydrafuran (THF)	11.2	2.0	12	3
Toluene	40	1.27	7.1	3
Xylenes	90	1.1[a]	7.1	3

[a]measured at 100°C.

The apparatus is to consist of a cylindrical metal cup made of copper, with hemispherical bottom and removable cover. The cup is to be $1\frac{1}{2}$ inches (38 mm) in internal diameter and $4\frac{7}{8}$ inches (124 mm) in depth, overall, and is surrounded at the sides and bottom by a metal jacket, with spacing of 1 inch (25.4 mm) between the exterior of the cup and interior of the jacket. The top of the jacket has inlet and outlet openings for a cooling medium liquid nitrogen, precooled ethylene glycol–water mixture, etc. and a vent tube of $\frac{1}{8}$ inch (3.2 mm) inside diameter. A tublar insert, $\frac{7}{16}$ inch (11.1 mm) inside diameter, extends from outside through the jacket and terminates in an opening 2 inches (50.8 mm) below the top of the cup for mounting a thermometer in the liquid sample within the cup.

The procedures for operating the special apparatus are also very specific(3). The apparatus, or cup, is filled to within 1 in. (25.4 mm) of the top with the test sample. The sample is to be precooled to a temperature at least 20°F (11°C) lower than the expected flash point. With the lid closed the sample temperature is increased at a rate of 1.8°F (1°C)/min. The rate is controlled by passing a cooling agent through the surrounding jacket. At higher temperatures a heating medium is used in place of the cooling medium. When the sample temperature has reached a level 10°F (5.6°C) below the expected flash point, the lid is removed from the test cup and a small gas flame, $\frac{5}{32}$ in. (4.0 mm) in diameter, is inserted in the liquid's vapor just above the surface and promptly withdrawn. The lid is replaced on the cup. This procedure is to be repeated at 3.6°F (2°C) temperature intervals, until a flash is noted when the small flame is inserted.

An estimate of the closed-cup flash point of hydrocarbons can be made by use of the following equation(4)

$$T_F = 0.6813 T_B - 71.7$$

where T_F = closed-cup flash point (°C)
T_B = boiling point of test sample (°C)

The use of flash points for planning and designing safe usage of liquids must be done with some precautions. The need for these precautions centers around the obvious fact that flash points deal with vapors above the liquid. There are situations where mists form as a result of spraying or foaming. These mists can ignite at temperatures below the listed flash points. Another potential hazard is that the flash points are determined with fuel–air mixtures. Any environment which would cause an increase in oxygen concentration above that of normal air lowers the flash point value. Pressure also is a determining factor in the value obtained for flash points. Table 6.4 includes the flash points of several commonly used chemicals. A more complete listing can be found in the *Handbook of Hazardous Materials*(5).

Ignition temperature, or fire point, is a term often confused with flash point. Ignition temperature is the minimum temperature to which a vapor–air mixture must be raised to produce self-sustained combustion. That is, it is the lowest temperature to cause a sustained burning of the fuel. Since it is affected by several variables, the ignition temperature is not a fixed value. These variables include the shape of the container, pressure, concentration of the vapor–air mixture, and the effects of any catalysts that may be present.

Three other terms that are crucial to safe handling of flammables are lower flammability limit (LFL), upper flammability limit (UFL), and flammability range. They can best be understood by examining a graph that records the change in flash point of varying vapor–air mixtures as temperature is changed (see Fig. 6.1). Figure 6.1 shows the changes that occur in the

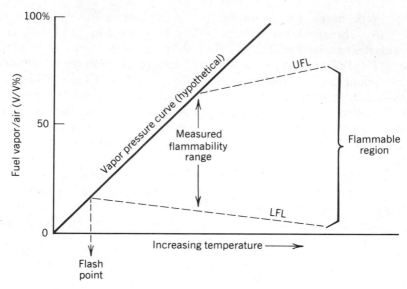

FIGURE 6.1.   Fuel–air mixtures and flash point relationships.

vapor pressure of a fuel, and the resulting change in flammable conditions as the temperature is increased. Until the temperature reaches the flash point the vapor concentration is too low to achieve a flash when the ignition source is inserted. Also note, however, that increasing the temperature of a vapor–air mixture will permit the ignition of progressively leaner fuel–air mixtures. The line connecting the increasing temperature and decreasing fuel–air ratio constitutes the LFL. Correspondingly, the value of the limiting fuel–air ratio, above which ignition is not possible, increases as the temperature is increased. The line connecting these values on the graph is called the UFL. Any point on the graph that lies above this UFL represents a ratio that is too rich to ignite.

The area on the graph that is enclosed by the LFL, UFL, and the vapor pressure curve constitutes what is normally called the flammable region. Any combination of temperature and fuel–air ratio that falls within this region offers the potential for a fire or explosion. In fire prevention we try to avoid such conditions.

The flammability range is normally expressed as the LFL and UFL values at 25°C. This introduces another insidious hazard, working with a fuel–air mixture at an elevated temperature, greater than the one where the flammability range was measured. Since the flammability range widens as the temperature increases, that elevated temperature may place that fuel–air mixture within the flammable region.

Also, we need to remember that the LFL and UFL values are determined under controlled laboratory conditions, which may differ from the actual workplace. In this case, the difference is that the fuel–air mixtures in the laboratory measurements are uniform, whereas, in the work situation there may exist spots of higher concentrations. As a consequence, it is becoming common practice to limit work conditions to a value $< 20\%$ of the published LFLs.

Another term whose understanding is important to fire prevention is autoignition temperature (AIT). The AIT is the temperature at which the vapors of a fuel ignite spontaneously without the introduction of an external ignition source. Spontaneous combustion of oily rags stored with poor ventilation is the classic example of where a slow oxidation raises the temperature to the temperature where autoignition occurs. The AIT of a substance will vary depending upon the concentration of the vapors, the presence of catalytic material, and environmental factors such as pressure and volume.

A catalyst is a substance that increases the rate of chemical reaction. Catalysts exert their influence on reaction rates by lowering the activation energy needed to make the reaction occur. In the case of burning, this means that the flash point and LFL are lowered—another insidious hazard.

Explosion damage is the physical result of rapidly expanding gases. Such expansion can be caused by very rapid oxidation (burning) of a fuel. If the oxidation reaction causing the explosion spreads to the unreacted fuel at a speed less than the speed of sound the explosion is called a *deflagration*. If the oxidation reaction spreads to the unreacted fuel at a speed greater than the speed of sound the explosion is then called a *detonation*. The term *explosible* is normally used when discussing the rapid burning of dusts. Regardless of what they are called, explosions are to be avoided and fire prevention is one way to avoid them.

6.3 RATE OF BURNING AND RESULTS

As you read the previous section you should have become aware that the rate of burning has a strong influence on the potential damage that a fire can produce. In one case a *slow* fire will simply cause thermal destruction of the fuel, where the fuel could be any solid, liquid, or gas that will burn. A more rapid burning can, in addition, introduce the physical damage brought on by the force of an explosion. Obviously the nonexplosive fire presents less risk to those involved, whether they are present at the time of initiation of the fire or whether they are involved in extinguishing the fire. An understanding of factors that affect the rate of burning of a fuel becomes most useful when trying to prevent or extinguish fires.

6.3.1 Nature of Fuel

One very important factor is the chemical and physical nature of the fuel itself. Thus, gaseous fuels such as hydrogen, methane, and acetylene, can react with an explosive rate if the conditions are not controlled. Of course, controlled burning of these fuels gives us useful fires. The gas range (stove), and hydrogen and acetylene torches are examples. On the other hand, the rare gases such as helium and neon do not burn but they can release explosive force when a compressed gas cylinder containing one of these gases is exposed to a fire situation.

Gasoline can burn with explosive force in an internal combustion engine because of its ability to readily convert from liquid to vapor and mist. Heavier, or more dense fuels such as heavy oils are not easily vaporized or made into mists. As a result they are less likely to burn at an explosive rate.

6.3.2 Surface Area

A large chunk of coal does not ignite when a match is held to its surface, yet coal miners have a very healthy respect for the possibility of a coal dust explosion. Similarly, those involved with the handling of dust-producing fuels such as flour, seed hulls, and many metals take great precautions to avoid these dusts because they can burn explosively.

Ignited fuel materials burn at explosive rates when in the gaseous state or are mists or dust-sized particles because under these conditions the two chemicals (fuel and oxygen) are in close physical contact with each other over an extensive surface area. The amount of surface area of the fuel exposed to the air has a direct influence on the rate; increase the surface area of the fuel and the burning rate also increases. The log sliced into small pieces offers more surface area to the air; powdered coal and the flour disturbed as dust offer more surface area, the gasoline and heavy oils converted from liquid to vapor do the same thing to increase the rate of burning. The fuels that exist as gas at room temperature have already converted to the smallest possible particles, molecules, and thus are ready for the maximum amount of contact with the oxygen in the air. This direct relationship between fuel surface area and the rate of burning is the driving force for the implementation of dust abatement programs in facilities where flammable solids are used or produced. Failure to control dust has led to many mining and industrial fires and explosions, which caused loss of life and millions of dollars of property damage. An example of such an industrial dust explosion and the consequences are well demonstrated by Fig. 6.2.

6.3.3 Effects of Catalysts

As previously mentioned, catalysts can increase the rate of chemical reactions by lowering the energy, for example, the temperature needed to cause the reaction. Extremely small quantities of catalyst suffice to produce explosive rates. As a result, extreme care must be exercised to avoid undesirable catalytic effects. The explosive decomposition of picric acid can be catalyzed by the presence of small amounts of the oxides of copper or lead. The

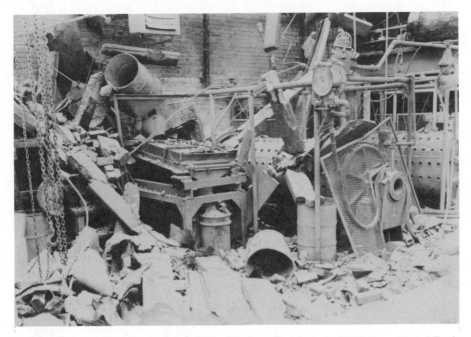

FIGURE 6.2. The result of a metal dust explosion. Copyright, © 1986 Factory Mutual Engineering Corp. All rights reserved. Reproduced by permission.

explosive decomposition of concentrated hydrogen peroxide has been *encouraged* by small amounts of dust. The unexpected presence of a catalyst in process equipment, or elsewhere, may lower the energy requirements to such a value that a fire can occur even though the temperature is well below the normal AIT value.

Fuels burning at an explosive rate cannot be adequately discussed without including the term BLEVE. These result from the puncture, or rupture, of a fuel container with a resultant release of gases or vapors and a sudden explosion upon ignition of the fuel. The ignition source can be several feet distant from the source of the vapor. The vapors travel until they contact the ignition source where they ignite and by a phenomenon called *flashback* the flame travels back the vapor trail to the original source of the vapor and then involves the source in the fire. All of this can happen in a matter of seconds and can result in very damaging explosions when the original source is ignited. Of course, the farther the vapors have to travel to reach the ignition source the less chance there is of a BLEVE occurring. This is true because as the vapors travel toward the ignition source they will be undergoing a dilution effect by mixing with the air. If these vapors undergo enough dilution with the air the resulting mixture will be below the LFL of the fuel when it reaches an ignition source. On one occasion in Fertile, Minnesota, in 1975 the ignition source was closer. During a train derailment a tank car containing liquid propane gas (LPG) developed a leak and the spilled fuel was ignited to form a BLEVE. The tank car was hurled nearly 700 ft through the air. During its ensuing crash landing it hit one house, ruined an automobile parked adjacent to the house, and proceeded to ruin a neighboring house, finally stopping on top of a second automobile(6).

BLEVEs in laboratories usually are the result of containers of flammable liquids being exposed to fires. The heating of the container simultaneously causes boiling of the liquid and a weakening of container walls. The resulting vapors generate an internal pressure, which causes the weakened wall to rupture.

A BLEVE can occur with any size or shape of container that contains a flammable liquid or gas. The container may be a 55-gal drum or a 20-mL bottle. The force of such an explosion will by necessity be smaller but even a 20-mL BLEVE can deliver significant material and personal damage. A BLEVE is characterized by the occurrence of a fireball. High's equation permits an estimate of the size of the fireball produced by a given quantity of fuel(7):

$$D = (3.86W_F)^{0.32}$$

where D = diameter of the fireball in meters
      W_F = quantity of fuel in kilograms

Thus, 20 mL of a flammable liquid could produce a fireball with a diameter approximately 0.4 m (assuming a fuel density of 0.83 g/mL).

Such a calculation could be useful when doing planning and design work as well as determining emergency evacuation distances. As we work with flammables we must always be mindful of the potential of a BLEVE. This means keeping the minimum quantities of flammables in the workplace. Quite obviously we want to minimize the possibility of having any container exposed to a fire situation. Consequently, the proper storage of flammables takes on additional importance.

6.4   THE FIRE TRIANGLE (TETRAHEDRON)

Professional and volunteer fire fighters are made familiar with a principle called the fire triangle. Understanding the fire triangle is valuable in fire prevention and fire fighting. The basis of the principle is that three essentials (sides) are needed in order for a fire to begin. Three essential ingredients of a fire triangle are fuel, oxidizer, and ignition energy. See Fig. 6.3. Only when all three are brought together (and the fuel–oxidizer ratio is appropriate) can a fire be initiated. When fighting a fire the principle is reversed and one of the sides of the triangle, usually the fuel and/or oxidizer, is removed and thus the fire is extinguished. Some fire fighters more recently have

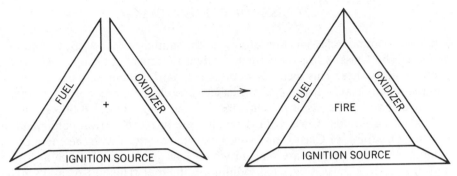

FIGURE 6.3. Fire triangle. The three elements of a fire are fuel, oxidizer, and ignition source. Only when all three are brought together can a fire be initiated.

expanded the fire triangle to a tetrahedron (four-sided) when discussing fire fighting. The fourth side involves the chemistry of continued burning. Once a fire is ignited the fuel proceeds through a series of chemical reactions until it is converted to the final products of burning.

As an example, the chemical equation for the complete oxidation (burning) of methane gas (CH_4) is as follows:

$$CH_4 \;+\; 2O_2 \;\rightarrow\; CO_2 \;\;+\; 2HOH$$

methane   oxygen   carbon dioxide   water

An examination of this equation reveals the burning process must be more involved than it first indicates. The methane molecule has four hydrogen atoms attached to the carbon atom, whereas the product carbon dioxide has no hydrogen atoms attached to the carbon atom. This indicates that before oxygen can attach to the carbon atom the hydrogens must be removed from the carbon. This hydrogen removal can be indicated as follows:

$$CH_4 + energy \rightarrow CH_3 \!\cdot\; + H\cdot$$

The $CH_3\cdot$ is called the methyl free radical. (The dot represents an unpaired electron.) This free radical then proceeds through a series of reactions, forming other free radicals, which eventually result in the production of carbon dioxide and water with the release of energy. The released energy then generates more free radicals from the remaining fuel, thus sustaining the fire. The removal of the free radical from the burning process would also extinguish the fire, as would the removal of the fuel or oxidizer. The possibility of interfering with the burning process by removal of the free radicals is the fourth side of the fire tetrahedron when considering possible methods of extinguishing a fire. How to accomplish this removal of free radicals as well as fuels and oxidizers will be discussed in Section 6.6.

6.5   CLASSES OF FUELS AND FIRE

The most common classification of fires is the familiar Class A, B, C, and D method. The Class A fire is one that involves the burning of solid fuels that leave behind ashes. This includes wood, cardboard, rubber, and plastics. As a mnemonic, Class A fires—"A" for ashes. The class B fire involves burning liquids or gases. It is easier to remember if you think of class B fires involving "b"oiling liquids. Class C fires involve energized electrical equipment. For this one think of Class C and your possible "c"ontact with the energized equipment. Class D fires are ones in which flammable metals, such as lithium, magnesium, potassium, and sodium are the fuel. The NFPA, in it's 704 system of labeling hazardous materials, classifies flammables according to the severity of the fire hazard(8). The NFPA-704 system uses numbers, 0–4, to indicate the severity.

4	Extremely flammable
3	Ignites at normal temperatures
2	Ignites when moderately heated
1	Must be preheated to burn
0	Will not burn

Table (6.4) gives the NFPA rating for several chemicals. Others can be found in the Alliance handbooks and the NFPA.(5, 9).

6.5.1   FUELS

A person can memorize the principle of the fire triangle and know the best fire prevention technique is to avoid having the components of a fire come together but that person can accomplish little in the realm of fire prevention if knowledge of what constitutes a fuel is lacking.

Carbonaceous fuels, those containing carbon, are the most common of fuels. They can exist as solids, such as wood, coal, paper, plastics, and so on, and as mentioned previously, create Class A fires. The carbonaceous fuels can also occur as liquids. Examples are alcohols, acetone, gasoline, and many of the organic solvents. These organic liquids are fuels for Class B fires. As suspected, the carbonaceous fuels can also exist in the gaseous state and cause Class B fires. Methane, ethane, and acetylene are the more common examples (the mnemonic fails us here).

When dealing with Class B liquid fuels one needs to remember these liquids vaporize to form an invisible vapor that is more dense than air. These denser than air vapors will flow along the ground or the floor in a manner similar to a liquid; vapor travel much more than 100 ft is possible. These are the vapors that can become involved in the flashbacks mentioned earlier. The severity of vapor flashback is highlighted in a report which states that 25% of all fires in educational institutions are initially fueled by flammable vapors(10). This should serve to highlight an often heard statement in fire prevention talks, "It's the vapors that burn, so vapor control is the real problem." The flashback phenomenon can make the task of separating the three legs of the fire triangle a difficult one that requires careful planning as the storage and use of flammable liquids is being contemplated.

Although the metals are involved in fire less often than the carbonaceous fuels, they still represent problems which make a metal-fueled fire a special problem and one to be treated with caution.

The first group of metals includes what we call the active metals. Sodium (Na), potassium (K), and lithium (Li) are the main metals of this group. They are called *active* because of the ease with which they undergo chemical reactions by losing electrons. One of these reactions is the metal reacting with oxygen. This reaction will occur simply by exposing the metal to atmospheric oxygen.

$$metal + oxygen \rightarrow metal\ oxide$$

In the case of potassium, potassium peroxide, an explosive, is formed also. This reaction with atmospheric oxygen to form the oxide or peroxide necessities that special methods of storage and handling be used.

The oxides are formed as a white solid which, if formed during an uncontrolled fire, can be spread in the atmosphere as a corrosive dust that presents a problem to all those who may come in contact with it. The problem centers around the following reaction:

$$metal\ oxide + water \rightarrow base\ (caustic)$$

The significance of this reaction is that the active metal oxides all can react with water to form the corresponding base, NaOH, KOH, or LiOH. All three are highly corrosive. When these metal oxide powders come in contact with the eyes or skin or are breathed into the respiratory system they react with the moisture present to form hydroxide, which then corrodes the victim. Potassium peroxide is a strong oxidizer and an explosive. To avoid its hazards store only minimum quantities, or none, and prevent all contact with air.

A second chemical reaction involving these metals with water also contributes to the fire hazards of the active metals.

$$metal + water \rightarrow hydrogen\ gas + base + heat$$

This reaction occurs violently, even when the water is cold, releasing the corresponding base, hydrogen gas, and energy. The hydrogen gas is itself a fuel and will ignite explosively at its room temperature LFL of 4% (< 4% in a fire). The heat released by the initial reaction is sufficient to ignite the

4% (or lower) hydrogen–air mixture. The total heat released changes some of the remaining water to steam—which carries with it some of the hot caustic. Clearly, these metals should not be stored in an area with a water sprinkler system; water should not be used as an extinguishing agent for metal fires.

A second group of metals, less active, includes aluminum (Al), magnesium (Mg), zinc (Zn), and zirconium (Zr). When hot, however, these metals will react with water to produce hydrogen gas and the corresponding bases. The reaction can be almost as violent although the corresponding bases, such as aluminum hydroxide, are less corrosive than those of the active metals. Note that in this context, *less active* refers to less active than, say, metallic sodium. Thus, an NFPA report states that the reaction of metal powders, such as Al and Mg, with water is "theoretically capable of producing a slightly more violent explosion than nitroglycerine"(11). This gives an indication of the hazard potential when working with these metal powders. In a nondust form, however, they are relatively difficult to ignite. When these metals are once ignited the extinguishing of the fire presents special fire fighting problems.

Magnesium fires are familiar to the public; older camera flashbulbs utilized burning Mg. The Department of Transportation's method of classifying the flammability of materials according to their flash points places Mg as *unclassified*. The classification method labels those with flash points < 104°F (40°C) as *flammable,* those with flash points between 104 and 199.4°F (40–93°C) as *combustible,* and those with flash points > 199.4°F (93°C) are labeled as *unclassified*. Magnesium with its dust ignition temperature of < 650°C would be labeled as unclassified. However, it does burn with a brilliant white flame that emits harmful ultraviolet (UV) radiation. When burning, it combines with atmospheric oxygen as follows:

$$\text{magnesium} + \text{oxygen} \rightarrow \text{magnesium oxide}$$

It also reacts with nitrogen to form magnesium nitride. Approximately 25% of the magnesium will react with nitrogen in the atmosphere and the remaining 75% will react with the oxygen of the atmosphere. This reaction with nitrogen adds complications if a magnesium fire is to be extinguished by allowing it to deplete the oxygen supply, or by adding normally inactive nitrogen.

6.5.2 Dust Explosions

In Section 6.5.1 we discussed the flammability of certain metals. When those metals, and almost all others, are dispersed as dust they can burn with explosive force, if a suitable ignition source is present. Exceptions to the last statement are the powders of copper, certain lead powders, nickel, some impure irons, and the noble metals. The powders of zirconium, magnesium,

magnesium alloys, aluminum, and titanium are considered to be the most hazardous. An intermediate group of metal powders are zinc, silicon, and carbonyl iron. The powders of cadmium, copper, chromium, lead, and milled iron have a low flammable and explosive hazard rating. The explosive potential of the flammable metal powders warrants careful consideration when the possibility of such powders exists. At a plant in Centerbrook, Connecticut, a magnesium powder explosion killed four workmen who were attempting repairs on a closed system for the manufacture of magnesium powder. The inert atmosphere of helium, which was kept in the system, had been partially replaced by air(12, p. 25).

Metal powders have been known to explode spontaneously. A violent explosion occurred in a drum containing partially wet, finely divided, and contaminated scrap zirconium. Two employees were killed and a third lost an arm as a result of the spontaneous explosion(13).

The fire hazards of other dust materials have been known for some time but they have taken on new importance as new materials, such as plastics, have been introduced and manufacturing facilities have grown larger.

Although we have discussed organic fuels they deserve another inspection as a result of organic dusts acting as explosible fuels. A spilled container of organic solvent will probably be cleaned up as quickly as possible to avoid a fire but a spill of a flammable powder, or dust, may go unnoticed, yet alone be cleaned up. It may be a fire hazard of explosible proportions just waiting for an event that can cause it to become airborne close to an ignition source. Bodurtha estimates that generally, the minimum explosible concentration (MEC) of finely divided solid organics in the air is about 20 mg/L(14, p. 137). A listing of some common dust problems and explosible concentrations are given in Table 6.5. Since most dusts can burn, a dust abatement program is an essential component of a good fire prevention program.

Much to the dismay of science and industry the solution to one problem often creates a new problem. Such is the case in pollution control. In an effort to remove particulate matter from discharges into the atmosphere and to obey water pollution laws, industry has moved toward the use of dry collection devices to remove particulate matter from stack exhausts. These dry methods may involve the collection of dry flammable dusts that have caused fires and explosions.

6.5.3 Oxidizers

We often fail in our efforts to keep fuels and oxidizers isolated from each other because of an incomplete understanding of what an oxidizer is and just what chemicals can act as oxidizers. Of course, if we do not recognize a chemical as an oxidizer we can very well place it close to, or even mix it with, a fuel. My personal experience has allowed me to question thousands of chemical workers on the topic of oxidizers. Very few of them could define an oxidizer and even fewer could name an oxidizer other than oxygen.

Table 6.5. Common Chemical Dust Problems and Explosible Concentrations

Chemical Dust	Explosible Concentration $(oz/1,000 \ ft^3)$
Aluminum	45
Alkyl alcohol resin	35
Cellulose acetate	25
Coal (low volatile)	125
Coal (high volatile)	55
Methyl methacrylate	20
Magnesium	10
Polyamide	30
Polyethylene	20
Polystyrene	15
Tin	190
Titanium	45
Zinc	480

In a chemical definition we could define the burning process as one where oxidation–reduction (redox) has occurred. Redox involves the transfer of electrons from one chemical species to another. For example, in the burning of sodium (Na) metal in oxygen each Na atom loses one electron to oxygen. The Na has served as a reducing agent and the oxygen has been the oxidizer, or oxidizing agent. Likewise, in the reaction of methane (CH_4) with oxygen (O_2) the carbon of the methane molecule transfers, or loses, electrons to the oxygen to give the following reaction:

$$CH_4 + 2O_2 \rightarrow CO_2 + 2HOH$$

Atmospheric oxygen is a common oxidizer but certainly not the only one. Many oxidizers do not even contain oxygen. Chlorine gas is a strong oxidizing agent; it does not contain oxygen. An oxidizer must be able to accept electrons from the fuel, the reducing agent. Therefore, for this discussion, let us broadly define an oxidizer as any chemical that reacts with a fuel by accepting electrons. This definition permits us to classify water as an oxidizer when mixed with metals, particularly the active metals. This definition is not far from that generally used by fire officers: An oxidizing agent is any substance that reacts with combustible materials with the evolution of heat(12, p. 63). We consider this to be our operating definition.

A closer examination of the reaction of a metal with water will reveal that the water has accepted electrons from the metal. As strange as this may seem, water is an oxidizer in this case and quite obviously is not to be mixed with the fuel as explained previously. The exclusion of water as an oxidizer could possibly be based upon the argument that no fire was produced by the metal and water reaction but rather it occurred as a result of another oxidization reaction, the burning of the hydrogen gas produced by the first reaction. Regardless of how you look at it, there has been a fire produced.

Oxidizers occur as compounds in all three physical states and many have uses which make them common to most chemical laboratories. Some of the more common ones are listed in Table 6.6.

Most of our experience with oxygen as the oxidizer has involved it as atmospheric oxygen at a 20% concentration. As a result we tend to forget the rate of a chemical reaction is related to the concentration of the reacting chemicals, in this case oxygen. In an atmosphere of 40% oxygen a lighted cigarette or even its hot ashes can ignite clothing. Cloth material that has had liquid oxygen spilled on it will absorb oxygen; it remains oxygen rich for several hours during which time cigarette ash can still ignite it. Liquid oxygen in contact with combustible materials can cause spontaneous fires to erupt. When liquid oxygen soaks into relatively noncombustible asphalt the combination is a high explosive, sensitive to heat or shock.

Now that we understand the need to separate fuels and oxidizers we should consider the hazards of the next topic: Chemicals whose composition includes both a fuel portion and an oxidizer portion in the same molecule. That is, each molecule is both the fuel and oxidizer. It may come as no surprise that this type of compound is one of the leading causes of industrial fires and explosions. Some examples of this type of compound are organic nitro compounds, organic nitrates, organic chlorates, and organic chlorites. Table 6.6 shows all of these compounds as being oxidizers, but for each of these, the organic portion of these molecules is a fuel. Ammonium compounds that contain these oxidizers may be explosive. Thus, ammonium nitrate is used as an explosive in the coal mining industry. One of the

Table 6.6. Examples of Commonly Occurring Oxidizing Agents

Gaseous	Liquid	Solid
Oxygen	Hydrogen peroxide	Ammonium nitrate
Fluorine	Nitric acid	Ammonium nitrite
Chlorine	Perchloric acid	Perchlorates
Ozone	Bromine	Peroxides
Nitrous oxide	Nitric acid	Chromates
Oxygen difluoride	Sulfuric acid	Dichromates
		Picrates
		Permanganates
		Hypochlorites
		Bromates
		Iodates
		Chlorites
		Chlorates

world's largest nonintentional explosions involved ammonium nitrate (Texas City, 1947). This combining of the fuel and oxidizer in the same molecule presents safety problems that require isolation of that compound from all other chemicals. It is also desirable to have a cool storage location. Since many of these compounds may be shock sensitive, it is inadvisable to use "chipping" as a method of breaking up lumps of these compounds.

Some fire extinguishing agents can act as oxidizers. Thus, nitrogen (N_2), carbon tetrachloride (CCl_4), and other halogenated hydrocarbons have been used. (Carbon tetrachloride is no longer an acceptable extinguishing agent because of its toxicity and its ability to form even more toxic chemicals when used on a fire.) In the case of a lithium fire nitrogen would react as an oxidizer as follows:

$$6Li + N_2 \rightarrow 2Li_3N$$

Magnesium reacts similarly; nitrogen would be a poor fire extinguishing agent for such metal fires.

Halogenated hydrocarbons (RX) can react with active metals to produce carbon and the metal halide (MX).

$$RX + M \rightarrow C + MX$$

It should be noted that the carbon so produced is a fuel material also. All of the halon fire extinguishing agents can react in this manner. They should not be used on metal fires.

Carbon dioxide is a commonly used fire extinguisher but it also has limitations, acting as an oxidizer when in contact with burning metals such as sodium, lithium, potassium, and aluminum. With the first three it reacts to form metal oxides, such as sodium oxide, and carbon. With aluminum it forms aluminum oxide and carbon monoxide. Carbon monoxide is a toxic substance as well as being another fuel. Sand (SiO_2) is sometimes used to extinguish sodium and potassium fires but should not be used on lithium fires because of the redox reaction between the two chemicals:

$$SiO_2 + 4Li \rightarrow 2Li_2O + Si$$

Again, the intended extinguishing agent is an oxidizer for the burning metal to which it is being applied. The topic of extinguishing metal fires is discussed further in Section 6.6.4.

6.5.4 Ignition Sources

As you will remember, the third side of the fire prevention triangle is the ignition source. Also, we mentioned earlier that the fuel and oxidizer molecules must come in physical contact before they can react. But not all contacts between these two lead to a chemical reaction. There also must be enough energy present to break the old chemical bonds in the fuel and oxidizer before the new chemical bonds of the product can be formed. When the molecular collision does not result in a reaction, the molecules did not have the prerequisite amount of *activation* energy to break the old bonds. If no molecules possess this required energy then no bonds will be broken.

The role of the ignition source then is to provide the required activation energy. If the required activation energy is not available then the fuel and oxidizer can coexist without burning. This explains why we say fires can be prevented by eliminating the ignition source from the fire triangle.

Many fires and explosions occur because people either are not aware of just what energy sources can act as ignition sources, or they become careless with common, everyday items such as matches and lighted cigarettes, or they use open flames for heating flammable chemicals. They can also fall victim to a series of events, each of which may not lead to a fire but when combined an accidental fire can occur. One absurd example that occurred in a college biochemistry laboratory involved an open bunsen burner flame at one end of a lab bench, a beaker of ether at the other end of the bench, and a mouse. For unknown reasons a student dropped the mouse into the beaker of ether. The mouse jumped out of the beaker and proceeded to run, wet with ether, along the bench top and toward the open flame where it caught fire. This was an unusual manner in which to bring the three sides of the fire triangle together but some of the things that occur in other laboratories can be almost as strange and also lead to more serious fires.

A somewhat insidious event can occur when fuels and oxidizers are brought together in the presence of a catalyst, such as platinum black. A common material, ferric oxide powder (rust), acts as a catalyst to lower the autoignition temperature of many combustible organic compounds(15). The catalyst can reduce the required activation energy to a point where the fuel ignites at its existing temperature. Table 6.7 is a list of some sources that are capable of causing ignition.

6.5.5 Pyrophorics

A related topic that needs to receive more attention than it usually receives is that of pyrophoric chemicals, those chemicals that rapidly undergo air oxidation and spontaneously ignite in air (see Section 7.3.9) Such chemicals thus do not require an external ignition source. All such chemicals should be stored in tightly closed containers and when being transferred they should be kept under an inert atmosphere or liquid. A partial listing of pyrophoric chemicals appears in Table 6.8. Again, failure to properly store and handle these chemicals can lead to severe and unexpected fires.

Table 6.7. Some Ignition Sources

Spark from an electric device, such as piezoelectric generator, switch, socket, motor, bare conductor, and so on

Spark from friction, such as flint lighter, metal grinding, and so on

Flame

Hot plate

Static electricity

Cigarettes and cigarette ash

Hot object, such as pipe, radiator, lightbulb, heater, chemical reaction vessel, glowing ember, overheated motor, and so on

Laser

Heat from friction

Catalyst

Table 6.8. Partial Listing of Pyrophoric Chemicals

Alkali metals
Calcium and magnesium
Metal alkyls and aryls
Metal carbonyls
Powders of aluminum, cobalt, iron, and manganese
Metal hydrides
Boranes and arsine
White phosphorus

6.5.6   Hypergolics

Another group of chemicals that also can lead to unexpected fires are those called hypergolic mixtures. These are pairs of chemicals which, when mixed, create enough heat of reaction to cause ignition to occur. Again, an ignition source is not needed. Fire prevention becomes more than just avoiding ignition sources, it entails the task of making certain the hypergolics are not mixed.

There are many examples of such mixtures. Perchloric acid and magnesium powder produce a white-hot flame within microseconds of mixing. The same is true of nitric acid and phenol. Acetone, plus 85% nitric acid, constitutes a hypergolic mixture, as does concentrated nitric acid and triethylamine. Red fuming nitric acid forms such a mixture with many different aromatic amines. Divinyl ether ignites after 1 ms when mixed with 96% nitric acid which contains sulfuric acid at concentrations $> 5\%$. A 90% solution of potassium permanganate in red fuming nitric acid will ignite when mixed with alcohols. Ketones, esters, and liquid alcohols form hypergolic mixtures with solid potassium permanganate.

The short time interval from when hypergolics are mixed until they ignite justifies spending time before mixing to discover if you are working with such chemicals. For a more complete discussion of this topic refer to Chapter 7.

6.6   PREVENTING AND EXTINGUISHING FIRES

In previous sections of this chapter we covered information that made you more aware of the chemical nature of fire, what is needed for one to start, and special situations such as pyrophorics and hypergolic mixtures. In this section we shall apply this knowledge to prevention and extinguishing of fires.

6.6.1   Storage of Flammables

Chapter 11 of this text covers in detail the topic of storage. This section addresses some specifics essential to the proper storage of flammable materials. To emphasize the importance of proper storage of flammables let us consider the fact that the average laboratory is more likely to be storing a wide variety of flammables than is a production facility. The quantities stored in the laboratory are usually larger also. This helps to explain why 25% of fires in academic laboratories are initiated by flammable vapors(16). Some of the keys to proper storage are to isolate and keep separate, store minimum quantities, and provide adequate ventilation.

The first key, isolate and separate, is one we have already discussed. Keep the fuels and oxidizers as far apart as possible. The new color-coded labels that are being used by many suppliers of chemicals can be of tremendous assistance in accomplishing this task. A frequently found work habit of laboratory personnel is that of storing their chemicals close to the area where they are to be used. From an efficiency point of view this may be a reasonable practice but if it places incompatible chemicals, such as flammables and oxidizers, close to each other then it may be a practice that needs to be eliminated. Another practice, one used by workers who are more inclined to be orderly, is to store their chemicals alphabetically. Using this method of storage also can place flammables in close proximity to oxidizers, for example, acetic acid close to ammonium nitrate. A mixture of these two chemicals will ignite when warmed. Aluminum metal powder and antimony trichloride could be neighbors if stored alphabetically. Aluminum burns in the presence of antimony trichloride vapor. Another example of incompatible chemicals which could be stored close to each other are aluminum powder and carbon disulfide. These also will spontaneously burst into flame if mixed. To minimize fire hazard potential, alphabetical storage of chemicals is to be avoided.

It takes very little flammable chemical to start a fire, which can quickly spread to cause huge losses of property and life. Our nature as workers in laboratories tends to make us forget this fact and thus to accumulate quantities of flammables that are in excess of what is needed to be a functional laboratory and far more than can be stored safely. Just what constitutes the minumum safe quantity depends on several factors. Among these are such things as the flammability of your chemicals, storage facilities, toxicity of your chemicals, and rate of use. If your storage area is limited you must exercise control over the quantities purchased. Otherwise, you will create storage problems.

Various codes and regulations have been written to insure that we store only safe quantities in our laboratories. The Occupational Safety and Health Act (OSHA) makes definitive statements concerning the construction of inside storage rooms for flammables(17). The walls, ceilings, and floors are to be constructed of materials with at least a 2-h fire resistance. There should be self-closing Class B fire doors(17). The standards apply to the storage of Class II and IIIA combustible liquids and flammable liquids. All flammable storage rooms are to have mechanical ventilation, which is controlled by a switch located outside the door and should be equipped with explosion-proof lighting and switches. It is recommended that the ventilation intake be located within 12 in. of the floor and have an exhaust capacity of one cubic foot per minute per square foot of floor area. Regardless of the room size there should be a minimum flow rate of 150 ft³/min. More information on the complete details of this can be found in refs. (15) and (18).

The codes and regulations also deal with how the flammables will be stored inside the room. The NFPA requires that safety storage cabinets be used for the storage of Class I and II liquids if the quantity exceeds 10 gal/100 ft² of floor space. However, the 100-ft² floor area is not fixed but rather varies, depending on the construction and the fire protection available in the laboratory. The literature should also be consulted to find the standards for the cabinet and room construction. The NFPA standard (No. 30)(2) gives information concerning quantities that can be stored outside an approved cabinet.

In one recent publication it is recommended that "Whenever feasible, quantities of flammables greater than 1 liter should be stored in metal containers"(19). These metal containers are more commonly called safety cans and have certain structural features that make them especially suited for the storage of flammable liquids. They are constructed of stainless steel or coated steel and are equipped with spring-loaded lids. The spring is such that it will vent when the internal pressure of the can becomes 5 psig. This is enough spring tension to prevent liquids or vapors from escaping at normal temperatures. Of course, there is always the possibility that the spring can weaken, so safety cans need to be checked for tightness. This can be done by filling the can and holding it on its side over an appropriate vessel. If the liquid escapes at a drop rate greater than four drops per minute it does not meet the requirements of the Factory Mutual Insurance Company. Excess leakage can be the result of a faulty gasket as well as a weak spring. If so fitted, the flame arrestor's function is to act as a heat sink to prevent a flashback from reaching the inside of the safety can. Both the flame arrestor and spring-controlled lid can present frustrating situations when the can is being used. Because of this, laboratory personnel need to understand the purpose of each and not remove them to make use of the can a little easier. Also, if the liquid level is high enough so as to immerse the flame arrestor, it cannot act as a heat sink. For this reason care must be exercised to be certain the safety can is not filled to the flame arrestor. Contrary to the popular myth the flame arrestor is not a screen to filter the liquid. Safety cans that are no longer suitable for the storage of flammable liquids should be painted a color other than red if they are to remain in service for the storage of other liquids. Another potential problem with safety cans is a tendency of workers to not label them adequately, especially if the can is to be used for waste flammables. If flammables must be stored in glass containers instead of a safety can, put the container in a sturdy, flexible polyethylene container or other type of bottle jacket.

6.6.2 Vapor Control

As we have already discussed, it is vapors that burn. This, coupled with the concept of flashback, justifies further examination of the topic of vapor control. In the previous section we mentioned the requirements for ventilation in a storage room. This, of course, was to aid in vapor control in the storage room. Now let us examine the control of vapors in the laboratory. It should be obvious that the faithful use of an adequate fume hood system is a good way to control the vapors of flammables and keep them below their LFL value. This is not to mention the value of controlling them because of their toxic nature as well (see Chapter 16).

Those workers who insist on leaving containers of chemicals open, or working with large volumes of flammables in open containers need to be made aware of the fire hazards associated with such practices. Usually, their lack of understanding of the dangers of vapors is the cause of their carelessness.

Chemical spills can be the initial source of the vapors which cause laboratory as well as industrial fires(20). An explanation of this lies in the fact that we often do not prepare for accidents until after they occur. In this case we are not prepared to clean up a flammable liquid spill until it has happened. This is often too late. The knowledge of what is a suitable cleanup material is usually not readily available and the correct material is not available close to the area of the spill. While a search is being conducted to locate the proper material with which to clean up the spill, the vapors are "pouring" rapidly along the floor or ground. If they contact a suitable ignition source a flashback can occur, and uncontrolled fire can result.

The recent trend among chemical suppliers to print on the label the name of a suitable spill cleanup material is to be commended and encouraged. That material should be readily available before the spill occurs. Be certain that the spill cleanup material will control the vapors as well as absorb the liquid. Many materials that are often used, such as diatomaceous earth, or kitty litter, do not control the vapors. As a result the vapors continue to be released to the atmosphere, where they are the source of fuel for a fire.

A relatively new and insidious source of vapors has been introduced as a result of our concern to protect the ozone layer of the atmosphere. This concern has brought about the substitution of low molecular weight hydrocarbon compounds (such as propane) for freon as the propellant in aerosol cans. Since these hydrocarbon propellants are flammable gases, care must be exercised in how and where these aersol cans are used. Their disposal, whether they are full, partly full, or supposedly empty, also can cause harm to personnel close by.

6.6.3   Avoiding Ignition Sources

Many ignition sources are easily detected and thus avoided. Among these are open flames, welding, matches, sparks from tools, and cigarettes. Those that are more insidious are the ones that require special attention. The chief culprit of these probably is static electricity. Its generation usually goes unnoticed and its presence is usually discovered as an electrical discharge which contains enough energy to ignite the vapors of flammable liquids. The more energctic sparks are capable of igniting dusts as well as vapors. Because of this insidious nature it is essential to understand how it is generated and how its discharge as a spark can be avoided. For a more detailed discussion of this topic refer to NFPA 77(21).

Static electricity is generated by the contact and separation of an electrical conductor and a nonconductor or of two nonconductors. Examples are the movement of a liquid in its container during shipment, the movement of liquid through a hose or spout, and the conversion of a liquid into a mist or vapor. The rubbing of certain cloth fabrics and the movement of machinery are also sources of static electricity.

Since the generation of static charges is insidious in nature it becomes essential that we take steps to control it. Bonding and grounding are the main methods of avoiding the hazards of static electricity. A spark occurs only if there is a difference in the electric potential of two objects. The equalization of potential is accomplished by bonding. Since the two objects which are bonded can still have a potential with respect to the earth it is also necessary to ground the objects as well. Grounding allows the static charge to leak into the earth and thus eliminates a potential from developing. Number 8 or 10 AWG wire is the minimal acceptable size for grounding.

When making bonding and grounding connections it is vital to assure that metal-to-metal contact is made. Too often workers fail to penetrate a layer of paint or rust and thus fail to make electrical contact because neither

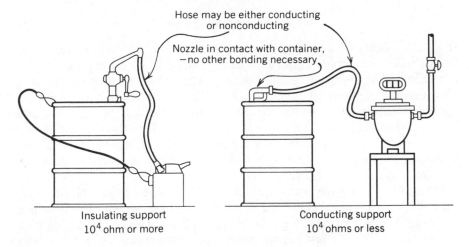

Hose may be either conducting
or nonconducting

Nozzle in contact with container,
—no other bonding necessary

Insulating support
10^4 ohm or more

Conducting support
10^4 ohms or less

Bond wire necessary except where containers are
inherently bonded together, —or arrangement is
such that fill stem is always in metallic contact
with receiving container during transfer

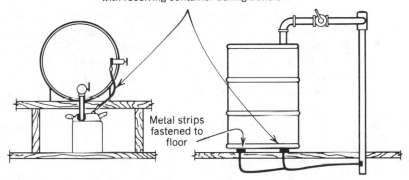

Metal strips
fastened to
floor

FIGURE 6.4. Bonding during container filling permits discharge of any static electricity generated. Reproduced with permission of Alliance of American Insurers.

substance is an electrical conductor. Figure 6.4 illustrates how bonding can be accomplished.

We all probably have experienced the discharge of static electricity as we touch a metal door knob after walking on carpeting on a cold, low-humidity day. The presence of high humidity permits a thin layer of water to form on the surface of the carpet and thus makes it more conductive. On a day of low humidity the carpet is less conductive and thus a charge builds up. The same theory applies to other nonconductors, such as concrete, plastics, and paper. Increasing the humidity of the work place then is one way of avoiding the generation of static electricity. A relative humidity of ~ 60 to 70% is required, to be of much value (14, p. 38).

6.6.4 Extinguishing Fires

If all our efforts to avoid having an unwanted fire fail, then we are confronted with the unpleasant task of extinguishing it. Referring back to the fire triangle, only now let us call it the fire tetrahedral, we soon see we have four general methods of fighting a fire. These are, cool it to below its ignition temperature, remove the fuel source, remove the oxidizer, and interfere with the chemical process of burning. Which method we use should be influenced by such factors as the type of fuel, that is, what class of fire we have, and the nature of other fuels that might become involved in the fire if the wrong extinguisher is used.

However, a more important question than the one of the type of extinguisher that is required is the one of whether to even attempt to extinguish the fire. In other words, "fight or flight." Most experienced fire fighters recommend that your first actions should be to sound the alarm, to secure the room by closing all windows and doors, and to vacate the area. After that, you may decide to attempt to extinguish the fire while waiting for the firemen to arrive. Only when a fire presents an immediate life-threatening situation should the first steps be to fight the fire.

In a situation where an extinguisher is to be used, then the questions of what type and size extinguisher should be available become more important. The fact that the extinguisher needs to be readily available cannot be overemphasized. Of almost equal importance is the question, do you know how to use the fire extinguisher? Fire fighting is no place for on-the-job training! All personnel should be trained on the proper use of fire extinguishers. The decision as to what type of extinguisher needs to be available should be based upon the materials in the laboratory; Class A, B, C, or D. The common agents for fire fighting are water (Class A), carbon dioxide (Classes B and C), dry chemical powder (e.g., potassium or sodium carbonate) (Classes B and C), dry chemical powder (e.g., monoammonium phosphate) (Classes A, B, and C), and halons (Classes B and C). Metal fires (Class D) are usually extinguished with sand, sodium chloride, or MET-L-X (trade name).

6.6.5 Water Extinguishers

Water's ability to extinguish Class A fires rests on its ability to cool the fuel and to exclude oxygen. That is to say, water can remove two legs of the fire tetrahedron. It excludes, or dilutes, oxygen by the formation of steam. Its high heat capacity accounts for its cooling ability.

Water has several advantages as a Class A extinguisher. Among them are its low cost, easy availability, and the relatively long distance it can be sprayed. The more distance you can keep between you and the fire, the safer you are. Some of its disadvantages are, it should never be used on Class B or C or D fires. Its use on Class B fires can lead to a worsening of the fire situation because most flammable organic liquids have a density less than that of water. As a consequence the fuel floats on the water and is spread as the water flows to lower levels. While floating on the water's surface the fuel's oxygen supply is not disturbed nor is the fuel cooled. Water has been used on Class B fires as a fog and in copious quantities. Water cannot be used on Class C fires because the water, with its dissolved solids, will act as an electrical conductor and thus endanger the life of the firefighter. Its use on Class D fires is to be avoided because of the chemical reaction which can occur between water and the metal (see Section 6.5.3). Another disadvantage of water in cooler climates is its relatively high freezing point. This difficulty can be circumvented by the addition of a soluble salt to lower the freezing point. The addition of such an electrolyte salt makes the danger of using water on a Class C fire even more severe. Water also suffers because of its lack of ability to cling to vertical surfaces. Adding thickening agents can alleviate this problem.

Portable water extinguishers usually have a maximum capacity of 2.5 gal. Such portable extinguishers will have a range of 30 to 40 ft and the discharge rate will be approximately 3 gal/m. This short operational time emphasizes the importance of having had "hands-on" training with such an extinguisher before the fire. In general, the operation is as follows; pull out the ring pin, grasp the nozzle, aim directly at the fire, and squeeze the lever. The fuel should be completely soaked after the fire has been extinguished to prevent a flare-up of the fire from smoldering fuel.

The effectiveness of Class A extinguishers has been standardized by the Underwriters Laboratories (UL). To do this the UL calls a fire that requires 1.25 gal of water to be extinguished a 1-A fire, and a fire that similarly requires 2.5 gal of water a 2-A fire.

6.6.6 Compressed Gas (Carbon Dioxide) Extinguishers

Although any of the inert gases can be used to dilute the oxygen supply, carbon dioxide is the most often used. In addition to the smothering effect it also has a cooling effect on the fuel. It can be used on Class B or C fires. Due to the possible chemical reactions that might occur it should not be used on Class D (burning metals) fires.

$$CO_2 + 4Na \rightarrow C + 2Na_2O$$

A similar reaction will occur if the burning metal is lithium or potassium. If the burning metal is aluminum, then the reaction produces carbon monoxide instead of carbon. Carbon monoxide not only is capable of acting as a fuel but also is a toxic substance. The force with which the carbon dioxide is expelled can cause Class A fuels, especially paper, to be scattered.

There are several advantages of carbon dioxide. It is readily available and cheap, it leaves no residue, and it requires no propellant since it is delivered as a gas.

Two of the main disadvantages are its low efficiency in extinguishing fires and its short range. Both are related in that since it is delivered as a gas it is dispersed quickly in the air. Because of its low temperature it can also cause thermal shock to occur in electronic circuit boards and thus hairline cracks in the circuitry. It is also an asphyxiant. The low efficiency coupled with the fact it is an asphyxiant mandates that a carbon dioxide extinguisher should not be used in confined spaces.

The extremely cold temperature at which the gas is delivered can cause frost bite if it is inadvertently sprayed on the human body. This low temperature can also cause atmospheric water to condense, which could prove a hazard if a water sensitive chemical was involved (22).

To operate a carbon dioxide extinguisher, pull the ring pin, direct the delivery horn at the base of the flame, and squeeze the operating lever. (If the delivery horn is made of metal it is not safe to use on Class C fires.) Move the horn from side to side, keeping it directed at the base of the flame. Due to the extreme cold of the carbon dioxide be certain you grasp only the insulated hand grip. Table 6.9 gives a summary of the ranges and UL ratings of carbon dioxide extinguishers.

6.6.7 Dry Chemical Monoammonium Phosphate Extinguishers

The monoammonium phosphate (MAP) extinguisher is suitable for Class A, B, or C fires. It should not be used on Class D fires. This diversity contributes to the fact it is the type most often found in homes and cars. It functions by melting to form a sticky residue that covers the fuel surface and thus excludes oxygen from the fuel.

The MAP extinguishers are highly effective but they do present cleanup problems because of the large quantity of fine powder which is released. This powder is especially troublesome for electronic equipment. Since MAP extinguishers have no cooling effect on the fuel or other heated objects involved in the fire the usual practice is to spray the area with water after the fire has been extinguished to prevent a flare-up. Do not spray a Class C fire with water until you are certain it has been deenergized! See Table 6.10

Table 6.9. Characteristics of CO_2 Extinguishers

Capacity (lb)	Range of Stream (horizontal ft)	Duration of Discharge (s)	UL[a] Classification
2.5–5	3–8	8–30	A, 1–5 B, C
10–15	3–8	8–30	A, 2–10 B, C
20	3–8	10–30	A, 2–10 B, C

[a]Carbon dioxide (CO_2) extinguishers with a metal delivery horn should not be used to fight electrical fires.

Table 6.10. Characteristics of Dry Chemical Extinguishers

Capacity (lb)	Range of Stream (horizontal ft)	Duration of Discharge (s)	UL Classification
1–5	5–12	8–10	1 to 2A, 2 to 10B, C
9–17	5–20	10–25	2 to 20A, 10 to 80B, C
17–30	5–20	10–25	3 to 20A, 30 to 120B, C

for a summary of the ranges, discharge times, and the UL classification of the various sizes of these extinguishers.

The UL effectiveness rating of this and other types of extinguishers on Class B fires is based on the ability of a nonexpert to use the extinguisher to control a contained area of gasoline fire. Thus, a rating of 20-B indicates it can control a fire area of 20 ft^2. An extinguisher can be used on Class A and B fires that may have a UL rating of 4-A; 40-B: C. This means it can be used on a Class A fire that requires 5 gal of water (4 × 1.25) and on a Class B fire that covers a 40-ft^2 area.

The method of using a powder extinguisher is the same as for a carbon dioxide extinguisher except that the precautions to prevent contact with a cold object are not needed.

6.6.8 Dry Chemical Potassium or Sodium Carbonate Extinguishers

A second type of powder extinguisher contains either potassium or sodium carbonate as the extinguishing agent. It can be used on Class B or C fires. It also should not be used on Class D fires.

These two chemicals will cause saponification of oils and fats. Since this occurs at the fuel surface, vaporization is hampered and thus a reignition is less likely to occur. Extinguishers using these chemicals also owe their high efficiency to the ability of the chemicals to interfere with the burning process by serving as a free radical scavenger (23).

One chief disadvantage is the cleanup of the finely divided powder. Also, when used on electronic equipment it can present the same problems as the MAP extinguisher.

Table 6.10 gives a summary of the properties of this type of extinguisher. You will note that the range is greater than that of carbon dioxide but less than that of water. The method of use is the same as the MAP extinguisher.

6.6.9 Halon Extinguishers

Earlier, in Section 6.4, the production of free radicals was discussed. The removal of these free radicals is the fourth leg of the fire tetrahedron. The halon extinguisher operates by this method.

A *halon* is defined as being a halogenated methane; meaning that flourine, chlorine, and/or bromine have been attached to the single carbon atom of the methane molecule. This is done by replacing hydrogen atoms that were on the carbon atom. The two common halon extinguishers are either bromochlorodifluoromethane (CF_2BrCl) or bromotrifluoromethane (CF_3Br). Each is designated by a set of four digits, which represent the number of atoms of each element present. Thus, CF_2BrCl is designated as 1211 and CF_3Br is 1301. The zero indicates that no chlorine is present.

Both are highly effective extinguishers because of their ability to act as scavengers for the free radicals produced from the fuel. The halon is unstable in the fire and decomposes as indicated in the following reaction to form another free radical:

$$CF_3Br + heat \rightarrow CF_3 \cdot + Br \cdot$$

which then combines with the fuel free radical, preventing it from proceeding with the burning process. All of the halons and their decomposition products have some degree of toxicity associated with the possible exception of tetrafluoromethane (CF_4)(24). The CF_4 molecule is stable and resists the formation of free radicals, which could serve as free radical scavengers. Since it is noncombustible its vapors smother a fire. Tetrachloromethane (carbon tetrachloride) (CCl_4) will extinguish a fire but it also is toxic and will react in a fire to form phosgene, a highly toxic gas. Hence, CCl_4 is no longer used as an extinguishing agent. For the same reasons the compounds dichloromethane (CCl_2H_2) and trichloromethane (CCl_3H) are not suitable fire extinguishing agents. For further coverage of the toxicity of these chemicals see Fawcett and Wood(25).

A 2.5 lb 1301 extinguisher has a range of 5.5 ft, whereas a 1211 of the same size has a range of 7 ft. Both are "clean", leaving behind no residue and causing no thermal shock to sensitive electronic equipment. Because of the related toxicities neither should be used in small, isolated areas.

The UL has given the 1211 extinguisher Class B and C ratings. In the larger capacities, 9-, 16-, and 17 lb capacity, it also has a Class A rating(25). The 1301 extinguisher can be used on the first three classes of fires but is less effective on Class A fires because of its weakness in attacking fires located deep in a burning pile of class A fuel. Such fires require > 20% concentration of 1301 and thus it becomes time consuming and expensive to use 1301. The halons also are not suitable for burning metals (Class D) fires.

6.6.10 Extinguishers for Class D Fires

Class D fires are particularly troublesome to extinguish because of the intensity with which they burn and their ability to react chemically with normal extinguishing agents. As an example, carbon dioxide will not react with cold sodium but accelerates the burning rate of ignited sodium.

Fire fighting equipment and strategy should not include the use of water, foam, carbon dioxide, or halons on fires involving alkali metals (12, p. 93). Graphite, soda ash, and powdered sodium chloride can be used on alkali metal fires but care must be exercised to insure they are dry or the moisture can react with the metal. If wet graphite is used on a lithium fire, an insidious event can occur. The graphite can react with the lithium to produce lithium carbide, which can then react with water to produce acetylene gas, a fuel. Even with this potential, graphite is considered a suitable extinguishing agent for burning lithium (26).

Two special fire extinguishing materials, MET-L-X (trade name) and a ternary eutectic compound (TEC) are effective on all alkali metal fires except lithium. For other metals and compounds that require special extinguishing materials, refer to Bahme (12, Chapter 4).

My personal observation indicates that many facilities that have active metals do not have adequate or suitable Class D extinguishers.

6.6.11 Location of Extinguishers and Telephone

An extinguisher will be of little value if it is not located in a clearly marked site that is readily available. *Readily available* means free of obstruction, positioned at a proper height (no > 5 ft), within a reasonable distance of the potential fire site, and well marked. It is recommended that extinguishers be no more than 30 to 50 ft from the fire site. This distance should be even less if furniture and other objects are located so that a direct path is not open. A fire can quickly become uncontrollable so it is the time needed to reach the extinguisher, not the distance to it that is paramount. Do not place it so close to the potential fire site that it can be involved in the fire and thus be unavailable.

A telephone is normally not considered as a fire extinguisher but if your first line of attack is to call for assistance then it becomes an important part of your fire fighting equipment. Thus, its location is also important. Positioning the telephone should receive the same careful consideration as does the positioning of extinguishers. The flammability of your fuels could justify the installation of an explosion-proof telephone.

6.7 FIRE TOXICITY

Most fires produce toxic products; of these carbon monoxide is well known. Other toxic fire products include hydrogen halides, halogenated hydrocarbons, partially oxidized hydrocarbon, and alcohol derivatives such as acrolein and other aldehydes, benzene and other aromatics and their derivatives, and a host of other compounds. Since the products that are formed in a fire depend on the substances involved in the fire, it is obvious that toxic exposures to fire fighters and bystanders in a laboratory fire could be quite serious. Persons other than fire fighters equipped with respirators should be promptly evacuated from a laboratory fire and removed to a distant location, preferably upwind from the fire. In exceptional instances and then only when a life-threatening situation exists, should rescue be attempted by a person not equipped with a respirator; in such cases the rescuers should know in advance that their own life is in jeopardy if they must remain in the toxic atmosphere longer than they can, by inhaling deeply before entering, hold their breath while attempting the rescue.

6.8   SUMMARY

Fires take thousands of lives and cause hundreds of millions of dollars in losses annually, not to mention the many jobs that are lost forever. The best way to cope with unwanted fires is to prevent their occurrence. Fire prevention centers around good ventilation, proper storage, proper spill cleanup materials, and adequate personnel training.

Accidents do happen which can cause fires so the second line of defense is to be prepared with the proper fire fighting equipment. The telephone, to make a call for assistance, should be the first equipment used. Proper fire extinguishers must be available before the fire begins.

REFERENCES

1. *Code of Federal Regulations,* Title 49-Transportation, Parts 100–177, Sections 173.115(a),(b), and (d); and 173.150.
2. *NFPA Flammable and Combustible Liquids Code,* NFPA No. 30-1984, National Fire Protection Association, Quincy, MA.
3. Standards for Safety, UL 340; Tests for Comparative Flammability of Liquids, Underwriters Laboratories Inc., Northbrook, IL, p. 9.
4. *Handbook of Industrial Loss Prevention, Factory Mutual Engineering Corporation,* Chapter 42, 2nd ed., McGraw-Hill, New York, 1967.
5. *Handbook of Hazardous Materials: Fire, Safety, Health, Technical Guide #7,* 2nd ed., Alliance of American Insurers, Schaumburg, IL, 1983.
6. *Fire J.,* March, 52 (1976).
7. R. W. High, *Ann N.Y. Acad. Sci.,* **152,** 441–451(1968).
8. *NFPA, Recommended System for the Identification of Fire Hazards of Material #704,* National Fire Protection Association, Quincy, MA.

9. *Handbook of Organic Industrial Solvents: Technical Guide #6,* 5th ed., Alliance of American Insurers, Schaumburg, IL, 1981.

10. A. M. Stevens, Research and Development, February, 1984.

11. *A Technical Mystery,* National Fire Protection Association Quarterly, National Fire Protection Association, Quincy, MA, 1957.

12. C. W. Bahme, *Fire Officer's Guide to Dangerous Chemicals,* 2nd ed., National Fire Protection Association, Quincy, MA, 1978.

13. C. W. Bahme, Fire Officer's Guide to Dangerous Chemicals, 2nd ed., National Fire Protection Association, Quincy, MA, 1978, p. 26.

14. F. T. Bodurtha, *Industrial Explosions and Protection,* McGraw-Hill, New York, 1980.

15. C. J. Hilado and S. W. Clark, *Fire Technol.,* **8,** 218–227(1972).

16. A. M. Stevens, *Research and Development,* February, 1984.

17. *Code of Federal Regulations,* Title 29, Section 1910.106 (d)(4)(i).

18. NFPA Standard Code No. 45, *Fire Protection for Laboratories Using Chemicals,* National Fire Protection Association, Quincy, MA., 1986.

19. *Prudent Practices For Handling Hazardous Chemicals In Laboratories,* National Academy Press, Washington, DC, 1981, p. 227.

20. *Case Histories of Accidents in the Chemical Industry,* 4 vols., Chemical Manufacturers Association, 2501 M St., NW, Washington, DC, 1962, 1966, 1970, 1975.

21. *NFPA Standard Code No. 77, Recommended Practice on Static Electricity,* National Fire Protection Association, Quincy, MA, 1977.

22. Flinn Fax, 1, 3(1986), Flinn Scientific Inc., Batavia, IL.

23. A. C. Willraham, *J. Chem. Educ.,* **56,** A312(1979).

24. A. C. Willraham, *J. Chem. Educ.,* **56,** A311 (1979).

25. H. H. Fawcett and W. S. Wood, *Safety and Accident Prevention in Chemical Operations,* Wiley, New York, 1982.

26. S. J. Rogers and W. H. Eierson, *Fire Technol.,* **1,** 103–111(1965).

CHAPTER 7

Chemical Reactivity: Instability and Incompatible Combinations

Leslie Bretherick

Woodhayes, West Road, Bridport, Dorset, England

7.1 INTRODUCTION

To improve upon a particular situation, existing deficiencies must first be recognized and classified. Next come decisions on what needs to be done to effect the required improvements, and the selection of appropriate means from those available to achieve these ends. The chemical laboratory is no exception to this generalization, and the essential first step is a safety audit (Chapter 5) to establish the existing situation and its deficiencies.

With respect to chemical reactivity considerations, the audit might include the level of understanding of chemical reactivity matters among laboratory workers, the ready availablity of suitable information resources, and the ability of personncl to use these effectively. This chapter attempts to provide information relevant to the two former topics, and the matter of suitable training relevant to the third topic is covered in Chapter 3.

7.2 REACTION ENERGY AND REACTION RATES

All chemical reactions implicitly involve changes in energy to effect the transformation of the starting material or mixture of materials into the product(s) of the reaction, the overall course of the reaction usually being described by means of an equation. As an example of a decomposition reaction, heating crystalline ammonium dichromate converts it to amorphous chromium oxide, according to the equation

$$(NH_4)_2Cr_2O_7 \rightarrow Cr_2O_3 + N_2 + 4H_2O \qquad (7.1)$$

As an example of a double decomposition reaction, the very toxic dimethyl sulfate (DMS) may be destroyed by treatment with dilute aqueous ammonia, according to the equation

$$(CH_3)_2SO_4 + 2NH_4OH \rightarrow (CH_3NH_3)_2 SO_4 + 2H_2O \qquad (7.2)$$

These indicate the changes in structure that have occurred in the reactions, but do not show what changes in energy were involved. It is important to understand the implications of these energy changes in the context of laboratory safety. These implications are explained next.

7.2.1   Reaction Energy

For a reaction to begin, the energy of activation must first be supplied to the reacting species, usually as heat but sometimes as radiation, so that the necessary transition state(s) can be attained for the reaction to proceed. When the reaction starts, in most cases energy is released in the form of heat, and unless this is removed at the same rate it tends to increase the temperature of the reaction system. Such reactions are described as exothermic reactions, and most reactions are of this type. There is another less common type of reaction in which the reaction energy is absorbed into the structure of the reaction products rather than being released, and these reactions and products are both described as endothermic. In such reactions the absorption of heat energy tends to reduce the temperature, and heat energy usually has to be supplied to the reaction system to keep the reaction going. Because of their relatively high bond energy content, endothermic compounds tend to be less stable than the products of exothermic reactions.

7.2.2   Reaction Rates: Temperature as a Major Factor

For a given reaction on a defined scale of working, the total amount of energy that will be released in an exothermic reaction is constant, being determined by the thermochemistry of the system. However, the rate at which that amount of energy will be released is not constant, and if it is excessively high (as in a violent or explosive reaction), then damage to the surroundings will be caused.

The rate of energy release will depend on a number of factors, some of which are more amenable to experimental control than others. Undoubtedly the most important factor is that of temperature, because in the Arrhenius equation, which describes the relationship between reaction rate and temperature, the latter appears as an exponential term. In practical terms, this means that a rise in temperature of 10°C will often double or triple the reaction rate. Control of the temperature in a reaction system will therefore be important, sometimes critically so, as a consequence of the fact that most reactions are exothermic and the reaction heat will tend to accelerate the reaction and lead to a runaway if cooling capacity is inadequate. This factor is probably the one most often involved in reaction accidents, both in laboratories and in industrial scale reactors. An example of a laboratory incident caused in this way involved the addition of bromine to acrylonitrile to give 1,2-dibromopropiononitrile. Cooling was not continuous, but was effected by occasional immersion of the 5-L flask in an ice bath, then allowing it to warm to 20°C. Partway through the reaction, the temperature rose to 70°C, then the flask exploded (1). In the Bhopal industrial disaster, the refrigerated cooling arrangements for the methyl isocynate storage tank were not operational, so the exothermic reaction caused by ingress of water could not be controlled (2).

7.2.3   Other Factors Affecting Reaction Rates

Other factors that have a direct bearing on reaction rates are those of proportions or concentrations in solution of the reaction materials—this follows from the law of mass action. Many instances are known where changing the proportion or concentration of a reaction component, either deliberately or unwittingly, has transformed an established and uneventful procedure into a violent reaction. For example, if in Eq. (7.2) concentrated ammonia solution (30 wt%) is substituted for the dilute solution (10 wt%) normally used to destroy DMS, the reaction becomes explosive (3). It is therefore important not to use large excesses or too-concentrated solutions of reagents, especially when attempting previously untried reactions. For many preparations, 10% is a commonly used level of concentration where solubility and other considerations will allow. When using reagents known to be highly reactive, however, 5 or 2% may be more appropriate. Catalysts (which effectively reduce the energy of activation) are commonly employed at these latter or even lower concentrations.

Further factors that must be considered in detail, especially when increase in the scale of working is contemplated, include the following:

- Temperature control arrangements, with sufficient capacity for heating, and particularly cooling of both liquid and vapor phases.
- Purity of materials, absence of catalytically active species.
- Degree of agitation and mixing.
- Presence of diluents or solvents, viscosity of reaction mix.
- Control of rates of addition (allow for induction period), of reactor atmosphere, of reaction or distillation pressure, and other experimental variables.

As an example of the result of failure to attend strictly to such details, during the dropwise addition of phosphorus tribromide to 3-phenylpropanol to convert it to the bromide, the stirrer stopped, allowing a layer of the dense tribromide to accumulate below the alcohol. Manual shaking later led

to an explosion, caused by the sudden release of much hydrogen bromide
gas as the layers were mixed (4).

7.2.4 Extreme Reaction Conditions

When reactions have to be carried out under high pressure (e.g., catalytic
hydrogenations or pressurized oxidations), the potential hazards which may
arise from the reaction energy or other properties of the reagents themselves
may be greatly enhanced by the very high kinetic energy content of the
autoclave or pressure vessel resulting from the internal gas pressure. Further
problems may arise from the fact that such vessels must necessarily be
massive to withstand pressure stresses, and will therefore have a very high
thermal capacity and be difficult to cool quickly. This makes it even more
necessary to ensure that very tight control is exercised over the reaction
temperature and pressure in such systems, which must be remotely operated
and located behind blastproof structures.

Similarly stringent precautions, though of a rather different nature, will
be required when working with cryogenic systems, especially those involving
liquefied flammable gases under pressure.

An unusual example to illustrate the very significant effects of pressure
on reactivity, and also the great importance of the rate of energy release is
the thermal decomposition of ammonium dichromate, referred to previously
in Eq. (7.1). Many will have seen the spectacular but steady decomposition
occurring when a small pile of the orange salt is ignited in the laboratory,
producing sparks, green chromium oxide, gas, and steam. If, however, the
same is attempted in a closed system pressured to 100 bar, decomposition
becomes so fast that a pressure of 510 bar is attained in a few milliseconds.
The rate of pressure rise, measured as 68 kbar/s, is the highest so far seen
for any substance that is not a high explosive, though the energy of decomposition (41.6 kcal/mol, 174 kJ/mol) of the salt is not particularly high (5).
This dramatic change is not what would be expected from Le Chatelier's
principle, which would predict that the rate of a gas evolving reaction run in
a closed system should fall off as the pressure increased and tended to
suppress the reaction.

7.2.5 Reaction Preliminaries

It will be appreciated from the foregoing that effective measures to control
reaction rates are therefore an essential part of improving the safety aspects
of practical laboratory chemistry by minimizing potential hazards. Preliminary action towards this end, taken of course *before* starting laboratory
operations, will be to find out as much as possible of what is already known
about the particular reagents and reaction system from existing literature
resources, which are now reasonably comprehensive. Information on individual reaction materials is to be found in the material safety data sheets

provided by suppliers or published in various forms (6–8). Information on practical techniques suitable for specific reactions or particular types of reactions may be found in many of the textbooks, monographs (9), or series (10,11) dealing with these matters. The best source of very recently published information is of course *Chemical Abstracts*. Specific information on thermochemical values (12) and on hazardous aspects of chemical reactivity is also published in collected and classified forms (13–15).

7.2.6 Probing the Unknown

Most of the previous section is related to information about chemical reactions involving materials about which something at least is already known or can be inferred from compounds of analogous structure. If, however, you have to work with compounds about which nothing is known or can be deduced, then a different approach will be needed. If the structure of the material is known, some idea of the likely energy of decomposition or of reaction with other compounds may be gained by using a computer program to estimate those energies (16), or by instrumental examination of the thermochemistry of decomposition. If the structure is unknown, then recourse must be made to one or more preliminary very small scale experiments (with appropriate safeguards) to assess the level of energy release and any potential control problems under the particular experimental conditions, prior to any attempts to increase the scale of working.

7.3 CHEMICAL STRUCTURE, REACTIVITY, AND INSTABILITY

For ready understanding of this relatively complex subject, it would be desirable to draw some distinction between those chemical structures that show a high level of reactivity in combination with other materials (including atmospheric air and moisture), and those structures that exhibit instability (or self-reactivity) in the absence of contact with appreciable amounts of other materials. However, there are some materials which show both high reactivity and a measure of instability at the same time (such as nitroalkenes or diazonium salts), and there are thermally unspectacular reactions (notably endothermic reactions) which may produce materials of very limited stability (alkyl azides or alkyl perchlorates), which rather obscure the distinction. These two closely related aspects, reactivity and inherent instability, are therefore considered next in conjunction.

Detailed recognition of what structural features will confer suitable levels of reactivity upon potential reaction components is the art of the synthetic chemist, which is well documented elsewhere (9–11). It is possible, however, to indicate some general structural features which are often associated with high levels of reactivity and/or instability in single compounds, and these are given next, classified on the basis of the elements and/or bonding systems present in the various molecular structures.

7.3.1 Compounds Containing Carbon

C=C–C=C	Dienes	C=C=C	Allenes
C=C–C≡C	Alkenynes		
C≡C	Alkynes, haloalkynes	C≡C–C≡C	Polyalkynes

7.3.2 Compounds Containing Carbon and Nitrogen

C–N=N–C	Azo compounds	C–N=N–N	Triazenes
C=C–NH–N=N	Triazoles	N=N–NH–N=C	Tetrazoles
CH₂–CH₂–NH	Aziridines	CH₂N=N	Diaziridines
CN₂	Diazo compounds	CN₂⁺	Diazonium salts
C–N₃	Alkyl, aryl azides	N–C(=N⁺H₂)–N	Guanidinium oxosalts
C–C≡N	Nitriles	N≡C–C≡N	Dicyanogen

7.3.3 Compounds Containing Carbon and Oxygen

CH₂–CH₂–O	Oxiranes		
C–O–OH	Alkyl hydroperoxides	C–O–O–C	Dialkyl peroxides
CO·O–O–CO·	Diacyl peroxides	C–O–CO·	Dialkyl
		O–O–CO·O–C	peroxydicarbonates
(–CMe₂O–O–)₃	Trimeric acetone peroxide		

7.3.4 Compounds Containing Carbon, Nitrogen, and Oxygen

C–N=O	Nitroso compounds		
C–NO₂	Nitro compounds	C(NO₂)₂	*gem*-polynitroalkyl compounds
C–O–N=O	Alkyl nitrites	CO·O–N=O	Acyl nitrites
C–O–NO₂	Alkyl nitrates	CO·O–NO₂	Acyl nitrates
C=NOH	Oximes	–C≡N→O	Nitrile oxides, fulminates
C–N=N–O–	Arenediazoates, bis(arenediazo) oxides		

7.3.5 Compounds Containing Nitrogen and Oxygen

NO	Nitrogen oxide	N_2O	Dinitrogen oxide
NO_2 or N_2O_4	Dinitrogen tetroxide	N_2O_5	Dinitrogen pentoxide
H_2NOH	Hydroxylamine and salts		

7.3.6 Compounds Containing Nitrogen and Other Elements

N–X	*N*–Halogen compounds	$-NF_2$	Difluoroamino compounds
N–Metal	*N*–Heavy metal derivatives	$-N-S-$	Nitrogen–sulfur compounds

7.3.7 Compounds Containing Halogen, Oxygen, and Other Elements

$-O-X$	Hypohalites	$O-X-O$	Halites, halogen oxides
$-O-X-O_2$	Halates	$-O-X-O_3$	Perhalates, halogen oxides
$N-Cl-O_3$	Perchlorylamide salts		

Many of the types of compounds listed in Sections 7.3.1 through 7.3.7 that show a high level of multiple bonding are endothermic compounds, some with very low values of energy of activation. Such compounds are especially hazardous because they will decompose with evolution of heat and this will rapidly accelerate the rate of decomposition, often to the point of explosion. It may be broadly anticipated that compounds with several of the structural elements indicated will show greatly reduced stability, even if capable of existence under normal conditions.

To give an idea of what levels of energy release are associated with various types of violent decomposition, some tabulated numerical data are given in Tables 7.1 and 7.2. The former gives some measured values for the decomposition exotherms for individual chemicals for a range of instabilities, while the latter gives ranges of values measured for some series of compounds each containing the same unstable chemical grouping. The range of values shown for each grouping in Table 7.2 stems in part from the effect on stability of the remainder of the molecule, but is largely related to the molecular weight of the compound. In general this means that the smaller the molecular weight, the larger will be the energy of decomposition per gram of compound, so that small molecules are potentially more hazardous than larger ones of the same structural type.

Table 7.1. Decomposition Exotherms of Specific Compounds[a]

		Exotherm			
Compound	kJ/mol	(kcal/mol)	MW	kJ/g	(kcal/g)
Trinitrotoluene	1159	(277)	227	5.1	(1.22)
Dinitrobenzene	981	(234)	213	4.6	(1.10)
Nitrobenzene	365	(87)	123	3.0	(0.71)
Ethylene oxide	67	(16)	44	1.5	(0.36)
Benzenediazonium chloride	212	(51)	140.5	1.5	(0.36)
Dibenzoyl peroxide	336	(81)	242	1.4	(0.33)
4-Nitrosophenol	148	(35)	123	1.2	(0.29)

[a]Reference 17 Tables 1 and 5, Copyright, The Institution of Chemical Engineers, reproduced by permission.

Table 7.2. Decomposition Exotherms for Series of Unstable Compounds[a]

Compound and Number	Unstable Group	Range of Measured Values (kJ/mol)	(kcal/mol)
Aromatic nitro (30)	$C-NO_2$	220–410	(52–98)
Aromatic nitroso (4)	$C-N=O$	90–290	(21–69)
Oxime (5)	$-C=N-OH$	110–170	(26–41)
Isocyanate (4)	$-N=C=O$	50–55	(12–13)
Azo (5)	$-N=N-$	100–180	(24–43)
Hydrazo (5)	$-NH-NH-$	65–80	(16–19)
Aromatic diazonium (5)	$C-N_2^+$	130–165	(31–39)
Peroxide (20)	$-C-O-O-$	200–340	(48–81)
Epoxide (4)	$C-C-O$	65–100	(16–24)
Double bond (6)	$-C=C-$	40–90	(9–21)

[a]Reference 17 Table 2, Copyright, The Institution of Chemical Engineers, reproduced by permission.

7.3.8 Ammine Metal Oxosalts

This is a discrete but very large group of compounds, all of similar general structure and containing nitrogenous ligands coordinated to a metal, with coordinated or ionic oxidizing groups (nitro, nitrate, nitrite, perchlorate, permanganate, and so on) also present. Several of these compounds have been described as treacherously explosive. Examples are bis(1,2-diaminoethane)dinitrocobalt(III)iodate, tetraamminecadmium permanganate, and dipyridinesilver perchlorate, all of which will decompose violently when the energy of activation is supplied by heating, friction, or impact. If the ligand is a reducing agent, such as hydrazine or hydroxylamine, the stability is so reduced that explosion may occur before the compound is isolated and dried. Compounds containing both oxidizing and reducing functions in the same molecule are known as redox compounds, and are highly energetic compounds of limited stability, some finding use as rocket propellants. An obvious example is hydrazinium perchlorate, but less obvious is the explosive double salt of potassium cyanide and potassium nitrite (reducer and oxidant, respectively).

7.3.9 Pyrophoric Compounds

Those compounds that react so rapidly with air and its moisture that the ensuing oxidation and/or hydrolysis leads to ignition are termed pyrophoric. Ignition may be delayed or only occur if the material is finely divided or spread as a diffuse layer (titanium powder, and mixed tributyl phosphine isomers, respectively). On the other hand, ignition may be virtually instantaneous, the delay time being measured in milliseconds, as is seen with trimethylaluminum. Pyrophoricity is sometimes put to practical application, as in the use of ferrocerium for lighter flints, and injection of trimethylaluminum to reignite jet aeroengines operating at high altitudes.

Although pyrophoricity is fairly widespread, a few types of molecular structure are notable for this behavior:

- Finely divided metals (calcium, zirconium).
- Metal hydrides or nonmetal hydrides (germane, diborane).
- Partially or fully alkylated derivatives of the latter (diethylaluminum hydride, trimethylphosphine).
- Alkylated metal alkoxides or nonmetal halides (diethylethoxyaluminum, dichloro(methyl)silane).
- Carbonyl metals (pentacarbonyliron, octacarbonyldicobalt).
- Used hydrogenation catalysts (especially hazardous because of the adsorbed hydrogen present).

Whenever such materials are proposed in a reaction, use of an inert atmosphere and specialized handling techniques are essential for safety (18).

7.3.10 Water-Reactive Compounds

After air, the next most common reagent likely to come into deliberate or accidental contact with reactive chemicals is liquid water. A few types of general structure, which may react exothermically and violently with water, particularly if in limited amounts (so that there is no significant cooling effect due to the high specific heat of water), are

- Alkali and alkaline earth metals (potassium, calcium).
- Anhydrous metal halides (aluminum bromide, germanium chloride).
- Nonmetal halides (boron tribromide, phosphorus pentachloride).

- Anhydrous metal oxides (calcium oxide, cesium trioxide).
- Nonmetal oxides (sulfur trioxide, hot boron trioxide).
- Nonmetal halide oxides (phosphoryl chloride, sulfonyl chloride, chloro-sulfuric acid).

The considerable exotherms shown by contact of some concentrated acids or solid bases in contact with water (sulfuric acid, potassium hydroxide) are physical rather than chemical effects.

7.3.11   Other Criteria for Instability

It is possible to deduce meaningful conclusions on the likely level of stability of individual compounds in many cases by examination of their empirical molecular formulas. If there is an unusually high proportion of nitrogen and $N-N$ bonds in the compound, instability should be suspected (hydrogen azide 97.6% N, and hydrazine 87.4% are both explosively unstable, but not ammonia, 82.2%).

The concept of oxygen balance is also important in this connection. The oxygen balance of a compound is the difference between the oxygen content of a compound and that required to oxidize fully all the other elements present. If there is a deficiency of oxygen the balance is negative, and if a surplus, the balance is positive; this is illustrated in the following examples.

Negative balance   trinitrotoluene (-64%, its total oxygen content is 36% of that needed for balance)

$$C_7H_5N_3O_6 + 10.5O \rightarrow 7CO_2 + 2.5H_2O + 1.5N_2$$

Zero balance   ammonium dichromate (no oxygen needed, energy release maximal)

$$Cr_2H_8N_2O_7 \rightarrow Cr_2O_3 + 4 H_2O + N_2$$

Positive balance   dimanganese heptoxide ($+57\%$, giving oxygen for further reactions)

$$Mn_2O_7 \rightarrow Mn_2O_3 + 3O$$

Compounds of positive balance have been mixed with fuels to give powerful explosives, so such compounds must be stringently segregated from any combustible materials.

Oxygen balance is not, however, the sole criterion in assessing potential hazard; the question of the energy of activation of the decomposition reaction is also of considerable significance in determining the margin between potential and actual hazard of explosive decomposition. Thus TNT (above, -64% balance) has a high energy of activation and will not detonate when burned or under impact from incendiary bullets, but requires a powerful initiating explosive charge (detonator) to cause explosive decomposition. On the other hand, 4-azidocarbonyl-1,2,3-thiadiazole (-54% balance) has a very low energy of activation and is extremely explosive in the dry state.

An old but useful rule of thumb states that if the oxygen content of a compound approaches that necessary to oxidize the other elements present (with the exceptions below) to their lowest states of valency, then the stability of that compound is doubtful. Exceptions are that nitrogen is excluded (usually being liberated as the gas), and halogen will convert to halide if a metal or hydrogen is present. Sulfur, if present, counts as *two* atoms of oxygen.

The concept of oxygen balance is also useful in considering reaction systems, as well as individual compounds for which it was originated. Most of the reaction systems found hazardous in practice have involved reaction of an oxidant with a susceptible material or materials. So in oxidation reactions (also one of the commonest procedures in preparative chemistry), the technique adopted should be such as to minimize the oxidant–substrate ratio for as much as possible of the reaction period, consistent with a satisfactory yield of the desired product. This will maximize the negative oxygen balance and minimize the potential energy release and reaction exotherm. In practical terms, this could mean adding the oxidant at a controlled rate to the other reagents, rather than vice versa, and with appropriate cooling, stirring, and other reaction control measures.

It is important to ensure as soon as possible during the experimental procedure that the desired (exothermic) reaction has in fact started, as judged by the physical appearance or thermal behavior of the reaction mix, to prevent buildup of an excess of unreacted oxidant and the possibility of a sudden runaway. It may be that in some circumstances it will be necessary for practical reasons to add the substrate into part or all of the oxidant, in which case the additions of substrate should be portionwise and controlled by the rate of heat release, once it is certain that reaction has begun. In this latter connection, it should be remembered that some reactions are subject to an induction period, usually occasioned by the presence of an inhibitor (functioning as a negative catalyst), and steps may need to be taken to destroy the inhibitor to allow the reaction to proceed smoothly. As an example, oxidation of 2,4,6-trimethyltrioxane (paraldehyde) by nitric acid to glyoxal is subject to an induction period, which can be eliminated by the presence of a little nitrous acid. In the absence of this additive, oxidation may become violent unless addition is very slow until the inhibiting species has been destroyed.

Many further examples of hazardous compounds and reaction systems relevant to the 11 topics in the previous section will be found in the cited reference works (13–15), and newer examples are published from time to time as "Letters to the Editor" in *Chemical and Engineering News* and similar chemical periodicals.

7.4   INHERENT VERSUS PREVENTABLE INSTABILITY

The story so far has been concerned with fast reaction problems arising from the reactivity of chemicals, and from inherent instability because of chemical structure. We now turn to another type of instability that arises from the slow reaction of pairs of chemicals, one of which is often atmospheric oxygen, leading to subsequent problems of instability after a delay of months or years. This type of instability (mainly peroxide formation, and of major significance), is preventable by suitably specific storage and handling techniques that need to be accurately matched to the properties of the materials concerned.

7.4.1   Chemical Structures Susceptible to Peroxidation

These structures invariably contain an autoxidizable hydrogen atom which is activated by adjacent structural features and/or the presence of actinic radiation (not necessarily continuously), so that it reacts slowly under ambient conditions with atmospheric molecular oxygen, initially to form a hydroperoxide.

$$RH + O_2 \rightarrow ROOH$$

This may itself be highly unstable, though usually a further (also slow) reaction step involving addition, rearrangement, or disproportionation of the initially formed hydroperoxide is necessary to develop potentially hazardous peroxidic instability, which may only become apparent after heating or concentration by evaporation.

Susceptible hydrogen atoms are often

On a methylene group adjacent to an ethereal oxygen atom ($-O-CH_2-$ as in diethyl ether, THF, dioxane, diglyme).

On a methylene group adjacent to a vinyl group or a benzene ring ($C=C-CH_2-$ or $Ph-CH_2-$, as in allyl compounds or benzyl compounds).

On a CH group adjacent to two ethereal oxygen atoms ($-O-CH-O-$, as in acetals or methylenedioxy compounds).

On a CH group adjacent to two methylene groups ($-CH_2-CH-CH_2-$, as in isopropyl compounds, decahydronaphthalenes).

On a CH group between a benzene ring and a methylene group ($-CH_2-CH-Ph$, as in cumene, tetrahydronaphthalenes).

On a vinyl group ($-C=CH_2$, as in vinyl compounds, dienes, styrenes, or other monomers).

The autoxidation reactions of these various structural types do not all proceed at the same rate, so the shelf life (that period for which a material may be kept under normal and appropriate storage conditions before significant amounts of peroxidic impurities have formed) will vary from compound to compound. A very comprehensive practical guide to the control of peroxidizable materials (19) covers structures, examples, organizational procedures, distillation and evaporation, detection of peroxides, and their removal. The guide distinguishes three different classes of peroxidizable compounds, gives examples, and recommends suitable precautions for each.

Class A contains those materials that readily form explosive peroxides without concentration, many of which separate from solution, and all of which have caused fatalities. These materials (diisopropyl ether, divinylacetylene, vinylidene chloride) are recommended for red labeling, dating on receipt, testing for presence of peroxides at maximum of three month intervals after opening, and if the test is positive, for disposal. Two other materials that are included within this classification are potassium metal and sodium amide, though in contact with air the former forms the superoxide KO_2 rather than a true peroxide, and sodium amide is similar.

Class B contains materials that peroxidize but only become hazardous on evaporative concentration. Many of the ether solvents are included (diethyl ether, tetrahydrofuran, glyme, etc.) as well as various unsaturates (vinyl ethers, dicyclopentadiene, tetrahydronaphthalene). These are recommended for yellow labeling and dating, with the testing period extended to 1 yr maximum after opening, again with disposal if peroxide is significantly present.

Class C contains those peroxidizable materials that can also polymerize exothermically when initiated by the peroxide content. These monomers include methyl methacrylate, styrene, acrylic acid, acrylonitrile, vinyl chloride, and several others. Labeling, dating, and testing requirements are as for Class B.

The general precautions proposed for the materials covered in the guide are all aimed at preventing peroxide formation as far as possible, and include storage in opaque containers and exclusion of air, preferably by nitrogen atmosphere, except for several Class C monomers that contain inhibitors for storage and need *limited* access of air for effective inhibition. The precautions also stress the need to test for peroxides before any heating or evaporation of a peroxidizable material is attempted, and the need to remove peroxides from such materials, or from stored materials of Classes B or C giving a positive test but which it is necessary to retain for use. Practical procedures to cover these eventualities, including precautions in disposal of materials used to remove peroxides from solvents, are fully detailed. Particular attention is directed towards the stringent precautions required for storing other than very small amounts (10 g) of uninhibited monomers for period of 24 hr or longer, and to the need for accurate and comprehensive labeling of all peroxidizable materials (ref.17, Table 2). See also Section 7.7.2.

In another publication listing about 100 peroxidizable compounds, one third of them are identified as forming peroxides with ease; the labels for these should (but alas do not) recommend destruction of the contents within 1 yr after receipt or 1 month after opening without any testing for peroxide content (20).

It has been recognized with some surprise in recent years that secondary alcohols (isopropanol, 2-butanol) are also susceptible to slow peroxidation, but that presence of traces of a higher ketone than acetone greatly increases the rate and extent of photoperoxidation (these alcohols were previously stored in clear glass bottles), leading to violent explosions on subsequent distillation (ref. 14, pp. 379 and 466). Further information on peroxides in solvents, identity of peroxidizable compounds, and preventive measures has been collected elsewhere (ref. 14, pp. 1643–1647).

7.4.2   Other Slow Reactions in Storage

There are a few materials other than peroxidizable species that have led to incidents of various kinds because of their slow decomposition or other reactions in storage. Many of them contain labile halogen atoms (often as a halomethyl group) and slowly and progressively degrade with evolution of the corresponding hydrogen halide gas, which may eventually burst the containing bottle or flask. Such decomposition reactions may be initiated or accelerated by the presence of moisture or traces of rust, so it is a worthwhile precaution to store them under dry and clean conditions, with, for example, the bottle stored over a desiccant within its own polythene or polypropylene snap-lid container. It is important, however, to label both the bottle and its external container with the date of preparation and to examine the bottle regularly thereafter for development of internal gas pressure, by loosening the lid and listening for the telltale hiss. After checking the desiccant for exhaustion, the external label should then be redated to record the examination. Materials showing this limited stability include several 2- and 5-bromomethyl or -chloromethyl derivatives of furan or thiophene, benzyl chloride or bromide, and sulfenyl chlorides. Such reactive species should not, however, be stored over molecular sieves as internal desiccants in the liquids, because it is known that when benzyl bromide was stored over 4A molecular sieve, the latter acted as a Friedel–Craft catalyst leading to intermolecular condensation—polymerization reactions and evolution of hydrogen bromide which burst the bottle after 8 days (21).

Another instance of an unexpected hazard arising from the use of a molecular sieve as an internal desiccant involved storage of dried nitromethane over large-pore (13×) sieve for several weeks. When a fresh portion of sieve was added, the contents erupted and ignited. This was attributed to formation of the explosively unstable sodium *aci*-nitromethanide inside the large pores of the sieve by an ion-exchange process, and its decomposition by the adsorption exotherm caused by addition of the further portion of freshly activated sieve. Use of the smaller-pored 3A or 4A sieve would probably have prevented the problem by excluding the solvent from the pores where ion exchange occurs (22).

7.5   TOXIC BY-PRODUCTS OF CHEMICAL REACTIVITY

In many reactions volatile or gaseous by-products may be produced as well as the required liquid or solid products, and in some cases the off gases emerging from the top of the reflux condenser may be toxic or highly toxic. The long established habit of allowing all such materials to just disappear up the hood outlet to the atmosphere is not good practice. Some thought should be given to dealing with such laboratory effluents in a more socially responsible manner, particularly when materials of high toxicity and significant quantity are liberated.

7.5.1   Practical Solutions

In some cases it will be possible to trap or absorb the toxic effluent in another reagent which will neutralize its toxic effects. Acidic water-soluble gases, such as sulfur dioxide, hydrogen bromide, or hydrogen chloride are effectively absorbed in a water or alkali scrubber, the alkali being essential when hydrogen fluoride is released. Hydrogen sulfide is best absorbed in soda lime. It is essential to wet and seal the spent soda lime in a closed container for separate disposal after use, to avoid the possibility of ignition (or slow liberation of H_2S) if it is discarded into an open waste bin containing moist paper wipes (23). Other acidic gases of low water solubility are absorbed and neutralized by soda lime.

Basic off gases are best absorbed in a washbottle or scrubber containing diluted sulfuric acid, taking care that the inlet tube is of wide bore to prevent blockage if the base sulfate is of low solubility in water. Relatively small amounts of combustible toxic gases that are completely destroyed by combustion (but not arsine!) may be fed via a fine jet into the air hole of a lit bunsen or meeker burner. This is useful for nickel carbonyl (probably the most toxic gas ever likely to be encountered in laboratory work) and the flame is colored silver by even small traces of the carbonyl. Other suitable methods of absorbing or neutralizing toxic gases or vapors may be devised based on the known chemistry of the materials involved, but think ahead to the ultimate fate of the toxic moiety.

7.5.2   Toxic By-Products of Accidental Contact

Toxic gases may also be liberated as the result of accidental, rather than deliberate contact of chemicals, so it is as well to be prepared for the potentially hazardous products of such contact. Remember, too, that most

Table 7.3. Toxic Hazards from Contact of Chemicals; Selected Examples[a]

Column A	Column B	Product
Arsenic compounds	A reducing agent[b]	Arsine
Azides	Acids	Hydrogen azide
Cyanides	Acids	Hydrogen cyanide
Hypochlorites	Acids	Hypochlorous acid or chlorine
Metal halides	Sulfuric acid	Hydrogen halides (especially HF)
Nitrates	Sulfuric acid	Oxides of nitrogen
Nitric acid	Copper, brass, heavy metals	Oxides of nitrogen
Nitrites	Acids	Oxides of nitrogen
Phosphorus	Reducing agents or caustic alkalis	Phosphine
Selenides	Reducing agents	Hydrogen selenide
Sulfides	Acids	Hydrogen sulfide
Tellurides	Reducing agents	Hydrogen telluride

[a]Chemicals in the column A should be stored and used so that there is no possibility of accidental contact with the materials in column B, when the highly toxic products in the third column will be evolved.
[b]Arsine has been produced by contact of an aluminum ladder with sodium arsenite solution.

proprietary domestic products contain chemicals and that some combinations (e.g., bleach or swimming pool chlorine when mixed with toilet cleaners containing sodium hydrogen sulfate, evolve chlorine) may be hazardous. It is therefore important to bear such possibilities in mind when planning storage arrangements, with appropriate separation of potentially hazardous pairs of materials. Table 7.3 indicates some of the more dangerous possibilities, but it is by no means comprehensive. If it is suspected that such possibilities may exist for other materials within your particular storage area, take the trouble to try to find out what is known about the materials from your accessible information resources.

7.6   CHEMICAL REACTIVITY: IMPLICATIONS FOR STORAGE

Just as unexpectedly violent reactions may sometimes occur under the relatively controlled conditions in experimental reaction systems, the same may be expected to happen under the relatively uncontrolled conditions pertaining to accidental contact of chemicals arising from spillage, breakage, or more seriously, fire conditions in a chemical store. It is therefore necessary to adopt measures to ensure the segregation in storage of mutually reactive types of chemicals to minimize these hazards. There are so many possible individual combinations that it is impractical to attempt to list them all, but some general principles that can lead to an understanding of what is required have often been quoted (see Chapter 11).

Because of the potential for high energy release, it is obviously important to segregate oxidizing agents from reducing agents and combustibles. Then powerful reducing agents should be segregated from readily reducible substrates, and pyrophoric compounds must be stored quite separately from flammables, again for obvious reasons. Gas cylinders, because of their additional hazard potential from the kinetic energy of their compressed contents, must be stored separately from other containers of chemicals. Thermally unstable materials need special refrigerated storage, and peroxidizable materials should be stored cool and dark. However, when these apparently simple segregation requirements are examined in the light of the widely varying properties of the chemicals that are to be stored, because many of them fall into two or more of the categories to be segregated, it is seen that the problems are in fact exceedingly complex, and that some extra criterion for segregation is needed for sensible decisions.

7.6.1   Fire as the Major Hazard

Because fire is statistically the major hazard in a chemical store, a scheme for segregation based upon fire fighting principles has been developed and described in considerable detail, and the reader needing to take action regarding segregation of chemicals in storage is referred to that published source (24), which forms part of a volume devoted to the whole subject of safe storage of chemicals. Tabulated lists of incompatible chemicals have also been published elsewhere in various forms (15, pp. 76, 25).

7.7   DEALING WITH CONSEQUENCES OF HUMAN FAILINGS

In spite of all the sound advice and exhortation on good laboratory practice and on responsible behavior, both above and elsewhere in this text, it will occasionally be necessary to deal with the consequences of human failings in these respects. Each circumstance will be different, of course, which makes it difficult to give advice that will be universally applicable, but there are some common features of such failings related to potentially hazardous chemical reactivity that can be outlined here.

7.7.1   Unlabeled Containers

Perhaps the most common example of poor practice encountered is the unlabeled or partially labeled container, arising either from slackness at the time when the container was filled (or refilled with something else), or from storage under poor conditions, leading to degradation of the label or its information, and possibly of the contents, too.

First action to be taken by the laboratory supervisor is to try to identify the culprit, either by questioning colleagues or from its location in the laboratory and relation to adjacent containers. If there is a label, the next step is to mask around the label with black or thick brown paper and examine it briefly with a UV (chromatography) lamp. This may reveal the original information, but the UV exposure must be brief to avoid the risk of initiating polymerization or degradation of susceptible materials. If this fails, before proceeding further with examination of the contents, it is desirable to check laboratory stores or ordering records to determine whether peroxidizable liquids (especially diisopropyl ether) or polynitro aromatic solids (picric acid) are likely to have been issued to the laboratory in question. Unless the answers are clearly negative, assume the worst and send for specialized assistance from a disposal contractor or the emergency services. If the unlabeled container has a crystalline deposit, with or without a supernatant clear liquid, or if it contains a dry yellow solid, assume the worst and send for assistance.

Only when it is certain that these conditions do not apply and there is no label, the next step is to don PVC gloves and a face mask and, in a good fume hood, gently try to release the stopper. If successful, and the contents look wholesome, remove a small sample and run some simple qualitative tests to try to establish the identity of the unknown. If on opening the contents appear degraded, and it is not possible to determine the likely nature from its odor (care!—sniff a drop on a glass rod, not the whole bottle), or its behavior with water (small scale test), then test a small sample for flammability and arrange for disposal via the local system. In the event of the cap or stopper being stuck fast, proceed straight to disposal arrangements.

7.7.2 Old Peroxidized Stocks

Probably the next most common problem of this type is that of old stocks of solvents or monomers that have survived well past their useful shelf lives, perhaps with labels but no dates and which are shown by one of the standard tests to be substantially peroxidized. Unless the material in question is so valuable or rare that disposal is completely out of the question, disposal is the route to follow, accompanied by appropriate documentation of its current state. If reprocessing is essential to recover the material and remove the peroxide (the level of which should be determined quantitatively), then a careful assessment of the options available (19) for deperoxidizing the particular material must be made, and safety measures adopted appropriate to the determined level of peroxide content.

Note that testing for the presence of peroxidized material can itself be hazardous, particularly if it is necessary to remove the test sample from a container that is corroded or has a damaged (air-leaking) closure. In such cases, the access to air and the evaporation of volatiles will both enhance the formation of peroxides and increase their concentration in the residual contents. Thus, if peroxide is present and has deposited in or near the threads of the closure, unscrewing this cap may cause detonation. Or, if deposited elsewhere in the container, the mechanical action of disturbing the container in order to move it to a testing location may be itself sufficient to cause detonation. Hence, if it is determined that peroxidizables should be tested for peroxides, it should also be the policy to make the first test for the presence of peroxide long before there is any possibility of peroxide formation and subsequent testing should also be frequent enough to preclude the possibility of a positive result. Any peroxidizable materials in containers showing corrosion or damaged or suspect closures must be disposed of immediately without being opened and with great care.

7.7.3   Poor Storage Practices

Another common failing, which can have far-reaching implications for the shelf life of reactive chemicals is to put them after preparation into containers that are unsuitable for their long term stability, making due allowance for the way in which the materials are likely to be used. For example, it would be self-defeating to store an air- or moisture-sensitive reagent in an ordinary screw-capped bottle with a card wad, since the slow but steady diffusion of both air and moisture through the wad would lead to progressive deterioration of the reagent, even if the bottle were never opened for use, and the latter operation would have to be performed in a glovebox or dry bag. If the reagent were normally used fairly regularly and in small amounts, it would be much more effective and convenient to seal it into a serum type bottle, capped with a PTFE-faced elastomeric seal, so that the small portions could be withdrawn by syringe without exposing the main bulk to the atmosphere. For exceptionally sensitive reagents, special serum bottles with an additional external screw cap with another PTFE-faced seal are also available from laboratory suppliers. Much may be learned about containers and caps that are suitable for chemicals of many different types and degrees of stability by study of the containers and caps used by reputable chemical suppliers, or of laboratory supply catalogs. The use of appropriate containers and labels (as usually supplied commercially), coupled with strict adherence to any special external storage conditions specified on the labels of reactive materials, will avoid many of the problems normally associated with storage of reactive chemicals. These and other related topics have been considered in great detail in Chapter 11 and elsewhere (ref. 24, Chapters 1 and 3–8).

7.8   PROTECTION OF PERSONNEL

The protection of personnel from the effects of undue reactivity or instability of chemicals and their reaction systems should be one of the primary considerations of laboratory management in the present context, so this topic has been left until last in this chapter, both to emphasize its importance and to maximize understanding of the relation between those effects and what needs to be done.

Probably the best way to try to protect personnel is to invest time and resources in ensuring that they understand their particular role and responsibilities in the overall laboratory organization, and the reasons behind the various procedures and points of technique that they are expected to implement in their laboratory work. This approach could well involve adding some explanatory background information to written instructions for standard procedures in a routine laboratory, and making time for some discussion of matters of both general safety interest and specific points of technique during safety review periods in a nonroutine or research context. This continued involvement of laboratory personnel in developing safety considerations as an integral part of laboratory work will eventually lead to an improved attitude and a diminution, though never to the disappearance, of the need for close supervision at all times.

This approach must be accompanied by the provision of a suitable range of readily available and easy-to-use protective equipment, so that when the need for such personal protection is perceived, preferably by the laboratory workers themselves, it may be readily brought into use. Care in matching equipment accurately to the technical and personal needs is vital, as uncomfortable personal protective equipment will not be worn. The range of equipment necessary will depend entirely on the type and scale of operations. It may extend from the absolute minimum of eye protection, through the use of a good hood, possibly supplemented by additional safety screens for normal laboratory operations, up to that maximum level of protection afforded by isolation cells of blast-proof construction with remotely operated equipment (26) when the potential energy release, for example, in high pressure process research at up to 2000 bar, demands it.

REFERENCES

1. Manufacturing Chemists Association, *Accident Case History No. 1214,* Washington, DC. (1966).
2. *Chem. Eng. News,* **63** (12), 4 (1985).
3. H. Lindlar, *Angew. Chem.,* **75**, 297 (1963).
4. D. A. H. Taylor, *Chem. Br.,* **10**, 101 (1974).
5. L. Bretherick, *Chem. Eng. News,* **64** (22), 2 (1986).
6. L. H. Keith and D. B. Walters, Eds., *Compendium of Safety Data Sheets,* VCH, Deerfield Beach, FL, 1985 (867 data sheets in 3 vols.).

7. Dutch Association of Safety Experts, *Handling Chemicals Safely 1980,* 2nd ed. (in English), Dutch Safety Institute, Amsterdam, 1980 (874 data sheets).

8. S. Templer, Ed., *Laboratory Hazard Data Sheets,* Royal Society of Chemistry, Nottingham, UK, monthly series from mid-1982 (published in RSC *Laboratory Hazards Bulletin* but available separately; No. 48, June 1986).

9. G. Brauer, *Handbook of Preparative Inorganic Chemistry,* (R. F. Riley, translation Ed.), Academic, London, Vol. 1, 1963, Vol. 2, 1965.

10. *Organic Syntheses,* Wiley, New York, Collected Vols. 1–5, 1944–1973, annual volumes thereafter.

11. *Inorganic Syntheses,* McGraw-Hill, New York, Vols. 1–23, 1939–1985, annual volumes thereafter.

12. R. C. Weast, Ed., *Handbook of Chemistry and Physics,* 60th ed., CRC Press, Boca Raton, FL, 1979, pp. D-61–84.

13. *Manual of Hazardous Chemical Reactions,* NFPA 491M, 5th ed., National Fire Protection Association, Quincy, MA, 1975.

14. L. Bretherick, *Handbook of Reactive Chemical Hazards,* 3rd ed., Butterworths, Stoneham, MA, 1985.

15. L. Bretherick, Ed., *Hazards in the Chemical Laboratory,* 4th ed., Royal Society of Chemistry, London, 1986.

16. *CHETAH, ASTM Chemical Thermodynamic and Energy Release Evaluation Program 4.3,* American Society for Testing and Materials, Philadelphia, PA, 1986 (modified versions for use on microcomputers, and a considerably expanded mainframe version are in course of development).

17. T. Grewer and E. Duch, "Thermochemistry of Exothermic Decomposition Reactions," in *Loss Prevention and Safety Promotion in the Process Industries,* Vol. 3, A1–11, Institution of Chemical Engineers, Rugby, 1983.

18. D. F. Shriver, *Manipulation of Air-Sensitive Compounds,* McGraw-Hill, New York, 1969.

19. H. L. Jackson, W. B. McCormack, C. S. Rondevstedt, K. C. Smeltz, and I. E. Viele, *J. Chem. Educ.,* **47,** A175 (1970).

20. Manufacturing Chemists Association, *Accident Case History No. 1693,* Washington, DC (1971).

21. T. D. Harris, *Chem. Eng. News,* **57** (12), 74 (1979).

22. L. Bretherick, *Chem. Ind. (London),* **1979,** 532.

23. L. Bretherick, *Chem. Ind.,* **1971,** 1042.

24. L. Bretherick, "Incompatible Chemicals in the Storeroom: Identification and Segregation," in *Safe Storage of Laboratory Chemicals,* D. A. Pipitone, Ed., Wiley Interscience, New York, 1984, p. 47.

25. H. H. Fawcett, *Chem. Eng. News,* **30,** 2588 (1952).

26. B. R. Franko-Filipasic and R. C. Michaelson, *Chem. Eng. Progr.,* **80,** 65 (1984).

Note: References 1 and 20 are out of print. To obtain copies, apply to the Docket Office, Occupational Safety and Health Administration, 200 Constitution Ave., NW, Washington DC 20210. Accident Case History Nos. 1214 and 1693 are found in Volume 3, page 46, 1970 and Volume 4, pages 27–28, 1975, respectively, of Case Histories of Accidents in the Chemical Industry, in Docket No. H-150.

CHAPTER 8

How Chemicals Harm Us

Stephen T. Springer

Richardson-Vicks, Shelton, Connecticut

8.1 INTRODUCTION

This chapter is an introduction to the subject of toxicology. Toxicology is the study of chemicals that can cause harm or injury to living beings in any of several ways. This chapter treats the concept of harm to persons, the types of chemicals that cause such harm, and how they cause their damage. The concepts to be discussed are (1) harm, (2) exposure, (3) reactions, (4) signs and symptoms, (5) susceptibility and risk, (6) treatment, and (7) prevention. The purpose of this chapter is to give the reader enough information to be able to make intelligent decisions about whether or not a chemical can cause harm and how to prevent injury.

A glossary (Section 8.10) defines some of the words commonly used in discussions of toxicology. A bibliography at the end of this chapter suggests additional sources of information.

8.2 HARM

Chemicals can *harm* (cause injury) in several ways. For example:

1. Injury by force, such as an explosion
2. Injury by poisoning, such as eating a poisonous mushroom
3. Injury by destruction of tissue, such as an acid burning the skin
4. Injury by displacing the air to be inhaled, such as carbon monoxide from an automobile exhaust building up in a closed garage.

The point here is to be always on the lookout for potentially harmful situations that may arise, usually through misuse or improper handling of chemicals in the workplace. Chapters 6 and 7 dealt with the first item in the list, above. This chapter and following chapters, particularly Chapters 9, 12, and 13, deal with the other three items. In the workplace, hazards can take many forms, some not so obvious. All chemicals can cause harm, even water, for example, if it is misused or handled carelessly. What makes a chemical harmful or harmless is the subject of Section 8.3.

8.3 EXPOSURE

The concept of *exposure* demonstrates that the same chemical can be both harmful and harmless. There are three aspects of exposure:

1. *Dose* (how much)—a teaspoonful or a gallon.
2. *Duration* and *frequency* (how long and how often)—a minute or an hour, once or every day for years.
3. *Route* (how the victim is exposed)—drinking it, spilling it on the skin, or breathing its vapors.

Changing any one of these three aspects can change the effect. As an example, water is harmless under the following conditions:

1. How much—3 gallons.
2. How long—5 hours.
3. How exposed—poured slowly over the head.

But take the same chemical, water, and change the exposure conditions:

1. How much—3 gallons.
2. How long—5 hours.
3. How exposed—head held under the water in a bucket.

In the first example, no harm is done; in the second, the victim is dead. In the second example, if the route of exposure was changed to "poured slowly into the mouth" instead of "head held under . . ." there would be less harm. Or, in the second example, if the time was changed to 5 seconds instead of 5 hours, even if the head was under water, there would be no harm at all.

8.3.1 Dose

In general, the larger the dose, the shorter the time it takes for an injury to occur. Conversely, it takes longer for injury to result from smaller doses. Also, as the dose increases, the severity of the injury increases until a dose is reached that causes immediate systemic reactions (see Section 8.4) that can be so severe that death results. On the other hand, the dose may be so small that immediate local and systemic injury is not noticed, but longer term problems are created.

Another term that applies here is *threshold*. Threshold is the dose at which a chemical begins to be harmful. A chemical may have, for example, a toxicity threshold of 6 g such that drinking < 6 g is not harmful but > 6 g cause severe nausea. Section 8.8.1 discusses threshold as it relates to exposure limits.

8.3.2 Duration and Frequency

How long (duration) a person is exposed to a chemical directly affects how severe the injury will be. A chemical that is capable of damaging the eye will do less damage the quicker it is washed out, and conversely, the longer it remains in the eye, the worse the injury. Any chemical that can penetrate through the skin and cause an illness will have a less severe effect if it is washed off quickly from the skin because in the short time it is on the skin, less penetrates through the skin.

How often (frequency) one is exposed to a chemical can directly affect the type, time of onset, extent, and severity of the toxic effect. Some chemicals cause immediate injury after only one exposure (*acute toxicity*) while others cause injury only after long-term exposure (*chronic toxicity*). This is so because the body has the ability to get rid of certain chemicals that enter into it. For example, certain chemicals pass through the stomach and intestines with little or no absorption (see Section 8.3.3). Others may be absorbed into the blood and be filtered out by the kidneys and end up in the urine. The speed with which this elimination process occurs is different for different chemicals. A chemical may be cleared from the body faster than it is absorbed, or, it may accumulate in various organs (see target organs in Section 8.4) slowly over time. Thus, one exposure to a large amount of a chemical may not result in any toxicity, but repeated exposures to smaller amounts of the same chemical could cause injury because each time a tiny amount is absorbed and eventually builds up to a harmful level.

8.3.3 Route

Chemicals can take different *routes* (pathways) to get inside the body (called systemic exposure). Probably the most obvious is by mouth (called ingestion). Usually, this is accidental, but not unlikely. For example, a victim might have a chemical on their fingers and then put a finger in their mouth, or might then eat a sandwich or smoke a cigarette without first washing the hands. Or keeping sugar in a beaker in the lab can result in putting the wrong white crystalline solid, from a different beaker, into hot coffee.

Exposure by inhalation (breathing) is also obvious. Dusts, mists, gases, vapors, and sprays can all enter the lungs through the nose and mouth during normal breathing. This route provides the most rapid entry into the body and usually causes harmful effects most rapidly.

Chemicals can be absorbed into the body through the skin. Sometimes the victim is not aware of the absorption. Usually, skin absorption is a slow process through intact skin. When the skin is damaged by cuts, abrasions, and the like, chemicals can be absorbed much more rapidly. In either event, chemicals can be absorbed through the skin when clothing, shoes, or lab coat is contaminated with chemicals. Some chemicals can penetrate the skin very rapidly, even causing death within 30 min of the initial contact.

A less obvious route of entry is through the eyes. Chemicals are very quickly absorbed through the eyes into the bloodstream. Also, there are open passages between the eyes, nose, and mouth so that anything that gets into the eyes and/or nose can be swallowed. Chemicals can get into the eyes as liquids or powders by splashing or by being on the fingertips when the eyes are rubbed. Gases, mists, fumes, and dusts also can enter the body through the eyes, and be absorbed. There are many ways a chemical can enter the body, sometimes without the victim knowing that it is happening.

8.4 TYPES OF REACTIONS

The body's *reaction* (response) to exposure from a chemical can take many forms. To begin with, the reaction may be immediate, within seconds or minutes of contacting a chemical. For example, a skin burn from spilling acid is immediate, the result of a one-time contact (acute exposure) with the chemical. Reactions that occur only after several separate exposures over a long period of time (chronic exposures) are referred to as delayed reactions. These types of reaction are usually more life-threatening than acute reactions (though not necessarily) and can be insidious because the victim often is not aware of the hazard until it may be too late.

Reaction to a chemical exposure can occur either *locally* or *systemically*, or both.

8.4.1 Local Reactions

Local reactions occur at the site of chemical contact. These sites typically are the skin and eyes, but not always. Taking a chemical by mouth could cause local effects inside the mouth, in the esophagus, the stomach, or the intestines. Local effects usually occur after a single exposure, ranging from minor irritation to severe tissue destruction. But local effects may require repeated exposures to show up, such as a delayed reaction resulting in an allergic skin rash (actually, a local contact skin allergy is a systemic reaction with local effects). These examples of specific types of local reactions will illustrate:

1. *Eyes.* Local eye reactions occur as a result of splashing a liquid or solid in the eye, or from contact with a gas, mist, or dust. Reactions are usually immediate and result in pain and tearing, but not always. Some chemicals can produce an immediate loss of feeling (anesthesia) in the eye, so even though there is no pain, the eye is being damaged, sometimes severely. Also, delayed local reactions can occur in the eye. Loss of sight may not necessarily be immediate, but may occur slowly over several days or even weeks.

2. *Skin.* Local skin reactions occur as a result of splashing or spilling a liquid or solid on the skin. As is the case with the eyes, skin reactions can be either immediate or delayed and result in pain and obvious tissue destruction.

3. *Digestive Tract.* Local reactions can occur in the mouth, esophagus, stomach, and intestines. They occur as the result of swallowing a chemical. Local reactions of this type include burns of the mouth and esophagus and ulcers of the stomach and intestines. These local reactions can cause severe long term illness and even death.

4. *Nose and Lungs.* Breathing in certain gases, powders, mists, or fumes can cause burns inside the nose and lungs. Again, as in the case of the digestive system, this type of exposure can lead to death, in this case by destroying lung tissue.

A special type of local irritation, known as *phototoxicity*, occurs when a chemical on the skin is exposed to sunlight. Although the chemical may not be an irritant by itself, it is activated by sunlight to cause the typical burning sensation characteristic of irritation. Thus, one may come in contact with a phototoxic chemical without knowing it because there is no discomfort until the exposed skin is in the sunlight.

The most severe local reaction is a *corrosive* reaction. Corrosivity results in direct tissue destruction. Some chemicals can burn a hole through the skin down to the bone. Others can destroy the tissues of the eye or the lining of the stomach or lungs. Solids, liquids, and gases are capable of causing corrosive reactions. Liquids probably are the most significant concern because they are the most easily spilled or splashed. Table 8.1 lists examples of common corrosive liquids, including organic and inorganic acids, caustic solutions, and organic solvents. Corrosive solids (examples in Table 8.2) pose somewhat less of a hazard than liquids because their effects depend to some extent on how soluble they are in the tissue, how much moisture is present in the air, and on the tissue involved and how long they are in contact. Whereas liquid corrosivity is almost instantaneous, solid corrosivity is somewhat slower and, therefore, quick removal of the solid can limit the extent of the damage. An additional danger that solids often pose is the heat given off when they come in contact with moisture on the tissue. The heat can be great enough to cause tissue destruction in addition to the corrosivity of the chemical.

Table 8.1. Common Corrosive Liquids

Inorganic Acids	Organic Acids
Chromic	Acetic
Hydrochloric	Butyric
Nitric	Formic
Sulfuric	
	Organic Solvents
Other Inorganics	
	Dichloroethylene
Bromine	Perchloroethylene
Phosphorus trichloride	Methyl ethyl ketone
Sulfuryl chloride	Gasoline
Peroxides	
	Other Organics
Caustic Solutions	
	Acetic anhydride
Ammonia	Liquified phenol
Sodium hydroxide	Triethanolamine
Potassium hydroxide	

Table 8.2. Common Corrosive Solids

Alkalies	Organic Acids
Calcium oxide	Oxalic
Calcium hydroxide	Phenol
Potassium hydroxide	Salicylic
Sodium hydroxide	Trichloroacetic
Sodium phosphate	
	Elements
Other Compounds	
	Sodium
Tin chloride	Potassium
Phosphorus pentoxide	Lithium
Potassium chromate	Phosphorus
Mercury chloride	Iodine

Liquids and solids can be seen; one can be aware of their presence. But gases are usually invisible, so they can be an insidious danger. Corrosive gases such as ammonia can destroy the lining of the lungs, resulting in death or in a chemical pneumonia that leaves the victim sick for a long time. Fumes and mists from corrosive liquids are almost invisible and are equally destructive. Dusts from corrosive solids are also destructive.

Contact with corrosive chemicals is perhaps the greatest danger to laboratory workers in their workplace.

8.4.2 SYSTEMIC REACTIONS

In general, *systemic* (meaning inside the body) reactions are much more serious than local reactions. As described earlier, there are many routes available to a chemical for entry into the body and there is not necessarily a local reaction at the site of entry to act as a warning. Some chemicals can pass through the eyes, skin, or lungs without causing any local effects, but they can cause severe systemic reactions.

For a systemic reaction to occur, the chemical must get into the bloodstream. Once in the blood, the chemical may be transported to all organs of the body, affecting some organs and, interestingly, not affecting others. An organ that is susceptible to damage by a chemical is referred to as a *target organ* for that chemical. Systemic reactions are classified by the organ or organ system affected, such as the heart, liver, bladder, respiratory system, and central nervous system. There are hundreds of different systemic reactions. The following general principles and examples will illustrate how organ systems can react to chemicals.

Like local reactions, systemic reactions can be immediate or delayed. Some allergic reactions (called anaphylactic reactions) can occur within minutes of exposure. They are usually so severe that death will result in a short time if not treated. Other reactions develop slowly over many years. For example, certain cancers can develop 20 years after only a few exposures to a chemical, even though the victim has never come in contact with that chemical in the interval.

Chemicals that affect the reproductive system may prevent a victim from having children but otherwise have no recognizable effect on their general health. Some chemicals may also change the genetic material in the reproductive cells (sperm or egg) such that the victim's own general health appears normal and the apparent health of the victim's young child also appears normal, but when the child nears physical maturity, at about age 20, they then develop a serious illness. This can occur as a result of the mother or father having been exposed to a chemical before conception, or the mother being exposed during pregnancy. Exposure of pregnant women to certain chemicals during certain specific times of the pregnancy can cause the unborn child to die, or to be born with any of several birth defects, even though the mother was never sick. These examples of delayed systemic reactions illustrate how serious the exposures to some chemicals can be, especially because there are no immediate signs or symptoms of exposure.

A special type of systemic reaction is the *sensitization* reaction. Sensitization is the process whereby the body reacts against a chemical by means of an *allergic* reaction. An allergic reaction consists of a series of events that take place inside the body as the body tries to get rid of the chemical. Sensitization develops slowly because it takes several exposures over a few weeks before the body begins to recognize the chemical as foreign (sometimes, it could take years for this to happen). However, once this happens, the body never forgets it. Usually, there are no noticeable symptoms of exposure during the development period so the victim is unaware of the developing sensitivity to the foreign substance until their sensitivity is fully developed, and by then it is too late.

Once sensitivity has developed, there are two possible types of sensitization reaction. The *immediate* type can occur within minutes of exposure. The severity of reaction ranges from anaphylaxis, as mentioned previously, to hay fever, which is not life threatening. These sensitizing chemicals enter the body through the mouth, eyes, nose, and skin. The *delayed* type reaction takes several hours or even a day to develop. The area of skin that comes in contact with the chemical becomes red, swollen, and itchy. The reaction usually gets worse over the next several hours, even though the sensitizing chemical is removed, and may last for days. Reaction to poison ivy is an example of delayed contact sensitization. The contact sensitization reaction can be distinguished from a local chemical irritation reaction by its itching and persistence. The local irritation reaction usually stings or burns rather than itches, and once the irritating chemical is removed, the reaction does not get worse as time goes on.

A special type of sensitization is *photosensitization*. This is a delayed allergic reaction that occurs when certain chemicals are exposed to sunlight. The chemical by itself does not cause an allergic reaction. It must be in the skin at the time of sun exposure. However, the victim does not have to get the chemical directly on the skin as in contact allergy. It may be a photosensitizing chemical that entered the body through the mouth, eyes, or nose. After being absorbed into the blood, however, it is distributed to the skin, where the sunlight can reach it. Thus victims can get photosensitizers in their skin and have no reaction as long as they are wearing long sleeves, or go to and from home and work in the dark. But upon going out the next morning into the sunlight, they get a strong allergic skin reaction. The symptoms of this type of reaction are the same as those of ordinary allergic skin reactions.

There is a special disease associated with photosensitization called *persistent light reactivity*. Although rare, this disease is becoming more frequent. Essentially, persons with this disease have been photosensitized to a certain chemical; the entire body produces a severe allergic reaction, not just a local skin rash, which occurs everytime they are exposed to sunlight—for the rest of their lives. These persons develop other complicating medical conditions, must spend the rest of their lives in the shadows, and are really quite miserable.

Table 8.3 lists examples of local and systemic reactions and their descriptive terms. This list is by no means complete; it is, however, a guide for the commonly recognizable symptoms of toxic reactions. The glossary (Section 8.10) also lists toxic reactions, but of the type that are only detectable by a physician.

Table 8.3. Signs and Symptoms of Chemical Exposure Recognizable by the Worker[a,b]

Sign of Toxicity	Descriptive Symptom
Abdominal cramps	Painful spasms of stomach area
Alopecia	Loss of hair; baldness
Amenorrhea	Stoppage of menstruation (period)
Amnesia	Loss of memory
Anesthesia	Loss of feeling
Angina pectoris	Chest pain
Anorexia	Loss of appetite
Anosmia	Loss of sense of smell
Anuria	Lack of urination
Anxiety	Troubled feeling
Apathy	Lack of emotion
Aphasia	Inability to talk coherently
Areflexia	Loss of reflexes
Arrhythmia	Irregular heartbeat
Arthralgia	Joint pain
Asphyxia	Suffocation
Asthenia	Loss of strength or energy
Asthma	Difficulty breathing
Ataxia	Inability to walk straight
Athetosis	Slow writhing movements of fingers
Back pain	Aching of back area
Blackened teeth	Darkening of the tooth surface
Blindness	Inability to see
Blurred vision	Not in focus
Bronchitis/bronchospasm	Coughing; difficult breathing
Burn	Tissue damage
Cachexia	Wasting away
Cancer	Abnormal tissue growth
Cataracts	Progressive loss of eyesight
Changes in body/breath odor	Abnormal body/breath odor
Cheilitis	Inflammation of the lips
Chills	Shivering with cold plus fever
Chloracne	Reddish skin rash
Chorea	Rapid, jerky uncontrollable movements of limbs
Colic	Abdominal pain, usually due to intestinal gas
Collapse	Loss of ability to stand
Coma/comatose	Extreme unconsciousness
Confusion	State of bewilderment
Conjunctivitis	Inflamed and reddened eyes
Constipation	Infrequent/difficult bowel movements
Constriction	Binding or contraction
Convulsions	Violent body spasms
Coughing	Forceful expiration of air
Coughing blood	Forceful expiration of blood
Cyanosis	Bluish skin color
Dark urine	Discoloration of urine

Table 8.3. (*Continued*)

Sign of Toxicity	Descriptive Symptom
Dehydration	Excessive loss of body water
Delirium	State of mental confusion
Dental erosion	Loss of tooth surface
Depression, bodily	Decrease in activity
Depression, mental	Feeling of great sadness
Dermatitis	Inflammed, reddened skin
Diarrhea	Frequent, loose bowel movement
Dilated	Expanded; opened up
Disequilibrium	Inability to maintain balance
Disordered gait	Change in walking pattern
Dizziness	Feeling faint; light-headed
Drooling	Excess saliva from mouth
Drowsiness	Falling asleep
Dysarthria	Difficulty speaking clearly, as in stammering
Dysosmia	Impaired sense of smell
Dysphagia	Difficulty in swallowing
Dyspnea	Difficulty in breathing
Dysuria	Painful or difficult urination
Eczema	Inflammatory skin disease with itching and burning
Edema	Fluid Retention; swelling
Emaciation	Extreme low weight; skinniness
Emphysema	Difficulty breathing
Epistaxis	Nosebleed
Erythema	Reddened skin
Euphoria	Exaggerated feeling of well-being
Fasciculation	Muscle twitching under skin
Fainting	Loss of consciousness
Fatigue	Tiredness; sluggishness
Fever	Increased body temperature
Fibrillation	Rapid muscle contraction
Finger clubbing	Rounded, swollen fingertips
Fluorosis	Darkening of the teeth
Footdrop	Dragging of the foot while walking
Frostbite	Freezing of tissue
Gangrene	Tissue death
Gasping	Difficulty catching breath
Giddiness	Dizziness; silliness
Gingival lead line	Dark line formed on gums
Glossitis	Tongue swelling
Halitosis	Foul-smelling breath
Hallucination	A sensing of things that are not real
Headache	Pain in head or neck area
Hemiparesis	Paralysis of one side of body
Hemorrage	Bleeding
Hyperkinesis	Excess activity or motion
Hyperpigmentation	Excess coloring of skin
Hyperthermia	Elevated body temperature

Table 8.3. (*Continued*)

Sign of Toxicity	Descriptive Symptom
Hyperventilation	Sudden rapid breathing
Hypothermia	Lowered body temperature
Icterus	Tissue discoloration
Impotence	Loss of sexual desire
Incoordination	Inability to accurately move a limb
Inflammation	Swelling; redness; warmth
Inflexibility	Rigidity; inability to move
Insomnia	Inability to obtain normal sleep
Involuntary defecation	Uncontrollable bowel movements
Involuntary urination	Uncontrollable urine passage
Irritability	Quickly becoming annoyed
Itch	Skin sensation causing scratching
Jaundice	Yellow discoloration of skin or eyes
Keratosis	Horny growths on skin
Labored	Not easy or natural
Lacrimation	Excessive eye tearing
Lactation changes	Decrease in amount of breast milk
Lassitude	Sense of weariness
Lightheadedness	Dizziness
Malaise	Uneasiness; discomfort; ill-feeling
Malnutrition	Inadequate diet
Melena	Black tarry vomitus or stools
Menstrual changes	Change in menstrual cycle (period)
Metallic taste	Taste in mouth resembling metal
Miosis	Pupil contraction
Miscarriage	Loss of baby by pregnant women
Myotonia	Temporary muscle rigidity and spasm
Narcosis	Stupor or sleep
Nasal ulceration	Perforation of nasal tissue
Nausea	Feeling of need to vomit
Nervousness	State of unrest/uneasiness
Nocturia	Excessive urination at nighttime
Numbness	Loss of feeling; prickly feeling
Nystagmus	Rhythmical movement of eyes
Ocular opacity	Loss of eyesight
Ochronosis	Dark spots on skin
Oliguria	Decreased urination
Opisthotonos	Spasms with body arched from head to heels
Pallor	Paleness of the skin
Palpitations	Forceful heartbeat
Paralysis	Loss of ability to move limbs
Paresthesias	Abnormal sensation; tingling
Paroxysmal	Sudden recurrence of disease
Perforation	Opening through a tissue
Pharyngitis	Sore throat; hoarse voice
Phlebitis	Swollen, painful vein
Photophobia	Inability to tolerate light
Photosensitization	Allergic reaction to light

Table 8.3. (*Continued*)

Sign of Toxicity	Descriptive Symptom
Phototoxicity	Irritant reaction to light
Pigmentation	Coloration
Prostration	Marked loss of strength; exhaustion
Ptosis	Drooping of upper eyelid
Pyorrhea	Swollen, bleeding gums
Pyuria	Pus in urine
Red blood cells in stool	Blood in bowel movement
Respiratory distress	Difficulty breathing
Rhinorrhea	Excessive nasal discharge
Salvation	Discharge of saliva
Scotoma	Blind spot in field of sight
Seizure	Convulsion
Sensitization	Allergic reaction
Shock	Depression of all bodily functions
Somnolence	Prolonged sleepiness
Spasm	Convulsive muscular contraction
Stomatitis	Swelling of the mouth lining
Strabismus	Lack of coordinated eye movement, crossed eyes
Stupor	Unconsciousness
Sweating	Excessive moisture on skin
Swelling	Enlargement
Tachycardia	Abnormal rapid heartbeat
Tenderness	Painfulness to pressure/contact
Tetany	Intermittent muscle spasms
Tick	Skin twitch
Tinnitus	Ringing in the ears
Tracheobronchitis	Coughing; difficulty breathing
Tremors	Shaking; trembling
Tumor	Swelling or growth
Ulceration	Tissue destruction
Unconsciousness	Not awake
Urticaria	Skin eruption
Vertigo	Feeling of whirling motion
Vesiculation	Blisters
Visual disturbance	Abnormal eyesight
Vomiting	Forceful expulsion of stomach fluid
Vomitus	Expelled stomach contents
Weakness	Lack of normal strength
Wheezing	Noisy breathing
Wrist drop	Inability to extend hand at wrist

[a]These are signs of toxicity most commonly found on Material Safety Data Sheets (MSDS). The words on the left are technical terms. On the right are the descriptive symptoms *you might recognize* if you were exposed to a chemical that caused the corresponding toxicity listed in the left column. This list is not meant to be all inclusive, but covers most of the major symptoms.
[b]There are many more signs and symptoms of toxicity from chemical exposure that are recognizable only by a physician. These are not listed in this table. Some may be found in the glossary (Section 8.10).

8.5 RECOGNITION OF EXPOSURE

How can a person tell if they have been exposed to a harmful chemical? Quite often the first sign or symptom is given by the senses. The victim may feel it—a liquid on the skin feels wet or warm or cold, or it hurts. The victim may smell or taste it (some gases or vapors in the air can be smelled or tasted). The chemical may be seen on the skin. The eyes or nose may begin to water or sting. Sometimes there is no sense stimulation; instead the victim feels dizzy, starts to cough, or has difficulty breathing. The exposure may cause a mild or severe headache, a feeling of nausea, or the victim may act as though they were slightly or severely intoxicated. A victim may also fall down, unconscious or dead.

Except for the last mentioned fatal effect, all of these are signs of exposure that provide an *early warning* of exposure and of possible injury. Unfortunately, these signs are not foolproof; as mentioned earlier, some chemicals may induce a local anesthesia, a loss of feeling. Other chemicals with strong odors can cause anosmia (loss of the sense of smell) very soon after exposure, so the victim thinks the exposure is over when it really is still present.

Table 8.3 lists additional examples of signs and symptoms of chemical exposure, mostly long-term exposure symptoms. A person must be continuously aware of how they feel on a daily basis and must be aware of any changes in their normal bodily functions. Some symptoms of exposure are tricky, they come and go and come again later. One day the victim does not feel right, the next day they feel fine. Then a week or so later, the victim again does not feel right, but this too passes. None of these warning signs should be disregarded. They are to be recognized as possible signs of exposure, whether they are persistent, strong, mild, or seem to come and go. Persons who suspect that they may have been exposed to a harmful chemical should consult with a physician promptly.

8.6 SUSCEPTIBILITY AND RISK

Not all people are affected to the same degree by the same chemical. Each has different levels of *susceptibility*. Many factors affect an individual's susceptibility, including:

1. *Age* Very young and very old people are usually more easily affected and by a lower dose than are people between the ages of 20 and 60.
2. *Sex.* Some chemicals specifically harm only the male or female reproductive organs. Also, a general harmful effect may be worse in one sex than in the other, because of differences in body chemistry (hormones) and metabolism (the chemical reactions that occur in the body).

3. *General Health.* People who are ill or very run down or tired seem to be more susceptible to injury than those who are more healthy and well rested.

These are only three of the many factors that affect susceptibility to injury. Some of these factors can be personally controlled, such as diet or the general state of health; other factors such as age cannot be controlled.

Risk can be defined as *the probability of being injured.* This probability, or risk, is influenced by the exposure aspects: How much, how long and how often, and by the route of exposure. Risk is also influenced by personal susceptibility factors: age, weight, sex, inherited characteristics, general health, and many others. It is not a simple matter to predict the risk of harm because it is influenced by personal susceptibilities. Not only do these differ from one person to another, they may vary for the same person from one day to the next, and they may vary without that person being aware of a change.

8.7   TREATMENT OF EXPOSURE

The author of this chapter is not a physician; this chapter is not intended to give specific medical advice, only recommended general first aid procedures.

The general principle in all first aid procedures is to *dilute* the chemical as much as possible, as quickly as possible, and to get professional medical care immediately. The following suggested procedures apply not only when someone else is accidentally exposed, but also when you, the reader, are exposed. In many cases, fellow workers may not know how to help when a victim is exposed to a chemical, or may panic at the sight of an accident. A person then may in fact save their own life by telling someone else how to help.

8.7.1   Acute Exposure

For local reactions, first remove the chemical from the site of exposure. That is, wash the exposed area immediately with copious running water (e.g., a safety shower delivering 50 gal/min) for at least 15 min. While washing, remove any contaminated clothing, including shoes, socks, wrist-watches, belts, and so on. Do not attempt to wash off the chemical with a solvent. In many cases this will only serve to drive the chemical into the skin more rapidly.

Eyes should also be rinsed immediately in copious running water for at least 15 min (e.g., at an eyewash station). While the eyes are being washed, hold the upper and lower eyelids away from the eyeball and move the eye up, down, left, right, continuously as the water washes the eye.

While the washing procedure is going on, a fellow worker should call for emergency assistance, doctor, hospital, or ambulance, informing them of the details including the chemical involved. Only when specifically directed by a physician (not a nurse or paramedical person) should the 15-min minimum washing be terminated prematurely and the victim removed for other treatment. Almost always the most important immediate action is to wash the chemical away; only a 15-min minimum washing with copious flowing water is known to be generally effective.

Under all circumstances, no matter how minor the exposure may appear to be, someone other than the victim should take the victim to a hospital or other suitable location for further treatment.

First aid treatment for ingestion is not the same for all chemicals. Unless the label or material safety data sheet (see Section 8.8) specifically recommends that vomiting be induced (by drinking two glasses of water and tickling the back of the throat), *do not induce vomiting*. Even if vomiting is reliably recommended, never give anything by mouth to an unconscious victim. Always, take the victim to the hospital promptly. Again, the victim should not go alone and the hospital should be alerted in advance that the victim is coming and the chemical should be identified, if possible.

Inhalation of a chemical should be treated similarly to local and oral exposure: dilute the chemical. In this case the chemical is diluted with fresh air. Get away from the contaminated area immediately and breathe fresh air. If there is difficulty in breathing, a fellow worker should be informed so that artificial respiration can be maintained on the way to the hospital. If a victim is overcome and is unconscious, remove to fresh air immediately and if not breathing, begin artificial respiration on the spot. Meanwhile, the emergency services should be alerted.

The most important thing is to remain calm and not panic (easier said than done). Do not shrug off an accidental exposure as not being serious enough to report or not serious enough to seek medical attention. All exposures should be reported and all exposures should receive medical attention, with no exceptions. Above all else, *a victim should never assume that they are able to treat even a small exposure themselves*.

8.7.2 Chronic Exposure

Persons who think they have been overexposed to a chemical recently, or even some time ago, should see a physician even though they may now feel fine. A physician specializing in industrial medicine is preferable. They should be prepared to identify the chemical(s) they may have been exposed to, when, and for how long, and bring along a copy of any Material Safety Data Sheets (see Section 8.8) for those chemicals. If a person is not sure they have been overexposed to a harmful chemical but "just don't feel right lately," they should see a physician. Some chemicals do not have any immediate or local effects, but cause harm after a period of repeated low-dose exposures. Persons who work with chemicals should have annual medical examinations to establish baseline medical values that can be compared from year to year. In this way, the physician may be able to pick up a change in a person's medical condition before the victim feels ill. The earlier an exposure is recognized, the better the chance of successful treatment.

8.8   PREVENTION OF HARM

The best way to prevent being harmed by a chemical is to prevent being overexposed to it. Laboratory workers should learn all about the possible harmful effects of every chemical they work with or even work near. They should learn and use the proper procedures for protecting themselves from excess exposures and learn and use the proper first aid procedures for treating an exposure. This information should be known *before* any work with that chemical is begun. It is found in brief form on the label of the chemical container and in detail in the Material Safety Data Sheet (MSDS) for that chemical. See Chapter 2.

Federal and state regulations now require that all available hazard information on a chemical be available to some workers by their employers before they begin work with that chemical. This information is contained in the MSDS, which the manufacturer of the chemical is required to furnish to all who purchase the chemical. See Chapter 12.

8.8.1   Knowing the Dangers

The following discussion deals with the *health hazard data* section of an MSDS.

Section 8.3.1 discussed the concept of threshold. For most chemicals, there is a *safe dose,* an amount to which a person can be exposed without harm. This amount is usually very small and is listed in an MSDS as the threshold limit value (TLV®), or as the permissible exposure limit (PEL). In the discussion that follows PELs and TLV®s are considered to be equivalent. The average amount of a chemical in the air to which a normal person can be exposed for 8 hours a day, 5 days a week, for their entire adult life without being harmed is the TLV®/Time Weighted Average (TLV®/TWA). This value refers only to a person's exposure by breathing the chemical in the air; it has nothing to do with local contact or ingestion. The one exception is those chemicals that can be absorbed through the skin. These are identified by a special skin notation. Another term is the TLV®/STEL, which is the TLV®/Short Term Exposure Limit. It is the amount of a chemical in the air, greater than the TLV®, to which a normal person may be exposed for a short period, (no more than 15 min) without harm, provided there are no more than four such exposures a day. Another term is the TLV®/C, the TLV®/Ceiling Limit. This is the maximum amount of a chemical in the air which must not be exceeded.

The concept of a TLV® should be used wisely. Not all TLV®s have been established by scientific experimentation. Many are estimates based on experience with a chemical over many years or based on known information about other chemicals with similar properties. The TLV®s are listed in a paperback book available from the American Conference of Governmental Industrial Hygienists (ACGIH); see the listing at the end of this chapter.

The Health Hazard Data section of an MSDS also addresses the toxicity of the chemical, an indication of how much harm the chemical can cause. The level of toxicity (harm) is indicated by a value referred to as the Lethal Dose-50 (LD_{50}) or the Lethal Concentration-50 (LC_{50}). The LD_{50} is the amount of a chemical that kills 50% of the animals it is given to by any route other than breathing. The LC_{50} is the amount of a chemical in the air that kills 50% of the animals exposed to it. You may also see an LD_{lo} or LC_{lo}, which are the lowest doses or concentrations of a chemical known to kill at least one animal by the appropriate route of exposure. The TD_{lo} (Toxic Dose -Low) is the lowest dose of a chemical that causes any signs of toxicity (not just death) in animals or humans. The LD and TD values are usually expressed in terms of a dosage—the amount of a chemical per unit of body weight, that is, milligrams of chemical per kilogram of body weight of the test animal (mg/kg). In the case of chemical concentration in the air, the LC values are expressed in terms of the number of parts of a chemical per 1 million parts of air (ppm) or the amount of a chemical by weight per unit volume of air (mg/m^3).

Although the absolute LD or LC number for a test animal may not be the same as for a human, the relative toxicity will usually not change between animals and humans. That is to say, if a chemical is highly toxic to an animal, it will most likely be highly toxic to humans. Table 8.4 will serve as an approximate translation of LD_{50} values into everyday language.

Another important piece of information in the MSDS is the *contact hazard*. This is an indication of the local harmful effects of a chemical and describes just how irritating or corrosive a chemical is to eyes, skin, and the respiratory tract. Unlike the Health Hazard, the Contact Hazard is not

Table 8.4. Degree of Toxicity Based on Ingestion of a Chemical[a,b]

Animal LD_{50}	Approximate Amount Ingested by 150 lb Adult	Toxicity Classification
Up to 50 mg/kg	1 teaspoon or less	Extremely toxic
50–500 mg/kg	1 teaspoon to 1 oz	Very toxic
500–5000 mg/kg	1 oz to 1 pt	Moderately toxic
5–15 g/kg	1 pt to 1 qt	Slightly toxic
Over 15 g/kg	> 1 qt	Practically nontoxic

[a]Adapted from: R. E. Gosselin, et al., *Clinical Toxicology of Commercial Products*, 5th ed., Williams & Wilkins, Baltimore, MD, 1984.
[b]This table indicates the approximate amount of a chemical ingested by humans which corresponds to an animal Lethal Dose-50 (LD_{50}).

usually expressed in terms of numbers, but rather in descriptive word form, such as *highly irritating to eyes*. Sensitization may be listed as either a Health or Contact Hazard.

The Signs and Symptoms information in an MSDS will often be listed as *effects of overexposure*. Compare the description in the MSDS with those listed in Table 8.3 of this chapter to be sure you understand the possible consequences of being overexposed. When a word or phrase is used in the MSDS but is not in Table 8.3, it may be found in the glossary (Section 8.10). Also, readers may wish to consult a medical dictionary or their supervisor to find out what is meant. It is important to be sure of the meaning; guesses about the meaning could have serious adverse consequences.

When there is incomplete or even no health or contact hazard information listed in an MSDS, it is advisable to treat the chemical as if it were the most toxic chemical known. This means take every precaution to prevent any contact with that chemical (see Chapter 9). Such a decision should not create problems between employee and supervisor, but failure to take full precautions with a chemical of unknown properties could place the employee's life in jeopardy. Once exposed, a victim cannot be unexposed.

8.8.2 How to Prevent Overexposure

Most MSDSs contain a *personal protective equipment* section. When properly prepared, this section provides specific information on how to prevent or reduce exposure to the specific chemical. The information includes equipment to prevent eye and skin contact and respiratory exposure. Depending on the chemical, the recommended equipment will range from simple safety goggles and a lab apron to full body enclosure with a self-contained breathing apparatus.

Pay close attention to the details in this section. If specific eye protection is described, ordinary corrective eyeglasses will not do; use only the type described. Do not wear contact lenses when working with chemicals, especially corrosives. If a corrosive is splashed on the eye it can be trapped between the contact lens and the surface of the eye and make an injury more severe.

When impervious skin coverings and a certain type of gloves are listed, use only that type; only the specified type will prevent the chemical from penetrating through the covering to the skin; and even then, keep in mind that no such barrier can prevent penetration for more than a few hours.

If the information in an MSDS is vague, such as "use eye protection," "wear gloves," "use local ventilation," employees should ask their supervisor for specific information. If the supervisor does not know, contact the industrial hygienist. If no one who is competent knows, the risk of handling that chemical is likely to be very large. For further information, see Chapter 9. Beware of MSDS that do not give specific protective information!

Because appropriate protective equipment is specific for each chemical, it is beyond the scope of this chapter to give such information for all chemicals. In a competently prepared MSDS, the section on personal protective equipment is probably the most important section. By applying that information, the harmful toxic effects of chemical exposures may be prevented. By ignoring that information, life may be jeopardized.

8.9   CLOSING REMARKS

Overexposure to a harmful chemical is the result of improper handling. Improper handling occurs because of a lack of knowledge, because of carelessness in applying that knowledge, or a combination of these two. The consequences of overexposure to harmful chemicals can be severe. The ounce of prevention described in this chapter is indeed worth a pound of cure. Use chemicals wisely.

8.10   GLOSSARY

Acidosis: body acid imbalance.

Acute hepatitis: liver damage without jaundice.

Adrenal gland: organ attached to the kidney.

Aerosol: a suspension of very small particles of a liquid or solid in a gas.

Albuminuria: protein in the urine.

Alkalosis: increase in body alkalinity.

Anaphylactic: pertaining to an extreme allergic reaction.

Anemia: fewer red blood cells than normal.

Arteriosclerosis: hardening of the arteries.

Aspirate: to inhale liquid into the lungs.

Atrophy: to decrease in size or waste away.

Automatic nervous system: controls voluntary movements.

Bilirubinuria: bilirubin in urine.

Bone marrow depression: inactivity of blood-forming organ.

Calcification: deposition of calcium in tissues.

Carcinogenic: capable of causing cancer.

Carcinoma: cancerous growth (tumor).

Cardiovascular: pertaining to heart and blood vessels.

Central nervous system: controls involuntary bodily functions such as breathing and heart beat.

Cerebral: pertaining to brain.

Cholinesterase: chemical in the body that relays nerve cell signals.

Chloracne: a skin disease resembling childhood acne but caused by exposure to chlorinated aromatic organic compounds.

Chromosome: material inside a cell that carries the genetic information.

Cirrhosis: progressive disease of the liver.

Colitis: inflammation of the large intestine.

Cornea: transparent covering of the eye.

Cystitis: inflammation of the bladder.

Degeneration: deterioration; worsening.

Demyelination: destruction of sheath surrounding nerves.

Emetic: a chemical that induces vomiting.

Emphysema: debilitating disease of the lung.

Encephalitis: inflammation of the brain.

Encephalopathy: brain disease.

Endocrine gland: hormone-secreting gland.

Epileptiform fits: seizures.

Epithelium: outermost living layer of the skin.

Esophagus: tube connecting mouth and stomach.

Fibrosis: fibrous scars.

Gallbladder: organ that secretes bile.

Gastric: pertaining to the stomach.

Gastrointestinal: pertaining to the stomach and intestines.

Genotoxic: capable of damaging the genetic material.

Glaucoma: increased pressure inside the eyes.

Glycosuria: glucose in the urine.

Hematoma: swelling containing blood.

Hematopoietic: formation of blood cells.

Hemoglobinuria: hemoglobin in the urine.

Hemolysis: destruction of red blood cells.

Hemolytic anemia: loss of red blood cells resulting from destruction.

Hormone: a biochemical secreted by the body that exerts an effect on an organ elsewhere in the body.

Hyperemia: congestion of blood vessels from excess blood.

Hyperglycemia: high blood sugar level.

Hypertension: high blood pressure.

Hypertrophy: exaggerated growth of a tissue.

Hypotension: low blood pressure.

Intoxication: state of being poisoned by a toxic chemical.

Keratosis: inflammation of the cornea.

Laryngeal: upper throat area.

Larynx: voice box.

Lesion: diseased or damaged tissue.

Leukemia: cancer of the blood cells.

Lymph: clear, yellow fluid found throughout the body.

Lymph nodes: glands that produce lymph.

Lymphatic system: vessels that carry the lymph to the blood.

Malignant: very injurious or deadly.

Mammary tissue: milk-producing tissue of the breast.

Metabolism: the sum total of all the biochemical reactions that occur in cells.

Methemoglobinemia: type of blood disease.

Mucous membrane: tissue lining of nose, mouth, esophagus, stomach, and intestine.

Mutagenic: capable of producing changes in the genetic material.

Mutant: an organism that has undergone a genetic change.

Narcosis: state of stupor or unconsciousness.

Nausea: upsct stomach; feeling of need to vomit.

Necrosis: dead tissue.

Neoplasm: abnormal tissue growth.

Nephritis: inflammation of the kidneys.

Nephrosis: kidney degeneration.

Neurogenic: pertaining to the nerves.

Neurologic: pertaining to the nervous system.

Ocular: pertaining to the eye.

Olfactory: pertaining to the sense of smell.

Osteoporosis: a condition in which bones become very fragile.

Osteosclerosis: hardening of bone tissue.

Ovarian: pertaining to the egg-forming organ in the female reproductive system.

Pancreas: insulin producing gland.

Pancreatitis: inflammation of the pancreas.

Papilloma: type of tumor.

Periorbital: area surrounding the eye socket.

Peripheral nervous system: nervous system controlling the arms and legs.

Peripheral neuritis: inflammation of the peripheral nerves.

Peritoneal: pertaining to the body cavity that surrounds all the abdominal organs.

Pharyngeal: pertaining to the pharynx.

Pharynx: sac surrounding the mouth, nose, and esophagus.

Phlebitis: inflammation of a vein.

Photoallergy: allergic response to a combination of a chemical and sunlight.

Photosensitization: word used to describe either photoallergy or phototoxicity.

Phototoxicity: irritant response to a combination of a chemical and sunlight.

Pigmentation: coloration.

Plasma: fluid part of blood and lymph.

Pleural thickening: thickening of tissue surrounding the lungs.

Pleurisy: inflammation of the lung cavity.

Pneumoconiosis: degenerative respiratory disease.

Pneumonia: infectious disease of the lungs that impairs breathing.

Pneumonitis: inflammation of the lungs.

Polyneuropathy: disease of several peripheral nerves.

Proteinuria: protein in the urine.

Ptosis: drooping of upper eyelid.

Pulmonary fibrosis: fibrous tissue forming in the lung.

Reproductive effects: pertaining to birth defects, death of a developing baby prior to birth, inability to have children (both men and women), and so on.

Respirable: capable of being inhaled.

Respiration: inhalation of air; breathing.

Salivary glands: glands in the mouth that secrete saliva.

Sarcoma: type of cancerous tumor.

Sensitization: becoming allergic.

Silicosis: lung disease caused from inhaling silica.

Spleen: organ that filters blood.

Teratogenic: capable of producing birth defects.

Testicular atrophy: wasting away of male reproductive organs.

Testis: male reproductive organs.

Tetany: intermittent spasms.

Thrombosis: blood clot.

Thymus: organ that forms cells involved in the immune response.

Thyroid: hormone-producing gland in the throat.

Trachea: passageway from nose to lungs.

Transplacental: across the placenta from mother to developing baby.

Tumor: cancerous growth.

Ulceration: destroyed tissue.

Urinary system: kidney, bladder, and connecting tubes.

Urologic: pertaining to the urinary system.

Uterine: pertaining to the uterus or womb (part of female reproductive system).

Vascular thrombosis: blood clot.

Vasoconstriction: narrowing of the blood vessels.

Ventricular fibrillation: rapid contractions of the ventricles of the heart.

BIBLIOGRAPHY

Arena, J., *Poisoning: Toxicology, Symptoms and Treatments*. 4th ed., Thomas, Springfield MA, 1979.

Clayton, G. D., and Clayton, F. E., Eds., "Toxicology," Vols. 2A, 2B, and 2C of *Patty's Industrial Hygiene and Toxicology*, 3rd ed., Wiley-Interscience, New York, 1981/1982.

Documentation of Threshold Limit Values, American Conference of Governmental Industrial Hygienists, Cincinnati, OH, 1985.

Dorland's Illustrated Medical Dictionary, 26th ed., Saunders, Philadelphia, PA, 1981.

Doull, J., *Casarett and Doull's Toxicology*, 2nd ed., Macmillan, New York, 1980.

Gosselin R. E., et al. *Clinical Toxicology of Commercial Products*, 5th ed., Williams & Wilkins, Baltimore, MD, 1984.

Hamilton, A., *Industrial Toxicology*, 3rd ed., Publishing Sciences Group, MA, 1974.

Parmeggiami, L., Technical Ed., *Encyclopedia of Occupational Health and Safety*, 3rd ed., International Labour Office, Geneva, Switzerland, 1983.

Sittig, M., *Handbook of Toxic and Hazardous Chemicals and Carcinogens*, 2nd ed., Noyes, Park Ridge, NJ, 1985.

Threshold Limit Values and Biological Exposure Indices for 1985–1986, American Conference of Governmental Industrial Hygienists, Cincinnati, OH, 1986.

CHAPTER 9

Handling and Management of Hazardous Research Chemicals

Andrew T. Prokopetz* and Douglas B. Walters*

National Institute of Environmental Health Sciences, Research Triangle Park
North Carolina

9.1 INTRODUCTION

The importance of the safe handling and use of chemicals takes on meaning in today's world unlike ever before. Not only is this evident in every aspect of the news media, but the status of regulations and recommendations is constantly being expanded and updated to encompass ever broadening areas of concern. While chemical health and safety in the *research* laboratory is of concern to a large segment of the scientific community, few standards and regulations have been promulgated which apply directly to these work areas. This, however, may change significantly in the near future if the OSHA Laboratory Standard is finalized. Of course, the burden for the safe use of hazardous research chemicals will always ultimately fall on the organization and its employees.

9.1.1 Regulations

For several types of hazardous chemicals specific guidelines and regulations can be found with little effort. Examples of these types of chemicals include radiolabled materials, legally defined carcinogens, certain pesticides, flammable and explosive materials, and controlled drugs. In addition, specific standards sometimes exist for certain defined operations involving hazardous chemicals. These work practices include: shipping requirements, waste regulations, clean air and water standards, right-to-know and other OSHA regulations, and fire department requirements for the storage of flammable and explosive materials. In short, numerous groups, starting at the federal

*This chapter was written by Mr. Prokopetz and Dr. Walters in their private capacities. No official support or endorsement by NIEHS is intended or should be inferred.

level and proceeding through state government to local organizations, are involved in regulating hazardous chemicals. The majority of research chemicals, however, do not fall under a specific category and the safe use of these products is dependent on the product practices of the individual investigators and the organizational policies under which they work.

9.1.2 Definitions of Hazardous Research Chemicals

For the purposes of this chapter, hazardous research chemicals (henceforth in this chapter called *hazardous chemicals*) are defined as belonging to one of three groups:

1. Chemicals known to have undesirable biological effects, either acutely or chronically. Appropriate regard is given to the size of the dose, duration, and type of exposure, as well as the physical state of the chemical necessary to produce such effects.

2. Chemicals for which reliable toxicity information is not available, whether or not they are highly suspect because of their similarity in chemical structure or function to other known toxic agents.

3. Chemicals that are explosive or violently reactive. Certainly other categories of chemicals, such as combustible or flammable, as well as caustic and corrosive materials, can also be easily categorized as hazardous. The intent in this chapter, however, is not to describe safety concerns for the more commonly encountered chemicals found in research laboratories, but rather to discuss those materials for which adequate training or experience is not necessarily acquired.

9.1.3 Administrative Controls

Responsibility for the safe use of hazardous chemicals begins when the decision is made to enter into the type of research necessitating the use of these materials. It is at this time that a uniform, general commitment for the safe use of these products is required. It is best to formalize this policy in a health and safety plan, handbook, manual, or other similar document. The purpose of such a document is to address the administrative and policy controls which the organization intends to follow to minimize the potential risk of personal exposure to, or environmental release of, hazardous chemicals.

It is highly likely that hazardous chemicals may be involved in only a portion of the work being performed in an organization or at a particular facility within the institution. Therefore, the formulation of a set of minimum requirements is necessary. The purpose of the minimum requirements is to define, in general terms, the control technology used in the specific laboratories where the work is actually being done.

Finally, a safety protocol is necessary that describes in detail who, what, where, when, and how the work will be done and precisely which hazardous materials or classes of chemicals will be used. Therefore, it can be seen that safe management of hazardous chemicals consists of a hierarchy of three levels of administrative control:

The Corporate Health and Safety Plan

Laboratory Specific Minimium Requirements

Chemical Specific Safety Protocols

9.1.4 Industrial Hygiene Principles

The basis for these administrative controls lies in the application of the three fundamental principles of industrial hygiene. These principles are:

Recognition

Evaluation

Control

This three-step approach is necessary as a foundation for the effective management of workplace hazards. For the safety professional, the object is the solution of occupational safety and health problems. The first step in this process is the recognition of potential problem areas and conditions where hazardous research chemicals are present. Not only is it important to know the precise chemicals being used and their physical properties, but the technology used to control exposure is also required (see also Chapter 16).

Evaluation of the environmental conditions is also needed to assess exposure potential and the concentrations or levels of exposure with respect to time. Knowledge of the toxicological properties of the chemicals, both acute and chronic, is necessary to determine possible toxic effects.

Finally, the control technology used to restrict the spread of chemical contamination needs to be assessed. The first step in this process should be the possible substitution of the hazardous material, process, or apparatus that is responsible for creating the potential hazard. If this is not feasible, other measures such as isolation or enclosure should be considered. A more costly form of chemical containment is the application of engineering controls such as ventilation and treatment, for example, hoods, scrubbers, and filtration. The use of personal protective equipment should never be relied on as a primary method of control. Such equipment is always subject to breaches in integrity, that is, glove permeation, respirator breakthrough, equipment failure, and the general lack of 100% protection. Personal protective equipment always needs to be used in conjunction with other means of control such as engineering technology. In this manner the personal protective equipment serves as backup and establishes a redundant system.

9.2 HEALTH AND SAFETY PLAN

9.2.1 Purpose

The purpose of a health and safety plan is to describe, in detail, the philosophies, responsibilities, organization, policies, and procedures which constitute a specific company's health and safety program. Such a program should be committed to effectively controlling and minimizing employee exposure to occupationally hazardous agents and conditions, as well as to safeguarding the environment. The scope of the plan should include, in general terms, identification of pertinent chemical, physical, and biological hazards potentially present in the organization's various institutes, facilities, and laboratories. The plan should then describe corporate efforts to contain and control these hazards. In effect, the entire health and safety program, its organization, and how it is implemented, needs to be described in the plan.

9.2.2 Philosophy

Philosophically, such a plan begins by outlining the company's goal of developing, providing and maintaining facilities, conditions, and practices that provide a safe working environment. A major objective is to completely describe the corporate health and safety hierarchy and responsibility as well as the flow of authority and communication. In this respect, nothing suits such a document better than a personal letter from director level management outlining the purpose, objective, and goals of the health and safety program.

9.2.3 Responsibilities

Next, the responsibilities for initiating, implementing, and maintaining such an occupational safety and health program should be given in detail. This begins with a description of the health and safety obligations of upper management in the parent organization and flows through all levels of management and supervision concluding with the individual workers and their duties and responsibilities. Responsibilities are described on each management level, both in general terms and specifically where actual reponsibility resides with designated individuals or groups. It is imperative that every new employee be given a copy of the plan when they begin work. The immediate supervisors have the responsibility of designating particular sections that are to be read first.

9.2.4 Organization

This section includes a full explanation of all safety and health offices, organizations, committees, and their staffs. Efforts should be made to ensure that representatives from all appropriate employee groups, such as unions, are also included on safety committees. A key point in every program is the provision that safety organizations or offices must have a direct line of communication with, and authority from, director level management. The responsibilities, authorities, interactions, and relationships with director level management need to be precisely detailed. Organizational flow charts are particularly useful in this section, where applicable. Provisions also need to be made to ensure that the plan is formally reviewed and updated on a routine periodic basis, every two years at a minimum.

9.2.5 Policies

A description of the processes used in formulating and carrying out such policies and measures to provide and maintain a healthy and safe workplace environment is necessary. With respect to hazardous research chemicals, this means inclusion of all steps from the initial acquisition of the chemicals and the completion of appropiate health and safety protocols and standard operating procedures (SOPs), through safe storage and handling, to the final disposal of excess materials and contaminated waste.

It is important to once again emphasize that the health and safety plan describes the organization's entire health and safety program in detail, whereas minimium requirements explain policies for specific laboratories or groups within an organization SOPs and safety protocols explain precisely *how* a particular task will actually be performed. All three levels of administrative control are necessary.

The plan should delineate and explain all policies pertaining to the following:

New employee training
Updating employee training
Medical surveillance program
Biological monitoring program
Health and safety regulations
Restricted access areas
Eating and smoking areas
Precautionary signs and labels
Fire prevention and protection
Emergency procedures
Evacuation procedures
Chemical acquisition
Chemical storage
Chemical transportation and shipment

Waste disposal

General housekeeping practices

Engineering controls

Personal protective equipment and clothing

Respirator fit programs

Personal and environmental monitoring

Laboratory safety inspections

Record keeping

Safety audits

Wherever possible, schematic diagrams should be included as an aid to clarification and explanation. Examples of areas where this is particularly applicable include evacuation routes and location of safety showers, eyewashes, and fire control equipment. It is also important to include completed copies of required documents, such as safety protocols, for the users of the plan.

9.2.5.1 Record Keeping. The importance of keeping good health and safety records can not be overemphasized. Records should be kept in a secured area, archived for 40 yr, microfiched on a periodic basis to conserve space, and be made available to employees and placed in their personal files as necessary. In the case of medical records, current thinking indicates that employees and their personal physicians should be apprised of any and all findings as soon as they are available, whether or not the medical item in question is work related.

Examples of records which should be archived include:

Name of employee (include title and function)

Medical surveillance records

Biological monitoring results

Air sampling results, logs of ventilation system performance and maintenance

Waste disposal records

Respirator fit-testing records

Accident and incident reports

Training records

Material safety data sheets

Signed employee statements that they have read and understood their duties and the health and safety plan

9.2.6   Procedures

All laboratories that handle hazardous chemicals should be required to have written standard operating procedures (SOPs). The purpose of the SOPs is to specify precisely how routine and often repeated tasks involving hazardous chemicals are performed. The SOPs, in this respect, apply to all employees in a specific work location, that is, hazardous materials handling facility, weighing room, barrier facility, toxicology testing areas, and so on. In contrast, a safety protocol explains precisely how and where all aspects of a very specific research project with a particular chemical or class of chemicals will be carried out by a very limited group of people within a defined research area. Details on the preparation of a safety protocol will be given later in this chapter.

A list of SOPs that are usually necessary follows:

Restricted access policy

Visitors access policy

Employee training

Medical surveillance

Respirator training and fit program

Eye protection

Personal protective equipment

General housekeeping practices

Ventilation system maintainance

Storage and transportation of chemicals

Accident response

Emergency response

Evacuation

Spill clean-up

Waste disposal

Special procedures (use of radiolabeled and infectious agents, controlled substances)

9.3   MINIMUM REQUIREMENTS AND REGULATIONS

Once policy decisions have been formally confirmed by the corporation in a health and safety plan, minimum requirements need to be established to assure safe conditions. These minimum requirements encompass the administrative, operational, engineering, and personal protection controls appropriate for the particular research laboratory. They should be viewed as a whole and not out of context. One of the key maxims in health and safety is to build in redundancy to ensure adequate protection in case of failure at the primary control level. Whereas the health and safety plan sets the overall policy, the minimum requirements are tailored to the specific type of work involved. There may be several sets of minimum requirements within one corporation, depending on the amounts of hazardous chemicals involved, the duration of the research and its potential for long term exposure, and the type of research (e.g., *in vivo* or *in vitro* toxicological research, or synthetic preparatory chemistry). The researchers that are affected by the minimum requirements will appreciate the concern and attention in selection of the appropriate requirements, rather than an irresponsible enactment of safety requirements with no relevance to the work being performed.

The language used throughout this section is in terms of recommendations to the reader ("should") rather than in mandatory terms ("will," "must," or "shall"). This is a fine distinction but one that needs to be considered by the laboratory in writing its safety policies. Does the laboratory choose the latter terminology in order to have enforcement capabilities over its safety requirements (and also, in certain cases meet local, state, and federal regulations) or merely treat them as safety recommendations or guidelines?

It should be noted that most state worker's compensation statutes have a reduction in award to the injured worker if the worker was not following the laboratory's safety *requirements* (Some states require that the safety provision be submitted by the employer to the commission and approved by them *prior to* the injury by accident.) Thus, the laboratory workers may not be treated as negligently contributing to their own accidents, if by the laboratory's own admission the provisions were not important enough to make the workers follow them. An example of a set of minimum requirements developed for support laboratories conducting research with hazardous chemicals is included at the end of this chapter, as Appendix 9.1. This appendix may be used by readers as a basis for tailoring an individualized set of minimum requirements for their own laboratories.

Minimum requirements can be effective only if, first, they are designed correctly for the appropriate area of research and second, they are effectively monitored for deviations from corporate policy. The focal point may either be the researcher in charge or, more appropriately, the health and safety officer.

9.3.1 Health and Safety Officer

Depending on the nature and extent of the work and the possibilities for exposure, the research laboratory should have a health and safety officer directly assigned to the laboratory. This is necessary, even if only on a part-time basis, to review and monitor operations critically. The safety officer should not report directly to the researcher but to someone in the corporate structure with the authority to intercede in cases of unsafe work practices. This helps ensure an unbiased viewpoint and avoids conflicts of interest. The health and safety officer should be educated at least to the bachelor's level in a technical field appropriate to the laboratory's research area. The safety officer should also have actual work experience in occupational health and safety and be enrolled in relevant health and safety training every one to two years to keep current in the field. The safety officer should have experience in working with the requirements of state, local, and federal statutes relating to occupational health and safety, environmental protection, and monitoring.

This health and safety officer then becomes the central focus of the minimum requirements, in that he or she may be instrumental in designing the requirements and later in implementing them. This may involve identification of problem areas and execution of corrective actions, which include redrafting certain aspects of the requirements.

9.3.2 Administrative or Operational Controls

The first aspect of the minimum requirements to be designed for the research laboratory should be the administrative controls. Attention should be directed toward a review of the applicable local, state, and federal regulations which may have impact on the laboratory. These regulations may relate to occupational health and safety, transportation and handling, radioactive materials, controlled substances, and environmental protection. The local and state right-to-know laws should be reviewed for applicability. This will then set the framework for the laboratory's minimum requirements. Areas that are either not addressed or only partially covered by the regulations can be supplemented through the following recommendations.

The second aspect of the administrative or operational control portion of the minimum requirements is to incorporate the SOPs for the research laboratory. There should be SOPs to cover the following:

1. Emergency/evacuation procedures in the event of a spill or leak of a hazardous material.
2. Storage and use of emergency protective equipment.
3. Training format for handling hazardous materials.
4. Fire training procedures/alarm system briefing/emergency evacuation preplanning.
5. General safety procedures for the particular research laboratory (e.g., safety glasses, eating and drinking area restrictions, clothing requirements, dry sweeping restrictions).
6. Warning signs and labels for the particular research laboratory.

The next area of administrative controls is occupational medical surveillance. All personnel who will be working with hazardous materials in the research laboratory should have a baseline medical examination, preferably a preemployment physical. This can serve to protect not only the employee but also the employer in a number of ways. For example, many states follow the last injurious rule. If an employer hires an employee who has been injured while in the scope of his/her employment under a different employer, but the employee was never compensated, and the present employer contributes, however slightly, to the employee's injury, the present employer must compensate the employee for the full extent of his/her injuries. The employer should therefore be assessed of the potential consequences of what type of work this employee cannot safely perform. Follow-up examinations should be performed every 12 to 18 months and at the time the employee leaves the research laboratory. Depending on the specific chemical's safety and handling protocol (e.g., in the case of work involving cholinesterase inhibitors) or where probable exposure to a hazardous chemical is indicated, the frequency of follow-up may be more often. Provisions should be made to ensure that those employees required to wear respirators be approved by a physician for use of this equipment. While the general scope of the medical examination should be specified in the health and safety plan, specific tests should be left to the discretion of the occupational physician, once the physician is informed of the hazardous materials to which the employee may be exposed.

Medical surveillance records must be maintained for each employee and each employee's files should include all accident/incident reports and exposure monitoring results. Any evidence of employee exposure that is discovered upon examination must be transmitted to the employee promptly. It is of interest to note that recent court cases have held that once employers undertake a duty to perform medical exams, they become obligated to transmit *all* information to the employee, whether or not work related. Failure to do so has been found to be negligent. Most importantly, the physician–patient obligations and duties are not the only ones to require action. An exposure immediately necessitates identification, evaluation, and control of the laboratory practices that may have contributed to the exposure, to prevent similar occurrences to other employees.

Finally, the last area of administrative controls is the provisions for injury or incident reporting. Records should be maintained for any minor or major injury as well as probable exposure to a hazardous material. The report should detail the full description of the incident, the time and date, chemical involved, protective equipment worn at the time of the incident, medical care required and if follow-up is necessary, who the incident was reported to and when, what SOPs were applicable at the time of the incident, impact of the incident on employees, facility, and environmental, remedial actions taken, and any follow-up that is planned to prevent similar incidents in the future.

9.3.3 Engineering Controls

The primary containment control provision utilized in the research laboratory should be engineering controls. It is a general health and safety principle to place the most hazardous operations in the areas with the most negative pressure. The hazardous operations research laboratory should operate under a series of increasing negative pressures, so that the airflow is from areas of lower contamination potential to higher contamination potential. First, the research laboratory should be isolated from other laboratory facilities and have its air supply under negative pressure with respect to connecting laboratories and hallways. This will contain all major spills/accidents within this laboratory and not contaminate the surrounding facility. If the common building exhaust system is used to maintain this negative pressure it should be ascertained that the exhaust air is not recirculated to other building areas. This requires that the exhaust ducts not be located near to or upwind of the air intake system.

Operations involving unpacking, storing, weighing, diluting, packing, synthesizing, or reacting hazardous materials (hazardous materials as defined in Section 9.1.2) should be performed in an isolated laboratory, because each of these operations poses a risk of exposure and release of material to the surrounding facility. This isolated laboratory should be designated a limited access area (only those persons authorized by the research scientist in charge may enter) that is posted with proper warning signs. Records should be kept of all personnel entering and exiting. It is recommended that personnel who work in these limited access areas have provisions to shower out (or at the minimum, wash their hands and arms) before leaving the facility.

9.3.3.1 Hoods and Vented Enclosures.

The policy stated in the corporate health and safety plan should indicate that all work involving hazardous materials will be performed in a properly functioning hood or vented enclosure. The minimum requirements should then detail the research laboratory's specific provisions for carrying out this policy.

First, all weighing operations involving hazardous materials should be performed with the balance placed, at all times, in an effective laboratory hood or vented enclosure with a face velocity of at least 50 feet per minute (fpm). It is highly recommended that only the smallest quantity of hazardous material needed be weighed out and handled. If practical, the use of an analytical balance to enforce this policy should be considered. Plastic-backed absorbent matting should be secured inside of the hood or vented enclosure wherever hazardous materials are handled. This will help in the cleanup and decontamination steps, after each working session in the hood, when this matting should be disposed of as hazardous waste.

An effective exhausted enclosure or hood, that is vented to the outside, should be used for operations involving hazardous materials (e.g., unpacking, storing, weighing, diluting, packing, synthesizing, or reacting hazardous materials). The evaluation criteria utilized in determining effectiveness of the enclosure or hood should include a combination of velometer and smoke tube tests to determine if there are sufficient contaminant capture velocities. The value normally recommended is 100 ± 20 fpm, but it should be noted that there are certain chemicals regulated by OSHA which require higher hood face velocities. (See Table 9.1 in Appendix 9.2). A regularly scheduled ventilation system monitoring and maintenance program should exist for all hoods and enclosures where hazardous materials are handled. This should include daily visual inspections before any operation is performed in the hood and quarterly inspections with smoke tube and velometer checks. This monitoring program should be included as a part of the hood operator's education and training program. An overall annual maintenance program should also be established and documentation of the testing kept by the laboratory supervisory personnel. (See Appendix 9.2 for an example of a laboratory hood and local exhaust system evaluation program.)

It is highly recommended that air exhausted from neat hazardous chemical handling areas (as opposed to areas where only the diluted chemical is handled), which handle bulk quantities larger than those typically used in an analytical laboratory (i.e., generally greater than milligram quantities), be passed through HEPA (high efficiency particle arrestor) filters. If volatile chemicals are handled, charcoal filters should also be used down line of the HEPA filters. Depending on the amount and type of chemicals being handled, the use of charcoal filters may not be practical or safe, as some chemicals pose fire and explosion problems (e.g., due to a heat of absorption onto charcoal). Other treatment methods such as electrostatic precipitation, incineration, or reaction prior to the filtration banks should be considered as alternatives. It is important to include these filtration systems in a periodic monitoring and maintenance program to ensure effectiveness. Personnel performing this maintenance should wear the same protective clothing as recommended in Section 9.3.4. It should be emphasized again that reentrainment of the exhaust air into the supply air intake is to be avoided.

Effluent exhaust vapor from analytical instruments (e.g., gas chromatographs and atomic absorption units) should be vented to the outside of the building. (As for vacuum pumps, see Section 9.3.3.2.) The instruments can either be fully enclosed in a hood or, more practically, the exhaust vapors can be captured by a local exhaust setup. This can be accomplished with the use of "elephant trunk" tubing, with one end manifolded into a laboratory hood exhaust duct and the other end incorporated into a small canopy hood positioned over the exhaust port. This allows easy access to the instrument without jeopardizing the employee's safety.

9.3.3.2 General Facilities Engineering Controls. General facilities engineering controls should include, at the minimum, safety showers and eyewash stations located throughout the research facility as required by local, state, and federal regulations. Any vacuum pumps or lines that are used when working with hazardous chemicals should be protected with scrubbers or HEPA filters. If there is any chance of the material getting through the trap or filter, the effluent exhaust should be vented to the outside (e.g., through a laboratory hood). If the material is flammable, the use of an explosion proof vacuum pump is required. When transporting chemicals, a nonbreakable, secured secondary container should be used at all times. Double bagging of contaminated waste material is also highly advocated.

The facility and operations should comply with all applicable state and local fire and building codes. Various provisions can be made to ensure that flammable liquids are stored and handled in a manner that will reduce the risk of fire and/or explosion. First, all nonworking quantities of flammable liquids should be stored in a storage cabinet approved by Underwriters Laboratories or Factory Mutual, or in a designated flammable liquids storage room with fire protection, ventilation, spill containment trays, and electrical equipment meeting the requirements of the National Fire Protection Association (NFPA) Standard Code No. 45 (1). Regardless of storage type, flammable liquids must be segregated from other hazardous materials such as acids, bases, and strong oxidizing agents. If low temperature storage of flammable liquids is required, they should be stored in an explosion-proof refrigerator that is labeled as such. If flammable liquid transfer is performed in the designated storage room, all transfer drums should be grounded and bonded and equipped with pressure relief devices and dead-man valves. If transfer is performed outside of this storage room, it should be carried out over a tray in an effective fume hood. Safety cans should be used when handling small quantities of flammable liquids. Finally, whenever flammable liquids are stored or handled, ignition sources should be eliminated. This especially means *no* smoking.

Fire extinguishers should be selected with consideration not only for the type of hazards to be protected from, but also for the strength of the personnel who might use the extinguishers. For the majority of laboratory applications, water and aqueous film forming foam (AFFF) extinguishers normally have a capacity of $2\frac{1}{2}$ gal. Dry chemical, carbon dioxide, bromodichloromethane (Halon 1211), bromotrifluoromethane (Halon 1301), and foam extinguishers are typically 20- to 30-lb capacity. Fire extinguishers should be conspicuously located where they will be readily accessible and immediately available in the event of a fire. This means that there should be a maximum travel distance of 30 ft to an extinguisher, which is located along the normal paths of travel, including the exists from an area.

Fire blankets and/or safety showers should be located within 25 ft from every entrance to a laboratory where flammable liquids are handled.

9.3.4 Personal Protective Equipment

Personnel in neat hazardous chemical handling areas, who work with bulk quantities larger than those typically handled in an analytical laboratory, should wear a disposable full-body suit over work clothing. No street clothes should be worn in this chemical handling area. The disposable overclothing should be dedicated to the work area and not worn outside of the chemical handling laboratory. The work clothing should be removed when exiting the laboratory.

Other personnel who work in laboratories where neat hazardous chemicals are handled in smaller quantities should wear a disposable full body suit or a disposable laboratory coat over street clothes. This overclothing should be disposed of on a weekly basis but disposed of immediately after contact with any known hazardous chemical. Disposable sleeves in conjunction with a nondisposable laboratory coat may be substituted for this overclothing.

The following personal protective equipment should be worn in areas handling both bulk and smaller quantities of neat hazardous materials. First, personnel should wear two pairs of disaposable gloves of dissimilar material (e.g., PVC, natural rubber, latex), and dispose of both pairs after any known chemical contact or after 2 h of use. Unless the hazardous chemical has been tested for permeation through various glove materials, this procedure should minimize permeation and subsequent skin absorption. Of course, if glove permeation data are available for the neat chemical and the duration of the work falls within the acceptable limits of the impermeability of a particular glove material, then a single pair of this type material may be worn. This information is strictly to provide a minimum safe level of protection. This is an example where the chemical specific safety protocol described in Section 9.4 will supercede the minimum requirements depending on the chemical's hazardous properties.

Second, a NIOSH (National Institute of Occupational Safety and Health) approved half-face respirator equipped with a cartridge, specific for the class of chemicals that are handled, should be worn in the neat chemical handling area. Cartridges should be changed once a month or after 40 h of use and the date of installation indicated on the new cartridges. All components of a written respiratory protection program will be practiced for routine and emergency use of respirators. (See Appendix 9.3 for a recommended respiratory protection program.) Safety glasses should be worn with half-face respirators.

Finally, personnel should wear disposable shoe covers. In areas handling bulk quantities of hazardous chemicals, footwear dedicated to the laboratory should be used in addition to the disposable shoe covers.

For those laboratories not involved in handling neat test chemical (e.g., chemical analysis), a single pair of disposable gloves, a laboratory coat, and safety glasses must be worn.

If any street clothing becomes contaminated with the hazardous chemical (e.g., by a spill), then the clothing should be discarded as laboratory contaminated waste. The same disposal precautions should be taken with street shoes if they become contaminated.

9.3.5 Waste Disposal

All wastes generated by the laboratory handling hazardous research chemicals must be disposed of in accordance with applicable federal, state, or municipal waste disposal regulations. In general, any contaminated solid or liquid waste should be collected in a leak-proof container. In preparation for disposal the container should be closed and placed in a secondary leak-proof container. Disposal containers must be stored in an isolated and secured area in accordance with local, state, and federal regulations. Options for disposal include incineration in a manner consistent with federal (EPA) and local regulations, or in a licensed hazardous waste landfill. If a contractor disposer is to be used, complete information on the firm's licensing should be obtained.

9.4   CHEMICAL SPECIFIC SAFETY PROTOCOL

A safety protocol should explain precisely how and where all aspects of a research project involving a hazardous research chemical will be carried out. This includes listing all personnel involved and specifying the research laboratories where this work is to be carried out. The protocol should be designed as chemical specific.

The safety protocol needs the support of management in order to be effective. The health and safety plan should explain the corporate policy requiring submission of a hazardous chemical safety protocol to the safety office prior to start up of any work involving the use of hazardous chemicals. This explanation should emphasize that the protocol must be reviewed and approved by the safety office and a hazardous chemical review committee before any work may be initiated. Preferably, clearance must be provided before even the hazardous chemical is ordered, received, or stored at the facility. Enough time should be provided to review the proposed procedures adequately and to determine compliance with the minimum requirements. Information in the protocol should allow determination of the degree of hazard involved in the research and the type of containment needed to ensure adequate protection for the employees, facility, and the surrounding environment.

The safety protocol requires the interaction of several individuals in order to be effective. The person responsible for completing the protocol should be the research scientist in charge of the proposed study. The safety protocol should preferably be typed or clearly printed and then forwarded by the researcher to his/her superior for approval. If that superior approves of the manner and use of the hazardous chemical, the protocol is then forwarded to the safety office, which then disseminates it to the protocol review committee. It is the responsibility of the review committee to provide comments and suggestions and to raise questions concerning potential problem areas with regard to the planned research.

Areas of review should include determination of the adequacy of the following: (a) experience and training of personnel; (b) containment facilities and equipment; (c) procedures to minimize exposure; (d) decontamination or deactivation procedures; (e) waste disposal procedures; and (f) emergency procedures.

Turnaround time from the review committee should be a reasonable five working days to foster cooperation with the research staff. During this time the safety officer should inspect the proposed facilities to ensure the adequacy of the planned safety measures (e.g.: Are the hoods performing adequately for the task? Are the fire extinguishers appropriate?). The safety officer should ensure that all personnel involved have the appropriate training and expertise to safely perform the work and also have had recent medical examinations and respirator fit testing, if necessary. For example, if respirators are required, have the personnel been approved by the physician for use of respirators? Have they been fit tested with the respirators indicated in the protocol? Are any other nonroutine medical tests advocated or required for surveillance of the specific chemicals to which the worker may be exposed?

The safety officer should then coordinate his or her comments with the responses from the review committee into a subsequent approval, disapproval, or conditional approval pending revisions under which the research can be performed. This is submitted to the research scientist, who either then proceeds with the work or makes the necessary changes for resubmission to the safety office.

It is the responsibility of the research scientist to ensure that every person involved in the research is made aware of the hazards and the safety procedures that need to be followed to minimize those hazards. The research scientist should also see that each worker is given a copy of the protocol for their own informational use. Finally, it is the responsibility of the safety office to keep all protocols on file for future reference. Protocols may be needed during site inspections, during review of chemical monitoring results, or in consultation with the occupational physician.

Safety protocol formats may vary in the ordering of their organizational structure, but certain key elements need to be present in every safety protocol. The principal user's name should be the first piece of information on the protocol. This allows the review committee and the safety office clear notice of who is in charge of the planned research. A statement of prior training of the principal user should be attached to the protocol so that any deficiencies in experience with the proposed research may be highlighted to the safety office and review committee. Training may then be recommended before the research may be started. All personnel participating in the research should also be clearly indicated on the protocol and copies of their prior training record should be attached.

The full chemical name should then be listed in conjunction with the CAS number and several common synonyms. If any abbreviations are to be used in the following sections, the abbreviations need to be indicated in this section. Health hazard information should be the next priority for inclusion in the protocol. If there are OSHA Permissible Exposure Limits (PEL),

ACGIH Threshold Limit Values-Time Weighted Averages (TLV®–TWA or Short-Term Exposure Limits (STEL), or NIOSH recommended exposure limits (REL), this information should be indicated in the health hazard section. The possible routes of exposure should be delineated as well as the effects of overexposure, both chronic and acute effects.

In addition, the protocol should state the maximum quantity to be purchased or stored and if the material is radiolabeled, the maximum activity to be purchased or stored. The rooms where the material will be stored and handled should then be specified and the ventilation provisions present in these areas indicated. If a special weighing area is to be used, this needs to be emphasized as well as the weighing technique that will be used (e.g., closed cover weighing or use of a vented balance cover).

The experimental procedures should be outlined as follows: the expected amounts to be used per month or week; the principal solvents to be used; where the experiment is to be conducted (e.g., hood, glovebox); and if the experiment is performed in parts, where each part will be conducted. Organic solvents used for extractions or for analytical procedures are to be listed. A summary of the rationale for the experiment should also be included in this section.

There should be an explanation of the safety precautions that will be taken. This includes information on storage, handling, solution preparation, transporting the material, personal protective equipment, ventilation requirements, and any analytical methods available for monitoring possible exposure levels. Emergency procedures should then be described, including first aid procedures for eye or skin contact, and inhalation exposure and ingestion provisions. Spill and leakage cleanup requirements should be outlined with an emphasis on the personal protective equipment required for cleanup of each of the possible physical states of the material. For example, if the material is to be handled both neat and in solution, provisions for cleanup/decontamination of both forms should be described.

Waste management and deactivation/disposal procedures need to be considered in conjunction with the company policies and waste contracts. Chemical and physical property information should be obtained from the literature. If it is not available after thorough searching, *not available* should be indicated on the protocol. The pertinent chemical and physical property information includes: (a) chemical structure; (b) molecular formula; (c) appearance/odor; (d) physical state; (e) melting point; (f) boiling point; (g) specific gravity; (h) vapor density; (i) vapor pressures at the temperatures proposed for use; (j) solubility/insolubility in water, ethanol, acetone, ether, and any other solvents to be used; (k) flash point; (l) autoignition temperature; and (m) flammable limits in air (percentage by volume).

Hazards associated with the chemical resulting from its physical or chemical properties should be highlighted as follows: (a) stability; (b) incompatibility; (c) handling and storage requirements; (d) recommended fire extinguishant; and (e) any other special precautions. Finally, toxicity data should be listed, with information on the route of exposure, species, dose, and the reference source. It would also be helpful to the safety office and review committee if a list of the reference sources, which were used in the preparation of the protocol, was attached to the protocol.

As can be seen from the information required, the researcher preparing the protocol must carefully examine each step of the experiment for potential exposure. The experimental design and respective safety provisions should be explained in sufficient detail to allow adequate evaluation by the reviewers. If the protocol is incomplete, then it should be returned to the researcher for additional information. An inadequate safety protocol can defeat the entire safety management scheme.

9.5   CONCLUSION

The approach described in this chapter to the management of hazardous research chemicals is based on an interactive hierarchy of three levels of administrative control: (a) corporate health and safety plan; (b) laboratory specific minimum requirements; and (c) chemical specific safety protocol. An inadequacy in any one of the levels can jeopardize the safety of the personnel.

Because this scheme is designed in increasing specificity as to the chemical and type and level of work to be performed, the system depends on the solid general chemical health and safety principles developed in the first two levels. Thus, a thorough corporate health and safety plan and appropriate minimum requirements reduce the likelihood for exposure potential due to an inadequate safety protocol, if all the checks and balances, required in the first two levels, are in place. This management scheme should then provide a solid foundation for handling hazardous research chemicals. The key to making this system work is interaction between management and research staff during the initial design and implementation stages and later in the redesign and improvement phases.

Only through interaction can both management and research staff be made aware of the problems and concerns in safely handling hazardous research chemicals.

REFERENCE

1. NFPA Standard Code No. 45, *Fire Protection for Laboratories Using Chemicals*, National Fire Protection Association, Quincy, MA, 1986.

APPENDIX 9.1   EXAMPLE OF MINIMUM REQUIREMENTS FOR SUPPORT LABORATORIES FOR THE MANAGEMENT OF HAZARDOUS RESEARCH CHEMICALS

I. *Administrative Controls*
   A. *Regulations*
      1. All work shall conform to the applicable local, state, and federal statutes relating to occupational health and safety, transportation and handling, and environmental protection.
      2. Where not superseded by this document, guidelines provided by *NIH Guidelines for the Laboratory Use of Chemical Carcinogens* (1) will be followed.
      3. Controlled substances will be handled in accordance with applicable regulations.
   B. *Health and Safety Plan*
      1. There will be written general safety policies at each laboratory including (but not limited to) the following:
         (a) Use of safety glasses in chemical handling areas.
         (b) No mouth pipetting.
         (c) No open-toed shoes.
         (d) No dry sweeping for maintenance or for spill cleanup;
         (e) Eating, drinking, smoking, applying cosmetics, and chewing gum will not be permitted wherever chemicals are handled. The presence of food or smoking materials will not be permitted in these areas.
      2. Personnel who handle (receive, store, weigh, dilute, transport, package, or administer) test chemicals, and/or radiolabeled agents will be provided with written material and trained on the associated hazards of these agents. This training will be conducted by a qualified person and will be properly documented. Training for handling radiolabeled agents will be handled according to any appropriate local, state, and federal regulations. At a minimum, training shall include the recommendations for handling these agents found in the *NIH Guidelines for the Laboratory Use of Chemical Carcinogens* (1).
      3. Warning signs and labels will be used wherever test chemicals are used or stored (i.e., on primary and secondary containers, affixed to entrances to work areas, refrigerators, and on containers holding hazardous waste). These signs and labels will indicate the presence of suspected carcinogenic, mutagenic, and teratogenic hazards, as recommended by *NIH Guidelines for the Laboratory Use of Chemical Carcinogens* (1).
      4. All personnel will receive initial training in fire safety. Course material will include hazard awareness, proper techniques for the handling and storage of flammable liquids, and a briefing on the

alarm system and emergency evacuation preplanning. In addition, hands-on training for appropriate personnel on fire extinguishers in future fire training activities is encouraged.

C. *Standard Operating Procedures*

  1. A written set of emergency/evacuation procedures to be followed by all project personnel in the event of a spill or leak involving the test chemical and/or positive control will be developed and posted in each laboratory. Personnel will be instructed to call for appropriate help (e.g., in-house emergency group or poison control center) in case of an emergency. This plan will address the storage and use of emergency protective equipment.

  2. Emergency protective equipment will not be stored in the laboratory where test chemicals are stored or handled.

  3. The written set of general safety policies will include actions to be taken in case of fire and/or explosion. They will address personnel assignments, evacuation routes, and notification procedures. The NFPA Life Safety Code, No. 101 (2), and existing manual pullbox locations will be considered when establishing means of egress.

D. *Occupational Medical Surveillance*

  1. Medical examinations for personnel who will be working with test compounds or animals will be performed at the time personnel are assigned to the program, before they are exposed to potentially hazardous chemicals.

  2. Follow-up medical examinations will be performed at least every 12 to 18 months, and upon termination of an individual's participation in the project.

  3. The scope of the medical examination will be specified in the laboratory's health and safety plan. Persons who are required to wear negative pressure respirators must be approved by a physician for use of this equipment.

E. *Injury and Incident Records and Reports*

  1. A record will be kept of any incident resulting in minor or major personal injury, and/or probable personnel exposure to test chemicals. These records will include a full description of the incident, the chemical involved, the medical attention required, any remedial actions taken, and planned follow-up, if pertinent.

II. *Engineering Controls*

A. *Isolation and Access Restriction*

  1. An isolated laboratory (or laboratories) separate from other laboratory facilities will be provided for unpacking, storing, weighing, diluting, and packing of test chemicals and/or positive controls; and where necropsy, tissue trimming, tissue processing, embedding, microtoming, and staining is performed.

  2. This isolated laboratory (or laboratories) will have its air supply

under negative pressure with respect to connecting laboratories and hallways. For test chemical handling, this area will be a limited access area.

3. A record will be kept of all personnel entering/exiting any restricted access area(s).
4. Personnel covered under paragraphs III.A.1–6 who handle neat test chemical will shower out and wash hair after handling the neat chemical.

B. *General Facilities*
1. A nonbreakable, secured secondary container will be used for transport of any chemical between laboratories.
2. Vacuum lines used when working with chemicals will be protected with an absorbent or liquid trap and a HEPA filter.
3. Safety showers and eyewash stations will be located throughout the facility as required by local, state, and federal regulations.

C. *Hoods and Vented Enclosures—All Support Laboratories*
1. Weighing of the test chemical, positive controls, and hazardous chemicals will be done using the smallest quantity needed. Analytical balance will be used whenever possible to preclude the need for handling large amounts of chemical. This balance will be placed at all times in an effective laboratory hood, or in a vented enclosure with a face velocity of at least 50 fpm.
2. Plastic-backed absorbent matting will be secured inside of any hood or enclosure wherever chemicals (including dilutions) are being handled. After each working session in the hood, this matting will be disposed of as hazardous waste.

D. *Hoods and Vented Enclosures—Chemistry Laboratories/Repositories*
1. An effective exhausted enclosure or hood will be used for unpacking, diluting, packaging, analysis, and other handling operations involving test chemical, positive controls, or other hazardous chemicals.
2. An effective exhausted enclosure or hood for these activities will provide sufficient contaminant capture velocities (100 ± 20 fpm), as evaluated by a combination of velometer and smoke tube tests. These enclosures or hoods will be vented to the outside. Enclosures and laboratory hoods will be routinely monitored.
3. Effluent exhaust vapor from analytical instruments (e.g., gas chromatographs and atomic absorption units) will be vented to the outside of the building.
4. Air exhausted from neat test chemical handling areas will be passed through HEPA filters. If volatile chemicals are handled, charcoal filters will also be used. These filtration systems will be periodically monitored and maintained and personnel performing maintenance will wear the protective clothing described for neat test chemical handling in paragraph III.A.1–6.

E. *Hoods and Vented Enclosures–Pathology Laboratories*

1. An effective exhausted enclosure or hood will be used for necropsy, tissue trimming, manual tissue processing, and manual staining.

2. An effective exhausted enclosure or hood for automatic tissue processing and automatic staining machines with exposed solvent systems will provide sufficient capture velocities, (50 fpm minimum) as evaluated by a combination of velometer and smoke tube tests. Exhausted enclosures for automatic processing will be provided with a fire protection system and/or emergency power backup.

3. An effective exhausted enclosure or hood for these activities (except automatic processors) will provide contaminant capture velocities of 85 ± 10 fpm, as evaluated by a combination of velometer and smoke tube tests. These enclosures or hoods will be mechanically exhausted to the outside. Enclosures and laboratory hoods will be routinely monitored.

F. *Fire Safety*

1. The facility and operations will comply with applicable state and local fire and building codes.

2. Flammable liquids will be stored and handled in a manner that will reduce the risk of fire and/or explosion. This includes the following:

   (a) All nonworking quantities of flammable liquids will be stored in a storage cabinet approved by UL or Factory Mutual, or in a designated flammable liquids storage room with fire protection, ventilation, spill containment trays, and electrical equipment meeting the requirements of the NFPA Standard Code No. 45(3), Fire Protection For Laboratories Using Chemicals. In either storage arrangement, the flammable liquids will be segregated from other hazardous materials such as acids or bases.

   (b) In the event that flammable liquids must be kept at low temperatures, they will be stored in refrigerators that are explosion proof. All explosion-proof refrigerators will be labeled as such.

   (c) Flammable liquids transfer will be done in the designed storage room or over a tray within an effective fume hood (as defined previously). In the former location, all transfer drums will be grounded and bonded and will be equipped with pressure relief devices and dead-man valves.

   (d) Safety cans will be used when handling small quantities of flammable liquids.

   (e) Whenever flammable liquids are stored or handled, ignition sources will be eliminated. This includes the prohibition of smoking.

3. Fire extinguishers will be conspicuously located where they will be readily accessible and immediately available in the event of fire. There will be a maximum travel distance of 30 ft to an extinguisher, which will be located along normal paths of travel, including exits from an area. For the majority of laboratory applications, water and AFFF extinguishers will have a capacity of $2\frac{1}{2}$ gal. Dry chemical, carbon dioxide, bromochlorodifluoromethane (Halon 1211), bromotrifluoromethane (Halon 1301), and foam extinguishers will be 20- to 30-lb capacity. The specific type and size of extinguisher will be selected with consideration for the hazards to be protected and the strength of the personnel who might use the extinguishers.

4. Fire blankets and/or safety showers will be located within 25 ft from every entrance to a laboratory where flammable liquids are used.

III. *Personal Protective Equipment*

A. *Where the neat test chemical is packaged, stored, weighed, and diluted,* at a minimum, the following personal protective clothing will be worn at all times:

1. In chemical repositories and in laboratories performing dose preparation and development of mixing, storage, and handling procedures, personnel handling neat chemical will wear a disposable full body suit over work clothing. Disposable overclothing will not be worn out of the laboratory or repository where neat chemical is handled. Work clothing will be removed upon exit from the laboratory on a daily basis.

2. Other personnel in analytical chemistry areas where neat chemical is weighed or diluted will wear a disposable full body suit, or a disposable laboratory coat, disposed of on a weekly basis or disposed of immediately after any known chemical contact; or disposable sleeves, disposed of after each use, if a nondisposable laboratory coat is used.

3. Two pairs of dissimilar disposable gloves (i.e., PVC, latex, natural rubber, etc.): both pairs will be changed after any known chemical contact and/or after every 2 h of use.

4. An approved half-mask respirator equipped with a combination cartridge (this filter is specific for organic vapors, HCl, acid gases, SO_2, and particulates): these cartridges will be changed once a month or after 40 h of use; the date of installation will be marked on each new cartridge.

5. Safety glasses.

6. Disposable shoe covers. In chemical repositories footwear dedi-

cated to that space will be used in addition to disposable shoe covers.
B. *For laboratory operations not involving the handling of neat test chemical:* (e.g., chemical analyses, histology, tissue trimming, and necropsy), a single pair of disposable gloves, a laboratory coat, and safety glasses must be worn.
C. Any street clothing contaminated with neat test chemical or positive control (e.g., by a spill) will be discarded as laboratory contaminated waste. Any street shoes contaminated with test chemical will also be disposed in the same manner.
D. All components of a written respiratory protection program will be practiced for routine and emergency use of respirators.

IV. *Waste Disposal*
A. Any contaminated solid or liquid waste will be collected in a leak-proof container. For disposal this container must be closed and placed in a secondary closed container in the limited access area. Disposal containers must be stored in an isolated and secured area. All wastes generated by the laboratory will be disposed of in accordance with applicable federal, state, or municipal waste disposal regulations.
B. Residual or archival wet biological samples (tissues) will be double bagged.

REFERENCES

1. *NIH Guidelines for the Laboratory Use of Chemical Carcinogens,* NIH Publication 81-2385, May, 1981.
2. NFPA Standard Code No. 101, Life Safety Code, National Fire Protection Association, Quincy, MA, 1985.
3. NFPA Standard Code No. 45, *Fire Protection for Laboratories Using Chemicals,* National Fire Protection Association, Quincy, MA, 1986.

APPENDIX 9.2   EVALUATION OF LABORATORY HOODS AND LOCAL EXHAUST SYSTEMS FOR THE MANAGEMENT OF HAZARDOUS RESEARCH CHEMICALS

For laboratory operations involving chemical carcinogens, mutagens, and/or teratogens where exhaust ventilation is used for primary control of personnel exposures, a regularly scheduled ventilation system monitoring and maintenance program should exist. These programs should be designed and directed by health and safety or other qualified personnel. Because of the wide variety of laboratory hoods (chemical fume hoods, biological safety cabinets, vented enclosures) and other local exhaust ventilation (e.g., vented waste containers, refrigerators, etc.), the monitoring program for each piece of exhausted equipment should reflect the specific manufacturer's recommended operating practices. Each monitoring program should include, at the minimum, the following items:

I. *Daily Visual Inspections before Operation*
  A. *Exhaust Slots.* Check adjustable rear exhaust slots in laboratory hoods for proper position. When the exhaust inflow and down flow are unbalanced in biological laminar flow cabinets, there is potential for contaminated air to be pushed outside of the hood face.
  B. *Airflow Check.* When the exhaust system is operating (hoods, any vented equipment), perform a tissue paper check to ensure that the exhaust is functional. For a glovebox, tissue paper should be placed inside at the exhaust slots (but should not be allowed to escape into the exhaust system).
  C. *Pressure Gauges.* Check that any pressure gauges are indicating pressures that are within an acceptable range. All pressure gauges that indicate fan shutdown or filter loading should be marked to demonstrate a safe operating range.

   Slight variations in duct static pressures can have significant impact on face velocities or negative airflow in a glovebox. (Negative pressure in a glovebox should be at 0.5-in. water gauge.) This range should be initially determined by health and safety personnel.

II. *Quarterly Inspections*
  A. *Smoke tube tests* to evaluate irregular or turbulent airflow patterns should be performed on a regular schedule.
    1. *Chemical fume hoods* (with or without auxiliary air supply). The smoke should move from several inches in front of the sash directly to the rear exhaust slot, with the sash in its normal operation position. The smoke tube should be placed at and above the interior working space to locate any dead or turbulent spots.
    2. *Biological safety cabinets* (*laminar flow*). The smoke should move from several inches in front of the sash into the forward intake grill. The smoke at the working area should move toward either the forward or rear exhaust slots with a minimum of turbulence.
    3. *Gloveboxes.* The smoke tube should be placed at the outside glove gaskets and inside the rubber gloves to check for leaks into the glovebox.
    4. *Other exhaust equipment.* The smoke tube should be placed around the perimeter or the outer boundaries of the area to be exhausted (e.g., the edge of a square waste container vented along the opposite edge). The smoke should move directly to the exhaust slots.

B. *Hood face velocities* and laminar downflows should be measured with velometers on a quarterly schedule. These tests should be performed by qualified personnel using a properly calibrated thermal or mechanical velometer.

1. *Chemical fume hoods.* The average of the velocities in each grid (see Section 16.2.1) will represent the face velocity. Mark the hood sash frame to indicate the fully open, half-open and one-quarter open positions of the hood sash. Use a six area grid when the sash is fully open and a three area grid for the other sash positions. Determine the average velocity for each sash position. When these measurements are made, any auxiliary air should be turned off. If the hood is connected to other hoods, the face velocity should be measured under the maximum "worst" conditions, that is, with the sashes fully open on the other connecting hoods. The face velocity should be 100 fpm ± 20%.

   The OSHA regulates the face velocity of laboratory hoods when specific chemicals are used or handled, see Table 9.1. To comply with the regulations, a face velocity of 125 to 150 fpm must be strictly maintained.

2. *Biological safety cabinets (laminar flow).* In Class II Type A and B hoods the supply blower should be switched off for the face velocity measurement. This face velocity should be 100 fpm ± 10%. The vertical downflow of the supply blower should also be measured. The downflow at the work area should be

Table 9.1. Chemicals Regulated by OSHA That Require Specific Laboratory Hood Face Velocities

Name	Standard	Percentage by Volume or Weight above Which Chemical Is Regulated
4-Nitrobiphenyl	1910.1003	0.1
α-Naphthylamine	1910.1004	1.0
Methyl chloromethyl ether	1910.1006	0.1
3,3'-Dichlorobenzidine (and its salts)	1910.1007	1.0
Bischloromethyl ether	1910.1008	0.1
β-Naphthylamine	1910.1009	0.1
Benzidine	1910.1010	0.1
4-Aminodiphenyl	1910.1011	0.1
Ethyleneimine	1910.1012	1.0
β-Propiolactone	1910.1013	1.0
2-Acetylaminofluorene	1910.1014	1.0
4-Dimethylaminoazobenzene	1910.1015	1.0
N-Nitrosodimethylamine	1910.1016	1.0

approximately 50 to 80 fpm depending on the manufacturer's recommendations.

Several types of Class II, Type B, hoods are not designed for easy switch off of the supply blower for face velocity measurements. For these types of hoods, a combination of supply/working surface measurements and inlet smoke tubes tests should be performed by experienced personnel. If the inlet supply air is too great (e.g., > 80 fpm) and the smoke tubes indicate lazy inflow air patterns at the face, the supply/exhaust airflow may be out of balance, the HEPA filters could be overloaded, or the exhaust fan malfunctioning. Whatever the problem, further testing or maintenance may be required.

3. *Glovebox.* If the glovebox has a filtered air inlet with no blower, a velometer reading can be taken to determine the exhaust volume. (The velometer can be placed directly at the air inlet to measure the inlet velocity. This velocity times the inlet area will give the exhaust air volume.) This exhaust volume for most manufacturers' specifications is in the range of 30 to 50 cfm (cubic feet per minute). When a glovebox has a supply air blower (consequently the inlet air volume is not available) combination periodic smoke tube/leak tests (outlined previously) and annual exhaust velocity duct measures (discussed in next section) should be performed.

III. *User Training*

The daily visual inspections and smoke tube tests should be an important aspect of a hood operator's education and training. Before an employee begins using any laboratory hood, he/she should be trained in its proper use and monitoring programs. The results of these tests for all types of exhaust equipment should be recorded. These records should be kept in an accessible location so hood users can refer to them when suspected hood malfunctions occur. Examples of monitoring forms are shown in Fig. 9.1 and 9.2.

IV. *Annual Maintenance*

In addition to the periodic monitoring, an overall annual maintenance program should also be established. These programs should be performed by *qualified* personnel. Documentation of these tests should be maintained by laboratory supervisory personnel (see Figure 9.3).

In the following list are the general maintenance items that should be performed annually on exhaust equipment. The next section includes more specific information for special hoods and enclosures (e.g., the *National Sanitation Foundation Standard for Class II Biohazard Cabinets,* NSF No. 49).

(a) *Exhaust fan maintenance.* The fan manufacturer should recommend necessary maintenance including lubrication, belt checking, fan blade deterioration, and for speed check.

Room No.: _____   Date: _____

Location within room: _____

Sash: Yes: _____   None: _____

          Vertical movement: _____   Horizontal movement: _____

Damper: Yes: _____   None: _____

Switch: Yes: _____   None: _____

Apparent connection to other hoods:

Sketch: (Show position of hoods, benches, windows, doors, makeup air)

Face Velocity Measurements   (test with other hoods on):

Condition/Position	Top Left	Bottom Left	Top Center	Bottom Center	Top Right	Bottom Right	Average
Sash open, door open							
Sash open, door closed							
Sash $\frac{1}{2}$, door open							
Sash $\frac{1}{2}$, door closed							

Hood checked by: _____

Figure 9.1   Example of a hood monitoring form.

Lubrication of the fan and fan motor may be required on a more frequent basis depending on the operating conditions.

(b) *Duct work check.* All the duct work between the hood or exhaust opening should be checked for corrosion, deterioration, and buildup of liquid or solid condensate. Any dampers used for balancing the system should be lubricated and checked for proper operation. Unused duct work or old hood installations should also be removed.

(c) *Air cleaning equipment.* In-line exhaust charcoal or HEPA filters should be monitored for containment buildup. Mechanical or absor-

LABORATORY _____   MODEL _____

INVESTIGATOR _____   SERIAL No. _____

TELEPHONE No. _____   DATE _____

Supply blower speed
    control setting   _____   SUPPLY VELOCITY

Magnehelic gauge
    reading   _____

New _____   Used _____

High ____   Low ____   Average work
                                    surface _____

Air Curtain

Access opening   _____

Exhaust Velocity   Ducted _____
                        Not ducted _____
                        Size _____

High ____   Low ____   Average ____

cfm _____
        Calculated air intake _____

REMARKS: _____

_____

_____

_____

Testing Technician _____

Figure 9.2. Example of a biological safety cabinet velocity profile monitoring form (laminar flow).

bent filters not equipped with differential pressure gauges or audible alarms should be leak-checked annually. Adsorbent or absorbent filters for gas and vapors can be leak checked by the release of halogen gas inside the hood and a halogen meter monitoring the filter outlet airstream. The HEPA filters can be checked using the dioctylphthalate (DOP) method (see NSF No. 49).

DATE _____

UNIT IDENTIFICATION   UNIT LOCATION

   Type _____   Building _____

   Model No. _____   Room _____

   Serial No. _____   Other _____

TESTS PERFORMED

   ( ) Velocity profile
   ( ) Calculated inflow
   ( ) In-place leak testing of HEPA filters
        (single particle monitor – cigarette smoke method)
   ( ) In-place leak testing of HEPA filter (DOP method)
   ( ) Paraformaldehyde decontamination
   ( ) Halogen leak test
   ( ) Ground continuity – containment and intrusion
   ( ) Smoke pencil
   ( ) Noise level and vibration test

FINAL TESTING RESULTS

   ( ) Unit passed tests   ( ) Unit failed tests

        ( ) All leaks repaired   ( ) Requires new filters
        ( ) New filters installed   ( ) Requires frame repair
        ( ) Other   ( ) Other

REMARKS: _____

_____

_____

_____

                          Testing Technician _____

Figure 9.3.   Example of test summary form.

(d) *Velocity measurements.* As mentioned earlier, total exhaust Class II, Type B biological safety cabinets and gloveboxes require exhaust duct velocity measurements in order to verify measurements of the proper airflow in the enclosures. (In the total exhaust hood this supply volume (cfm) can be subtracted from the exhaust volume (cfm), and this value can be divided by the area of the face opening in order to calculate the face velocity). For the glovebox the exhaust duct volume should be in the range of 30 to 50 cfm.

When these maintenance procedures are done, extra precautions should be taken to protect personnel from any of the toxic contaminants in the hood, duct work, or filters. Any excess contaminated material or filters removed from the hood system should be disposed of according to the facility's approved toxic waste disposal practices.

BIBLIOGRAPHY

The following are additional sources that can provide more detailed information on ventilation monitoring programs for specific types of laboratory hoods and exhausted enclosures.

Caplan, K.J. and Knutson, G.W., "The Effect of Room Air Challenge on the Efficiency of Laboratory Fume Hoods," *ASHRAE Trans.* **83,** Part I, 141–156 (1977).

Fuller, F.H. and Etchells, A.W., "The Rating of Laboratory Hood Performance," *ASHRAE J.,* 49–53, October (1979).

Industrial Ventilation—A Manual of Recommended Practice, Committee on Industrial Ventilation, American Conference of Governmental Industrial Hygienists', 19th ed., Cincinnati, 1986.

National Sanitation Foundation Standard for Class II (Laminar Flow) Biohazard Cabinetry, NSF No. 49, The National Sanitation Foundation, Ann Arbor, MI, 1983.

Recommended Industrial Ventilation Guidelines, U.S. Department of Health, Education, and Welfare, HEW Pub. No. 76-162, The National Institute of Occupational Safety and Health Contract No. CDC-99-74-33, prepared by Arthur D. Little, Inc., Cambridge, MA, GPO, 1976-657/5543, January 1976.

U.S. Department of Labor, Occupational Safety and Health Administration, Occupational Safety and Health Standards, 29 CFR 1910, all parts (Sub part Z), U.S. Government Printing Office, Washington, DC, revised annually.

APPENDIX 9.3   RECOMMENDED RESPIRATORY PROTECTION PROGRAM FOR THE MANAGEMENT OF HAZARDOUS RESEARCH CHEMICALS

A. *Selection of the type of respirator.* The type of respirator worn by an employee must be matched to the chemical and physical properties of the substance and the degree of hazard associated with its use. The information necessary to make these decisions is assembled in the respirator decision logic used in the Joint NIOSH and OSHA standards completion program. Included in this decision logic are PELs of the substance, warning properties, eye and skin irritation, flammability, immediately dangerous to life and health (IDLH) concentration, vapor pressure, physical state, sorbent efficiency, and skin absorption. Any respirator which is selected will be approved by the National Institute of Occupational Safety and Health and/or the Mine Safety and Health Administration (NIOSH/MSHA TC approval number will be on the cartridge). Respirators for routine and emergency use can be chosen by utilizing these parameters and the known limitations (protection factors) of the various types of respirators.

B. *Medical Considerations.* Wearing of any type of respirator imposes physiological stress on the wearer. Air-purifying respirators (e.g., half-masks) restrict normal inhalation because the filter or cartridge resists free air flow. Consequently, any reduced lung or cardiovascular function will inhibit the wearer's ability to use a respirator. Before a respirator is used, any complicating medical conditions of the wearer will be known.

C. *Training.* A session to acquaint the respirator wearer with the proper fitting, maintenance, and limitations of the respirator is important. A discussion of engineering controls, respirator selection, the potential health hazards when a respirator is not used, and recognition and handling of emergencies will also be included.

D. *Respirator fit testing.* Individual fit testing of the specific types of respirator to be worn is necessary due to the importance of the facepiece–face seal in an air-purifying respirator. Negative and positive pressure fit testing is not sufficient to check a facepiece fit. Initially and at periodic intervals isoamyl acetate or smoke tubes will be used to check the facepiece seal. The proper cartridges will be installed to stop the test agents adequately.

E. *Maintenance of facilities and work practices.* In a laboratory setting the respirator is normally providing an additional safety margin for the wearer and is considered a secondary control method. An integral part of the selection and use of a respirator is the effectiveness of the laboratory hood and ventilation system. This primary control method and the individual work practices will be reviewed quarterly to ensure that the respirator can provide the necessary protection.

F. *Documentation and review of the respirator program.* A written record of the selection, training, maintenance of the respirators, and maintenance of the facilities will provide a program review mechanism and documentation that the program exists.

CHAPTER 10

Other Hazards

Jack J. Bulloff

New York State Legislative Commission on Science and Technology
Albany, New York

10.1 INTRODUCTION

Nonchemical, that is, physical, hazards also cause accidents in chemical laboratories. They can compound or be compounded by chemical hazards in a variety of ways. Physical hazards can change the nature or degree of the hazards with which they interact. This list of physical hazards demonstrates that most chemical laboratories have many of these to guard against:

Slippery spills, including solids that make footing insecure.
Ionizing radiations, including radioactive radiation.
Other electromagnetic radiations, including lasers.
Electrical currents and magnetic fields.
Electrostatic sparks, including coronas in copiers.
Vacuum, including Dewar flasks.
High pressure, including compressed gases.
Low temperature, including cryogenic systems.
High temperature, including undissipated heat.

10.2 SLIPPERY SPILLS AND SUBSTANCES

Some surfaces are inherently slippery. Spills can make other surfaces slippery. Thus flooring and shoes that are slippery should be avoided in the laboratory. So should smooth shelving that makes it easy to slide or knock off a bottle, Labware that is hard to hold or clamp can also cause spills and accidents. Any items that are not intrinsically slippery but which become so when wetted by water, solutions, or other liquids not commonly considered to be slippery themselves should generally be kept out of chemical laboratories. Use of aqueous solutions is so common and mopping of laboratory floors so frequent that flooring made slippery by water, soap, or detergent solutions should be shunned. "Wet Floor" warnings are inadequate safeguards against slips, which may in turn, cause a slippery spill to occur.

Failure to use spill-containing or spill-proof carriers, break-proof or shatter-proof reagent bottles, plastic bottles or containers, or containers safe to hold if they or their contents are hot, wet, or slippery may assure hazardous and costly dropping of equipment and spills and consequent added injuries. All those working in a laboratory, particularly newcomers and students, should be periodically checked to see if their training and experience are sufficient for their use of products and processes that are likely to pose slippery spill hazards. Emergency response teams that serve such laboratories should be alerted if it seems likely that hazardously slippery material use is planned

Slippery spills are usually liquid spills. There are many kinds of intrinsically slippery liquids that can make almost any flooring or support surface slippery. Some liquids, whether or not they are initially intrinsically slippery, may make a surface that has irregularities in it slippery by either solution or dissolution of the irregularities, thus making the surface smoother and slipperier. As well as dissolving the irregularities, a solvent may swell or soften the surface to make it slipperier; or in dissolving the surface material, the solvent may thicken and become syrupy or jellylike and, again, slippery. Several of these augmenting effects may go hand in hand to make for a very slippery spill. Of the many mechanisms that make for a spill some others are lubrication, as by an oil or a frictional-flow-facilitating fluid, and saponification.

But solids may also be slippery, for example, soft flakes like those of graphite, of boron nitride that is isomorphous with graphite, of the disulfides or diselenides of molybdenum or tungsten, or of the numerous layer silicates like common talc or mica. A quite different, but equivalently disastrous footing hazard (or floor slipperiness) is provided by spills of rolling or rollable solids, for example, spheres like ball bearings, glass or quartz beads, glacier or Ottawa sands, or small short cylinders such as boiling stones, catalyst supports, or ceramic frits.

There is a vast difference between spills that cause injuries from a fall and those that cause concomitant traumas like burns, irritations, or poisonings. The substances spilled make the difference.

10.2.1 Slippery Substances

In general slippery substances are not thin watery fluids. Syrupy liquids, greaselike semisolids, and waxy, crayonny solids tend to be the slipperiest. To illustrate, in removing sulfuric acid from a low shelf, a young lady broke a 9-lb bottle. She tried to step out of the spreading spill. But she slipped in and into it, sitting down on the broken glass in the spill. Her fright made her speechless despite the pain of acid on the open cuts. As she thrashed about, unable to rise, she skidded and broke a second 9-lb bottle into the spill. That left her prone, silent, burning, and bleeding in 18 lbs of sulfuric acid. Her rescuer realized the situation as he came upon it fortuitously, and seeing that her hair was the only safe hand-hold, he used it to pull her out of the spill and to the nearby shower. She survived, and her pain, disability, disfigurement, and fright are still being paid for.

10.2.1.1 Slippery Acids. Concentrated sulfuric acid is the most widely encountered slippery acid solution in chemical laboratories. It is not the most hazardous water-miscible or aqueous acid. A solution of sulfur trioxide in sulfuric acid, fuming sulfuric acid, is an even stronger dehydrating and charring agent, a stronger oxidizing agent, and a more concentrated acid than sulfuric acid itself. Common laboratory cleaning solution, now shunned as a confirmed carcinogen, is a stronger oxidant than fuming sulfuric acid. Nitrating acid is a stronger and more oxidizing acid than any of the foregoing; it is a mixture of nitric and sulfuric acid. If either fuming nitric acid or fuming sulfuric acid is used to make nitrating acid, or if both fuming acids are so used, the resulting products are among the most hazardous slippery liquids known. Also, they can cause fires on almost any but stone-surfaced or cement-surfaced floors.

Other syrupy acids include phosphoric acid, perchloric acid, and the poly-acids formed by dissolving phosphorus pentoxide in either phosphoric or metaphosphoric acid (or by other numerous methods). The sodium salts of these, and other alkali polyphosphates, polymetaphosphates, polyhexametaphosphates, and ultraphosphates yield unusually and extremely thixotropic viscous solutions that can even flow from a lower to a higher beaker once the flow has been started. Some of these solutions are water greases. Sodium metasilicate solution, "water glass," is not as slippery as water greases, but it is more common in laboratories. Aqueous cream emulsions of fatty acids are more slippery; fortunately, these are not corrosive to the skin and underlying tissues.

10.2.1.2 Slippery Alkalis. The common alkalis, sodium and potassium hydroxide, can impart slipperiness to flooring or shoe-sole materials, which they can saponify or attack. Even their weak solutions saponify underlying fat and attack skin so that fingerprints disappear. The hands feel soapy, but there is no pain or sensation of burning, because nerves are destroyed or altered so that no pain signal or response occurs. For this reason, alkalis are more insidious hazards than acids. Victims often do not realize they have been burned until it is too late for first aid or even subsequent remedial assistance. Alkalis are another reason why there should never be bare feet in a laboratory.

Pellets of sodium or potassium hydroxide are extremely hygroscopic.

When spilled, they get wet and slippery in a short time. Cleanup should follow immediately after the spill. Other alkalis calling for safeguards are the salts of strong bases and weak acids that hydrolyze to what are essentially alkali hydroxide solutions. The alkali polysalts of phosphorus and silicon just mentioned and certain well-known thickeners like sodium carboxymethylcelluloses and carboxymethydextrans yield such strongly alkaline solutions.

There are less strongly, even weakly, alkaline but very strongly saponifying slippery materials, for example, pastes, pasty emulsions, and gelled hand cleaners, soaps, and detergents. Some are made more hazardous by the addition of strong alkalis in excess, for example, builders for detergents and soaps. It might be noted that some of these materials occur in laboratory medical or safety cabinets and others in the janitorial supplies used to maintain "good housekeeping" in the laboratory. It behooves the laboratory management to include maintenance as well as other laboratory staff in safety training, particularly for avoidance or minimization of slippery spills and for their much-overlooked amelioration.

10.2.1.3 Polymers and Other Slip Producers. It is well known that water can be thickened or even gelled by polymeric solutes (or macromolecular solutes like gelatin). It is little known that certain polymeric solutes have been used to reduce the friction of water flow; this is the mechanism that allows coal to be more cheaply transported as a slurry or a nuclear submarine to have its speed under water greatly increased. The same mechanisms can make water exceedingly slippery. In general, the thickening ability of a given polymer or polymer type is greatly increased by increasing either its degree of polymerization (the weight-average molecular weight), its degree of branching, or, most effectively, its degree of cross linking. These three mechanisms also apply to the thickening of other liquids. For flow facilitation, on the other hand, as in the examples just given for water, or in the formulation of multiviscous and long service motor oils, the solute polymers have to be as linear as possible, with the highest practical degree of polymerization. When both kinds of polymers are present, slipperiness is highest.

10.2.2 Slippery Spill Safety

If they have proven effective in the past, existing chemical laboratory safety programs for ordinary spills should be amended to include slippery spills. If there have been prior problems with online spill safety programs, they should be reformulated with adequate inclusion of slippery spill protection. In either case, solid as well as liquid spill prevention and response should be included. The possibility that one slippery spill can cause another and the fact that a spill can be caused by thoughtless grasping of hot, slippery, wet, or corrosive objects should not be overlooked. It must also be realized that safety materials may be slippery, as already discussed. That includes response materials like fluids, foams, and powders used in fire fighting, and certain sorbents thrown upon spills. Response teams should be aware of these points.

10.2.2.1　Slippery Spill Prevention.　Usually the preventive means applied for slippery spills are essentially those already applied to control and clean up spills of any hazardous laboratory chemical, provided that the preventive means in use have proven themselves effective in the past. Spill prevention safety precautions are well worth emphasizing:

- Use bottle carriers capable of containing all the liquid that can be spilled (and make sure the container material is compatible with the liquid carried).
- Use vessels either made of unbreakable (highly drop-resistant) material or else made of glass but coated with a plastic that can hold the glass and contents if dropped and broken.
- Find a substitute that does not give hazardous, or at least does not give slippery spills. Failing that, see if slippery substance use and transport can be decreased and made safer by replacing macroscopic scale with semimicro scale or micro scale operations.
- Find substitutes for nonslippery chemicals that make floors or shoes slippery. For example, hydrofluoric acid, which makes cement floors slippery, can be replaced for many uses by sodium hydrogen fluoride. For other uses hydrofluoric acid can be generated in a hood from safer materials.

Of course, all the precautions that apply to nonslippery chemicals and spills also apply here.

10.2.2.2　Response to Slippery Spill Accidents.　As already indicated, the victim of a slippery spill may not be able to call for help, so response plans should be so framed that there will be a response when needed even if there is no call. Part of the laboratory safety training should include notification of the safety team in advance of work to be performed with a hazardous substance. In turn, the safety team should not permit the work if the appropriate remedies and precautions are not in place. One of these, for workers in an isolated location, is the requirement that a safety team member shall look in on the site of the work as frequently as considered advisable for the hazardous product or process in use.

10.3　IONIZING RADIATION

Radiation that overcomes the forces within atoms and molecules causing those particles to become ions is called ionizing radiation. Ionizing radiation can be fatal in acute exposure. For chronic exposure no completely safe lower level of exposure has been or is likely to be established. One complication is that there is no zero level of exposure; we live in a universe of cosmic rays and on an earth in which there is natural radioactivity and natural nuclear reaction and fission. Further, the medical effects of radiation absorption are so complex that standards of exposure and dosage are still in flux. What seems more settled are the standards characterizing the activity or intensity of radiation.

The radiation most commonly encounted in chemical laboratories is α, β (including positrons), or γ emission from natural or artificial radionuclides. Less commonly, chemical laboratories use neutron sources; other sources include X rays from X-ray tubes and bremsstrahlung X rays from the surface of cathode ray tubes, and electrons and protons from synchrotons.

For radioactivity, the decay rate has traditionally been reported in curies (C_i or Ci). Now that SI (*Systéme International*) units are replacing cgs units, the Ci is being replaced by the becquerel (Bq), 1 disintegration per second (dps) or 2.703×10^{-11} Ci. A curie is 3.7×10^{10} Bq or dps, or 2.2×10^{12} disintegrations per minute (dpm). But a mere count of the number of material particles or electromagnetic photons emitted by a source is not a measure of its hazard or of the protection against it that is required. In addition, some radiations are directly ionizing and others are indirectly ionizing.

Charged-particle radiation, which can be considered as ionic to begin with, interacts directly, and therefore strongly, with extranuclear electrons. Its high linear energy transfer (LET) limits it to a short range in matter. Neutral particle and electromagnetic radiations interact so weakly with extranuclear electrons that they have to collide or interact with the nucleus to produce ionizing radiation. Such indirectly ionizing radiation has low LET and a long range in matter. It is penetrating radiation that is hard to stop and shield against, while high LET radiation is easy to stop and shield.

For X-ray and γ radiation, exposure is defined in roentgens, R. In air R is 2.58×10^{-4} coulombs (C)/kg (of energy), and a roentgen forms 1.61×10^{12} ion pairs/g of air. The SI exposure unit of 1 C/kg is 3.876 R. For any kind of radiation, absorption or dosage (hence radiation absorbed dose or rad) is defined in units of 100 ergs/g or rads. In SI units it is in joules per kilogram (J/kg) or grays (Gy). A gray is 100 rads. These are the quantity definitions that make it easier to evaluate and protect against ionizing radiation injury, but though they do so much better than curies or becquerels, they cannot be used directly either.

Roentgen equivalent man, or rem, is the unit that is used directly at this time (but the constants and methods of calculating dosage and damage are still the subjects of ongoing discussion and change). Since different organisms, organs, tissues, cells, and cell components, including different proteins, ribonucleic acids (RNA), and deoxyribonucleic acids (DNA) respond differently to dosage, dose equivalence defined in rems cannot be a simple quantity. The rem is the product of three quantities:

1. Absorbed dose in rads.
2. Quality factor, Q, which is solely LET dependent.
3. N, a fudge factor supposed to take care of response differences and nonuniformities of exposures, dosages, and dose equivalence (in short, a learning curve correction that brings observations and predictions into some sort of interconsistency).

Q is usually unity for X, β, and γ rays, 3 to 10 for neutrons, depending on their energy, 10 for fast protons and α radiation, and higher for slow protons and heavier ions. If gray is substituted for rads, then the product in SI units is sieverts (Sv): 1 Sv is 100 rem.

10.3.1 Regulation of Radiation and Radiation Sources

International and national recommendations in terms of the units just defined guide national, state, and local regulation of radiation and radiation sources and uses. Federal regulations are detailed in the *Code of Federal Regulations* (CFR) as updated by the daily *Federal Register* (FR). Of pertinence are CFR refs. 1–6. These federal regulations, the corresponding regulations that apply in some states, and the corresponding regulations that apply in other countries reflect the recommendations of the National Council on Radiation Protection and Measurement (NCRP), the International Commission on Radiation Protection (ICRP), and the International Commission on Radiologic Units (ICRU).

The NRC *Standards for Protection Against Radiation* (ref. 1) deal with the units discussed previously. A rem is taken to equal a roentgen of X rays or γ rays and so is a rad (and thereby β rays are included). For neutrons or high-energy protons a rem equals 0.1 rad. For low-energy protons it equals 0.05 rad. Based on action levels or ascertained needs, these suggested precautionaary procedures should be considered: surveys of radiation levels, personnel monitoring using dosage badges or portable counters, and use of caution signs, labels, signals, and controls. The caution signs differentiate radiative materials locations and the radiation flux and airborne radioactivity areas.

So ascertained, specific levels trigger specific controls. At 100 mrem/h visible or audible alarms and locks against area entry (but not against area egress) automatically operate wherever there is no direct safety escort provided. At 500 rem/h at one meter, replicate alarming, entry control, and check and backup devices are actuated. To assure that power failure cannot possibly lead to inadvertent exposure from failure of precautionary devices and their backups, design and construction provide multiple backup fail-safe operation. What inspired all this was that at a $^{60}_{27}$Co irradiation facility, when an electrician shut off power to make repairs, the warning light went out, the alarm buzzer stopped, the safety door opened, and an unwary technician entered to find himself staring at a huge source unretracted into its shielding water pool. But basically radiation protection is only a matter of optimally combining reduction of exposure time, increased distance from the source, and shielding the source and/or laboratory worker.

10.3.2 Ranges of Ionizing Radiations

Because alpha particles have 7340 times the mass of an electron, they have a high LET; the lower the energy of an α particle, the greater its LET. Hence, with such a short range, α particles cannot penetrate the dead outer layer of skin to enter live tissue; when their source is external to the body they cannot constitute an internal hazard. But if α-emitting radionuclides are inhaled or injected, they become a very serious internal hazard. Electrons have longer, but still short, ranges. With an increase of average β energies from 0.05 to 1.13 meV, the range in air increases from 1 to 29 ft and penetration into the body increases from 0.03 to 1.1 cm. The percentage of β rays that penetrate the layer of dead skin (0.007 cm) increases from 11 to 97. Slightly different, but generally corroborative data exist.

So very weak β radiation, like that of tritium, ^3_1H, present no external hazard. But energetic β radiation presents hazards that have to be shielded against or avoided in some other way. In contrast, γ rays (and X rays of comparable energies) have theoretically infinite ranges that become practically infinite for very large sources, that is, most of the rays go through the whole body. With increase of γ-ray energies from 0.03 to 1.20 meV, the body depth that reduces the γ radiation 50% increases from 2.3 to 11 cm. These thicknesses are the half-thicknesses for those γ-ray energies; a second half-thickness would stop only 75% of the entering radiation. So γ rays are both internal and external hazards. Again, slightly different but corroborative data exist.

In working with natural or artificial radioactivities where parent nuclides originate extensive chains of sequential radioactive disintegrations, every kind of radioactivity may be present. Gamma and X-ray emission may accompany the decays. Bremsstralung, or braking radiation X rays may be formed by the stopping of electrons. In this instance protection has to be assured against the most penetrating radiation and there has to be assurance that the most serious internal hazard will be neither inhaled, ingested, nor injected.

10.3.3 Hazardous Exposure Levels

The NRC sets 1250 mrem/quarter and 5000 mrem/yr as the limits for whole body occupational exposures. The NCRP recommends no more than 500 mrem/yr for students because they have such a long life yet ahead of them and because for most of it they may fall under the lower limits set for nonoccupational exposures. The ICRP advises and emphasizes *as low as reasonably achievable* (ALARA) acute and chronic exposures. Chronic exposures set today may effectively be millionths of LD_{50} levels. LD_{50} is about 650 R: it is 400 to 500 R for LD_{50} 30 days. Above 600 rads all exposures are probably fatal.

The body metabolizes and excretes toxins while it repairs and recovers from their damage. For radiation, as LET increases, effectiveness of repair diminishes. But, for a given LET, internal radiation levels drop with time— and they may become less damaging more quickly than might be expected from the half-life of the radionuclide involved. For there is a biological half-life for retention of the radioelement in the body. That is the time it takes to excrete half the internal radioactivity. The effective half-life is always less than either the radioactive (or physical) half-life or the biological half-life according to the equation:

$$\frac{1}{\substack{\text{effective}\\\text{half-life}}} = \frac{1}{\substack{\text{biological}\\\text{half-life}}} + \frac{1}{\substack{\text{physical}\\\text{half-life}}} \tag{1}$$

Thus, if the biological and physical half-lives were each 2 days, the effective half-life would be 1 day. That is a 50% reduction of exposure time.

For bone-seeking radionuclides the biological half-life is so long that the effective life is almost the physical half-life and there is no effective reduction of exposure time. For other radionuclides the effective half-life is much shorter than the physical half-life and there is an important reduction in exposure time. Effective and physical half-lives are, respectively, 11 and 13 days for $^{140}_{56}Ba$, 162 and 165 days for $^{45}_{20}Ca$, 8 and 8.05 days for $^{131}_{53}I$, 14 and 14 days for $^{32}_{15}P$, and 64 and 64 days for $^{85}_{38}Sr$. But they are, respectively, 12 days and 5730 yr for $^{14}_{6}C$, 12 days and 12 yr for $^{3}_{1}H$, 44 and 88 days for $^{35}_{16}S$, and 20 days and 212,000 yr for $^{99}_{43}Tc$.

10.3.4   Radiation Safety Measures

It is obvious that internal exposure levels have to be much lower than external exposure levels. So there should be minimization of entry of radionuclides into the body and maximization of elimination of radionuclides from the body and the environment and of avoiding, or shielding from, their presence.

- Airborne activity calls for the use of respirators and filters.
- Food and drink should neither be carried into nor consumed in a radioactivity or radiation area. Pipetting by mouth should be made impossible through removal of ordinary pipettes from the laboratory and the provision of other pipettes.

- Decontaminants that are optimum for the radionuclides in use shall be on hand at all times.
- Chelating prescriptions and other internal decontaminants shall be available for prompt use. So shall any other means that decrease the biological half-life of ingested radionuclides.

Radionuclides must always be treated as poisons with no safe level of body entry or proximity. Procedures already in place to deal with hazardous chemical spills, fires, toxicities, irritabilities, and corrosivities should be strengthened as is necessary to improve prevention of accidents, for there are slippery, flammable, reactive, toxic, and corrosive radiochemicals as well as radiotoxic chemicals.

10.3.4.1 Reduction of Exposure Time.   Training staff and providing equipment that reduces the length of time of exposure to radiation should be planned only for reductions that do not unduly hasten operations so much that there is a countervailing increase in the chances of having a spill or some other accident or mishap. A more salutary way of reducing exposure time than speedup and streamlining of working and transfer operations is to perform those operations either through remote manipulators or by use of robots. As good as glove compartment boxes have become, they can still be improved. But remote automation is better, especially for routines like titration of radioactive samples.

10.3.4.2 Increase of Distance.   For point, certain line, and some other sources, exposure decreases approximately as the square of the distance. If the source is a short-range source, the worker can remain out of range all the time, thus incurring practically no exposure time. Use of distance has permitted much work with very high LET radiations and rather large short-range radiation sources. To a lesser extent, it has minimized exposures to longer range radiation. If no other radiation sources are present to complicate the calculation, working five times as far away from a source leads to a 25-fold (96%) reduction in exposure. If it is safe to work 40 days a year with a certain source, working twice as far away makes it possible to work 160 days a year. It is obvious that combining distance, time reduction, and shielding can go a long way towards making radiochemistry and nuclear chemistry in chemical laboratories much safer.

10.3.4.3. Shielding.   Shielding of sources of highly penetrating radiation, especially large ones, must have coordinated fail-safe backup. Surveillance TV is advised for some cases, for example, a lone worker in the laboratory. Respirators have been mentioned for airborne activity. Where both workers and sources have to be shielded, heavy-metal shielding that gives rise to bremsstrahlung (X rays from braking of electrons) is hazardous. Low density shielding should be used, for example, plastics. For neutrons, hydrogen atoms are the best stoppers, and paraffinic materials the most practical.

10.3.4.4 Monitoring and Decontamination. Since radiation cannot be smelled, tasted, heard, or felt, monitoring must be continuous. Many kinds of counters are on the market, and the literature is extensive. When badges are used, they serve mainly for establishing average, long-term, and cumulative exposures; though they inspire worker confidence about chronic exposure, for acute exposures, they are essentially after-the-fact documentation and not preventive. Whatever monitoring is conducted has to coordinate with ongoing survey activity identified previously. There should be feedback from it and the surveys to continual safety improvement.

Where survey and continuous personnel monitoring shows any contamination, whether significant, borderline, or suspect, of people, clothes, materials, equipment, or facilities, it is mandatory that it be cleaned up immediately and effectively. Where this is not accomplished safety measures such as isolation, sign posting, and shielding, should be undertaken. For whatever internal or external contamination could have been foreseen, there should already be in or near the laboratory a stock of all the chelant medicines, detergent and chelating cleaners, and other appropriate agents. Most of these can be located through the references listed in Section 10.3.5. Others have to be identified in the work planned for update of existing contingency emergency response protocols as provided in the regulations referenced and in local health and sanitation regulations.

Where there is energetic, say α (or fission) decay, the kinetic energy of disintegration may be so great that the daughter nucleus, radioactive or not, recoils, and the recoil carries with it neighbor radioactive atoms that then recoil later. Thus, if a source like $^{210}_{84}$Po is left open, especially if it is a high surface area or finely powdered solid source, recoil will empty it progressively and spread radioactivity to neighboring surfaces and objects. Letting a very active solution of any high energy α emitter dry down from the heat of the radioactivity is an invitation to disaster. When such radionuclides are used, besides extensive monitoring to get assuredly adequate decontamination, masks should be worn for persisting airborne radiation. If such radiation will thereby not contaminate a duct adversely, use of a hood will limit recoil effects in the laboratory itself, but cleanup of the hood must be optimum. In general high energy and high activity workplaces and laboratories should each be surveyed and monitored with greater frequency and thoroughness than less hazardous sites.

As stated, film badges, pocket ionization meters, and particle counters are after-the-exposure means. But they are valuable for checking both permitted chronic exposure levels in workplaces and the conditions and effects of allowed exposures. They help assure the adequacy of external radiation protection. Most important, their use is a constant reassurance to those in the laboratory that there is indeed a safety program in effect.

Surveys and passive monitorings have other uses. The exposure records show that all too frequently too many radioactive sources have either been put into the laboratory by management or brought in by those working there. For portable sources the rule should be: Put the one (and it should be the only one) you are using away before you bring in a second source. For management the rule should be: Have so few sources in the laboratory that the inverse square law holds well for every user who is using distance to minimize exposures to ionizing radiation. The exposure records also expose belated disposals of radioactive wastes (which should be removed from the laboratory ASAP) and incomplete or even overlooked decontaminations. Methods and means for taking care of all the eventualities already listed are abundantly documented in health physics and radiation safety publications.

10.3.5   Information for Radiation Safety

This brief annotated bibliography suggests the kind of information that a chemical laboratory using radiation sources should have available.

Brodsky, Ed., A., *Handbook of Radiation Measurement and Protection,* 2 vols.(I. Physical Science and Engineering Data; II. Biological and Mathematical Data), CRC Press, Boca Raton, FL, 1979, 1982.
This book offers a basis for detailed implementation of many radiation safety measures, but is not well suited for the instruction of laboratory beginners, particularly students. It is better suited for radiophysics than for radiochemistry and nuclear chemistry laboratories.

Fairies, R. A. and Parks, B. H., *Radioisotope Laboratory Techniques,* 3rd ed., Wiley, New York, 1973 (cf. Faires, R. A. and Boswell, G. G. J., *Radioisotope Lab Techniques,* Butterworths, Stoneham, MA, 1981).
In contrast to Brodsky, these books are better suited for new employee and student training, and quick staff and managerial guidance. It organizes material didactically that otherwise tends to be labyrinthinely discussed in larger more advanced and specialized works.

Friedlander, G., Kennedy, J. W., Macias, E. C., and Miller, J. M., *Nuclear and Radiochemistry,* 3rd ed., Wiley, New York, 1981.
This is the classical introductory level textbook in the field. It keys nuclear data compilations compactly and most conveniently collocates much of the salient information in them. One fault in this excellent work is that there is as yet no mechanism for keeping track of all the information which is changing with increasing rapidity today. Even so, the references in this book are more than one can keep up with. But the other works listed here must be used as well.

Klement, A. W., Jr., Ed., *Handbook of Environmental Radiation,* CRC Press, Boca Raton, FL, 1982.
This book supplements Brodsky above and it identifies situations that Brodsky applies to but does not identify. It correlates with the following reference.

Lawrence Berkeley Laboratory, *Instrumentation for Environmental Monitoring,* Vol. 1, *Radiation,* 2nd ed., Wiley-Interscience, New York, 1983.
This updates, and to some extent supplements the works of both Brodsky and Klement. It should be noted that there are other volumes that follow. Since other ambient and indoor air pollution occurrences can spread, or be synergistic with, radioactive pollution, it may be advisable for each chemical laboratory to review other similar works to see if these too should be on its shelves.

Mettler, F. A., Jr., and Moseley, R. D., Jr., *Medical Effects of Ionizing Radiation,* Grune & Stratton, Orlando, FL, 1985.
Here is one of the most lucid and concise books in the field; it is recent enough to reflect current and emergent philosophies of radiation safety and protection. Its introductory correlation of radiation, physics, and biology is instructive and useful. Its ensuing discussion of radiation sources aptly backgrounds action information needed in chemical laboratories using radiations.

Nargowalla S. and Przybylowicz, E., *Activation Analysis with Neutron Generators,* Wiley, New York, 1973.
This remains a seminal treatise most valuable for chemical laboratories using neutrons and performing radioactivation analysis and tracing. It is also valuable to those that use radionuclides that are products of neutron irradiation. Some of the technical updating that may be needed is keyed here, but much may have to be sought in the periodical literature through *Chemical Abstracts* and other abstracting publications. Much of the scientific updating can be tracked through use of Friedlander, above. It is interesting to note that many of the service publications of the National Bureau of Standards (NBS) use this treatise as a starting point.

National Council on Radiation Protection and Measurement (NCRP), *Operational Radiation Safety Program* (Recommendations of the NCRP), NCRP Report No. 59, Bethesda, MD, December 15, 1978.
This keys the background thinking and activity on which NRC, NCRP, ICRP, ICRU, and participating organizations base their philosophies and methodologies of radiation safety and control. This report has bibliographies on

- Operational radiation safety programs
- Facility design
- Warning and security systems
- Monitoring and control programs
- Personal protective equipment
- Orientation and training
- General health physics

- Emergency planning
- Occupational medicine programs
- Government regulations

For other pertinent reports see refs. 7 and 8.

Shapiro, J., *Radiation Protection. A Guide for Scientists and Engineers,* 2nd ed.,
Harvard University Press, Cambridge, MA, 1981.
This is the revision of a classic and pioneering guide that has been cited
often in the literature. It has a selected bibliography of NCR, NBS, American National Standards Institute (ANSI), International Atomic Energy
Agency (IAEA), ICRP, ICRU, and NCRP reports, and of topical books as
well as a long list of references. Though the text is often so dense with
information that it makes for murky reading and difficult search, it has
insights and details that are of wider application than is usually appreciated;
these at least are well and pertinently presented.

Stewart, D. C., *Handling Radioactivity. A Practical Approach for Scientists and
Engineers,* Wiley, New York, 1981.
By contrast, this is clear, lucid, informative, and easily used, read, and
taught. It reviews nuclear literature, technical and governmental radiation
standards, and various areas of radioactive radiation management, with
good referencing and adequate documentation.

10.4   OTHER ELECTROMAGNETIC RADIATIONS

There are other ionizing radiations than those just discussed. High frequency (short-wave) vacuum or far ultraviolet (UV) light has photon energies
comparable with those of γ and X rays. It too can eject extranuclear electrons from atoms or break chemical bonds to form ions. But lower energy
photons can cause ionization if enough of them are focused on a small target
volume. It is well known that heating increases the vibrational, rotational,
and translational velocities and motions of atoms, ions, and molecules.
When the collisions become violent enough from the kinetic energies of
motion so imparted, chemical bonds are broken and odd-electron species,
composed of one or more atoms, are formed. These are called free radicals.
At higher temperatures the collisions are so violent that ions are directly
formed by ionization of atoms or free radicals, or by breaking of the chemical bonds in free radicals or molecules.

So we have thermal ionization, which can go farther, abstracting many
electrons from one atom and thus forming a plasma ultimately composed of
only electrons and atomic nuclei. Since the energy and thermal levels at
which free radicals are formed are lower than those at which ions are
formed, it is puzzling that copious occurrences of free radicals on the way to
ion formation and in decay of ions by abstraction of electrons from the

environment is so little mentioned in the vast literature on ionizing radiations. This nuclear preoccupation does not exist in studies of the effects of electromagnetic radiations, and there is copious reference in them to photoradicals as well as to photoions.

Ordinary light can cause thermal ionization. Use of a lens to focus sunlight is commonplace; "burning glasses" have long been used to start fires. With the advent of modern solar technology, extremely high temperatures have been achieved by tightly focused solar radiation. It is easier to get tight focusing with single-frequency or monochromatic radiation. But the easiest focusing and the most intense monochromatic radiation is coherent or laser radiation. The hazards of laser radiation have been well appreciated by the scientific and medical communities. The public has heard of them in connection with laser beams for military uses, such as cutting through armor, imploding pellets of thermonuclear fuel for use in fusion reactors, and, of course, Star Wars. But the latter uses X-ray lasers that begin with ionizing radiation later made coherent. Ordinary lasers begin with nonionizing radiation that is made thermoionizing by being made coherent and focused.

The public has also become concerned with cancer from tanning in the sun. This involves ultraviolet radiation that is of lower frequency and energy (and of longer wavelength) than far UV. This near UV is so called because its wavelengths are not very different than those of the violet end of the visible spectrum. It can be blocked by thin films of UV absorbers (or *sunblocks*). These absorbers may not block infrared (IR) radiation accompanying the UV that can also cause skin damage.

Because it can go through nondense matter and cause vibrations that heat water molecules, microwave radiation can be hazardous subcutaneously.

All in all, nonionizing radiation can pose various hazards in chemical laboratories that should be met by protective planning.

10.4.1 Laser Radiation

Use of lasers in medical procedures triggered federal and some state and local regulation of laser use and safety. The FDA, BRH, and OSHA all participate in this. The American National Standards Institute (ANSI) and the American Conference of Governmental Industrial Hygienists (ACGIH) are also involved. Lasers have been thereby classified into four kinds in order of increasing hazard:

1. Enclosed systems that do not emit hazardous radiation.
2. Lasers limited to visible light and safe for momentary viewing.
3. Lasers not safe for momentary viewing that require prescribed controls and protection equipment.
4. Potentially hazardous lasers that can cause fire, skin burn, diffuse reflection that is harder to control than direct reflection, and retinal exposure.

For each of these, specialized appropriate training programs have been re-commended and prescribed.

American Conference of Governmental Industrial Hygienists Publication No. 4 (9) and *Laser Safety Handbook* (10), offer a good start towards deciding what kind of training might be best to fulfill requirements for a given kind of laser. *The Biomedical Laser,* by Sliney (11) is a good summary of the field, with good reference to ACGIH, ANSI, and World Health Organization (WHO) publications. Goldman wrote the laser chapter in the CRC handbook of Laboratory Safety. *Perioperative Laser Nursing,* has an amazingly helpful introduction (12).

In summary, protection against laser radiation is simply a matter of the following:

> Housing or shielding of beams that cannot be viewed safely so they can-not be seen at all or else assuring that emergent hazardous beams or their reflections, direct or diffuse, are positioned so no inadvertent eye or body exposure can occur.
>
> Using appropriate signs optimally positioned to prevent exposures, along with other control means to assure that exposures do not occur.

In one laboratory photocells that cut off powering to the laser beam have been placed so that exposure is almost impossible, yet positioning for all kinds of work with the laser does not cut off the power. In connection with this system, the two-door feed-box technique has to be used.

10.5   ELECTRICAL CURRENTS AND MAGNETIC FIELDS

Electrical and magnetic power sources and devices, including house current, are not ordinarily viewed as common hazards in chemical laboratories al-though they are so viewed in many physics and electrical engineering labora-tories. But electrical sparks have started many volatile flammable solvent fires and have caused some explosions. In addition, chemical laboratories use and store chemicals that are corrosive to electrical systems and thereby often negate what little fire protection so-called explosion-proof motors offer. Further, many corrosive laboratory solutions are electrolytes that are conductors of electricity and quite dangerous to spill on or near electrical connections or on connected equipment (13).

Besides the overt hazards of electric shock and fire from sparks, there are the insidious hazards of electric power outages in chemical laboratories. Refrigerated items that are hazardous to store at room temperature because they are reactive, unstable, volatile, and toxic or flammable, or for some other reason can warm up dangerously. (The corroded "spark-proof" refrigerator motor or the waste heat of refrigeration itself may then serve to ignite or explode the contents, with the door of the refrigerator acting to hold in the pressure until it flies off.)

Not all the insidious hazards are thermal. In one instance when a lasting power outage was announced, someone disconnecting a whisker capacitator did not realize it was still charged after the power was off and was badly shocked.

A power outage may cut off the working of an electrical or magnetic stirrer that is being used to avoid catastrophic local overheating or superheating of a viscous reactant mixture. An electromagnet holding heavy steel equipment aloft may drop it onto a tabletop on which a flammable mixture is being refluxed. Controls throttling passage of a reactive solution into a cooled reactor will fail because of a power outage just as the cooling, also needed to prevent a runaway reaction, fails. Conversely when power goes on suddenly without prior notice, items being pulled and other items not yet pulled out of the circuit become damagingly "live."

Emergency response team members and laboratory workers should be trained in electrical safety. They should know the uses and locations of all switches that control laboratory lighting and operations. Especially if outages might be a problem, either backups or fail-safe lockups should be provided. All electrical equipment should be examined at established, regular intervals to determine whether it is surviving in the laboratory atmosphere or needs repair or replacement, particularly when located in areas subject to volatile laboratory corrosives.

10.6   ELECTROSTATIC SPARKS

When two unlike materials contact each other, an electrical potential (contact triboelectrification) may result. When the two materials are displaced or rubbed over each other, there will be both charge separation that increases voltage and recharge of some areas that have transferred their charge to the other material. This frictional triboelectrification builds up the charging and potentials of contact electrification greatly. If there is enough buildup, there may then be electrostatic sparking that can start a fire.

Triboelectrification is lessened by leakage of charge into the air. This leakage diminishes as the relative humidity decreases. In winter moisture is usually well condensed out of ambient air, and when such air is taken into a heated indoor area, the humidity is exceedingly low. Triboelectrification is vastly facilitated. All know that crossing a carpeted hall to put a key in a keyhole can result in an impressive shock or a spark half an inch or so in length. At Prudhoe Bay, where subzero temperatures and desertlike ambient humidity occur together, it is quite obvious that dressing, brushing hair, touching a door knob—anything—can generate a spark. So, essentially, some electrostatic spark hazards are seasonal and climatic.

There are electrostatic hazards that are less limited by humidity and temperature. Electrostatic copiers use imaging toners that are triboelectrified by carriers so formulated that humidity differences matter but little (thus erasing the effect of outside temperature). Toner that has been photo-imaged on the drum is transferred to copy paper by a 10,000-V corona. That corona can cause a fire anytime of the year. Copiers have no place in a laboratory or storeroom where there are volatile flammable solvents or flammable gases. It is ironic that in many places flammable spirit duplicating fluid is kept near the electrostatic copier. Other sources of hidden electrostatic sparks or coronas may be dust precipitators, air purifiers, clothing (often protective clothing is highly triboelectrifiable), and plastic wrapping films.

10.7   VACUUM AND DEWAR FLASKS

Laboratories handling gases frequently use vacuum systems of glass or silica. For insulation against thermal change and for cryogenic studies Dewar flasks are frequently used. Vacuum systems and Dewar flasks pose the special hazards of implosion. When collapse occurs, sharp fragments therefrom keep moving and spreading rapidly. In addition to the cuts they may inflict on skin and eyes, the fragments might imbed in the flesh in a way difficult to remove, and in the process they could be injecting into the body toxic, corrosive, or irritant products that they had previously been in contact with. Flying glass from liquid traps in vacuum systems or from filled Dewars are most likely to do this. The greatest hazards come from systems or flasks in which radioactive, or even just radiotraced gases have been used. Radiation that otherwise might only have been external may thereby become internal.

The fall of a Dewar or the breaking of a system line or vessel might initiate the implosion. Or it could come from thermal as well as from mechanical shock. Glass, including fused or sintered silica, can weaken by devitrification, with promotion by stress or corrosion. Improper clamping may also result in mechanical or thermomechanical stress that causes cracking and breakage. Whatever the cause, the determinants of damage are the pressure gradient, the size of the apparatus, and the nature of the contents. In systems that contain a pyrophoric gas like diborane or nickel tetracarbonyl, imposion may well be followed by fire and explosion.

Wrapping to confine fragments is a good idea. But if an oxidizing gas like oxygen or ozone is to be stored in a wrapped Dewar, fire may start in the wrapping before the flask is used or dropped—and most certainly if it is dropped after surviving filling. Organic wrappings, particularly those that have volatile plasticizers or adhesives, should be used if, and only if it is certain that they will not react with future flask or system contents.

10.8   HIGH PRESSURES AND COMPRESSED GASES

High pressure is usually associated with cylinders of compressed gases and their numerous chemical laboratory uses. But many laboratories go beyond the standard piped-in gas services—fuel gas, compressed air, or vacuum. Some, mainly to avoid use or storage of gas cylinders in the laboratory, keep the cylinders outside of the laboratory and pipe in the gas needed. Commonly that may be nitrogen, oxygen, or chlorine. For some laboratories that have special areas of work, the gases appropriate for use may be those that would be considered exotic. Thus a semiconductor chemistry laboratory finds much use for diborane, boron trifluoride, phosphine, arsine, silane, and silicon tetrafluoride.

With the advent of gas chromatography gas lines as well as cylinders have multiplied for hydrogen, acetylene, and other gases that may be used for sensitive chemical instruments. Biochemical technology and clinical chemistry have led to use of sterilizing agents like ethylene oxide in growing amounts. It has already been mentioned that radioactive and radiotracing gas use has grown. The immense variety of pure gases and gaseous mixtures commercially available requires that those concerned are better informed than is possible by what follows here.

Studies of reactions at high pressures can be divided into two pressure ranges. In the lower range, pressures of interest are either reached directly, by pumping, or by pressurizing with an inert or appropriately reactant cylinder gas. The apparatus tends to be macroscopic and often larger than ordinary bench size. The pressures, understandably, do not depart far from those of compressed gases. An indirect way of reaching study pressures in this range is to fill cold and then go to the thermal or hydrothermal temperature of interest by internal heating. This produces somewhat higher pressures, say 10,000 psig.

The second pressure range is orders of magnitudes higher. It extends from thousands of pounds per square inch gauge to millions of atmospheres. This is the realm of synthesis of diamonds and of isomorphous borazon (boron nitride), of laboratory Moholes, and of shock waves for implosive nuclear fusion. The highest temperatures attainable outside of nuclear reactions and exploding electric wires are used in this, the ultrapressure range. Beyond the hydrostatic or static ultrapressure range there are the vastly higher transient or dynamic pressures generated by shock waves. More work has been done in these fields than is usually realized.

10.8.1   Compressed Gases in Cylinders

Nearly 500 kinds of gases are available in commercial size cylinders, and many more kinds can be purchased from stock or on order in lecture-size cylinders. It is with the former that compressed gases are commonly associated.) Guides for makers and users of cylinders and gases are published by the Compressed Gas Association (CGA). These, among other things, cover

Precautionary labeling and marking of containers, (14).

Pressure relief device standards (15). See ensuing discussion.

Fire testing of DOT cylinder safety relief device systems (16, 17).

Disposition of unserviceable cylinders (18).

Safe handling and storage (19).

Standards for visual inspection (20).

The first part of this CGA publication (21) treats oxygen as a fire hazard that supports ignition and combustion, and one which intoxicates and calls for dilution; the latter part of the work deals with sudden releases of gases that call for the use of a self-contained oxygen supply to avoid suffocation.

The CGAs *Handbook of Compressed Gases* (22) and *The Matheson Gas Data Book* (23) are the bibles of the industry. See also ref. 24. Matheson also distributes brochures (25, 26). The Linde Division of Union Carbide Corporation has published a series of general specialty gas precautions booklets, and has distributed precautions for use of oxygen, nitrogen, argon, helium, carbon dioxide, hydrogen, ethylene oxide, and sterilisant mixtures (27).

10.8.2 Cylinder Safety Management

All compressed gas manufacturers provide cylinders, specific valves, gauges, and fittings, labels and markings, and other ancillaries as recommended by CGA and in compliance with DOT specifications as in CFR. The regulations cover materials and manufacture, testing and certifying, inspecting and dating, and filling of cylinders and of keeping records thereon. They cover, as approved by the Bureau of Explosives, Association of American Railroads, the making and use of safety devices for cylinders and for their different contents as described below.

Cylinders are tested under hydrostatic water pressure every 5 yr (there are a few exceptions). The date of testing is stamped on the cylinder clearly recognizable as the latest date of inspection. Cylinders to be used for liquefied gases also have a tare weight stamped on them. Nonliquefied gases are filled to the service pressure marked on the cylinder shoulder (DOT3A-2000 means: according to DOT specification 3A the service pressure is 2000 psig or 2015 psia at 70°F temperature). If the gas is nonflamable, a 10% overfilling of the cylinder is permissible. Liquefied gases are filled to a filling density (maximum weight permitted as a percentage of the water capacity of the cylinder).

These pressure values are relevant to laboratory safety and health. Typically, the interior area of a cylinder is approximately 10 ft^2. At an interior gas pressure of 2000 psig., the force on the interior walls of the cylinder, then, totals 28,800,000 lb! If the cap that protects the cylinder valve fixture has been removed, and if without this protection that fixture is broken off when the cylinder, unsecured, is knocked over, then that cylinder becomes a destructive, uncontrolled rocket. With the recent advent of lighter, weaker aluminum cylinders that can rupture by impact or bounce more easily than a steel cylinder, the hazard is greater.

If the escaping gas is oxygen released within the confines of a laboratory, not only might laboratory chemical bottles and containers be broken and the contents ignited by open flames or hot surfaces, but the escaped oxygen would further increase the fire hazard—and the likelihood of oxygen intoxication, to boot. If it were an inert gas that escaped into a confined area, the risk of suffocation is likely. If the escaping gas were toxic, or an irritant, that would add to the possibility of suffocation toxic or irritant effects. If the escaping gas were combustible or explosive, the results could be fatal. Similarly, if the escaping gas were toxic, corrosive, and a strong oxidizer such as ozone, chlorine, or fluorine the same situation could occur. The rule seems to be that any chance taken with a gas cylinder may be the last chance ever taken.

When a cylinder is delivered it should not be accepted if thereby a chance will be taken with someone's life. When a cylinder is received, it should be checked for a stamped hydrostatic test date within the past 5 yr, for a stenciled or labeled identification of its contents, a DOT label that matches the preceding and other cylinder identifications, and the presence of a valve protection cap firmly in place and in condition such that what is within the cap can be inspected through the ports in the cap. If the test date, identification, DOT markings, cap, or cap ports are not in order, or if there is any mismatch, or if through inspection of the valve and fixture visible inside the cap these appear to be dirty, rusted, or inoperable, the cylinder should be rejected.

There is a practical reason for carefully examining a cylinder before it is accepted. Once accepted, a cylinder that is defective should not be used; the contents become a waste, very likely a hazardous waste. Under the Resource Conservation and Recovery Act (RCRA), the cylinder supplier may possibly not qualify as a transporter of hazardous waste. The receiver in such a case must comply with both EPA and DOT regulations for getting rid of what has become an exceedingly costly cylinder. And more to the point, as described previously, a defective cylinder that is put into use in the laboratory may cause a serious injury or fatality.

Accepted cylinders should be carted on a hand truck, firmly secured to the hand truck. A cylinder should never be carried by manhandling, nor rolled, to move it from one location to another. Except when being moved, cylinders should be securely strapped to a sturdy immovable solid support. At no time should cylinders, even empty cylinders, be allowed to stand free, unsupported. Clearly, an empty cylinder cannot become a rocket—the reason for securing filled cylinders; the hazard in not securing empty cylinders lies in the not impossible likelihood that a filled cylinder will occasionally be misidentified as empty.

Gas cylinder storage is discussed in detail in Chapter 11. In the laboratory or in a storage area, ganged chaining or strapping may permit the ganged group or individual members to fall; as single cylinders are removed from a ganged group, the probability of toppling a cylinder increases. Even pair ganging is not as good as individual securement.

Once a cylinder has been used it should be clearly marked "empty" or "MT" and never be stored along with full cylinders. If, for example, by error an empty cylinder is attached to a manifolded pressure system, suck-back will contaminate that cylinder and it will be unsuited for refilling.

Cylinders should never be completely emptied. They should be used down to 25 psig. at most and then immediately be marked MT to prevent further use by someone else. Prompt return to "MT storage" is the best policy.

Cylinders should never be dropped or permitted to strike each other. The safety devices cylinders carry should not be removed or tampered with. When open flames, sparking or arcing devices, electric or gas welding or soldering equipment, or any ignition sources whatsoever have to be used near gas cylinders, the user should do so under surveillance by someone who knows enough about gas cylinder safety to anticpate and stop any unsafe actions; for example, to ascertain that there are no cylinder or gas line leaks, to prevent striking an arc against a cylinder.

Only after a cylinder has been safely installed at its use site and is about to be used should the cylinder cap be removed. Before removal there should be a reinspection as rigorous as that used in accepting it. After removal of the cap there should be another examination of threads on the outlets and of the head and valve to make sure no other preliminaries need to be completed before proceeding further. If there is removable dirt or rust, it should be removed. If there are problems, this is the point at which to decide that another cylinder is to be used and that the problem cylinder is to be exchanged or handled elsewhere. If the valve and outlets seem operable, use can start.

CGA and the American Standards Association (ASA) have standardized outlets and safety devices for four different hazard families of gases to prevent interchanges of regulator equipment between gases that are not compatible. Hence, there are four basic types of cylinder valves, each different in outlet type and safety device. If adapters are used to subvert this built-in safety practice, incompatible gases may be mixed under high pressure; their reaction usually is violent. (One of the incompatibles is the residual gas in the adapter or in the wrong gage that the adapter has been designed to accommodate; the other is the gas in the cylinder.) For example, the interchange of equipment intended for use only with oxygen may result in fire or explosion. Use of an oil-pumped gas, say nitrogen, or use of equipment that has been used with an oil-pumped gas and is now oil laden, with oxygen, chlorine, ozone, or fluorine may also result in a fire or explosion.

The DOT does not require safety devices for toxic gases so poisonous that their release is more hazardous than potential rupture, explosion, or other failure of the cylinder. Cylinders for such gases are generally required to have a higher safety pressure factor than other compressed gases so that if failure occurs it is likely to be in a fire area that has been evacuated already. But sometimes this thinking does not work. For other gases safety devices are incorporated in the cylinder valve, in plugs in the cylinder itself, or in both.

10.8.3   Cylinder Gas Management

For safe cylinder management it is necessary to know the gas it contains. If a gas poses a multiple hazard—say it is toxic, reactive, shock sensitive, corrosive, and flammable—its hazards at the site of use should be considered. Thus, in one situation carbon monoxide may be regarded as likely to burn or explode; it is being used at high temperatures (long feed lines and remote siting of the cylinder is suggested). In a different situation, the cylinder is being used within rather confined quarters; sudden release might cause anoxia and asphyxiation.

Some gases have flammability ranges so wide that they are fire and explosion hazards in almost any quantity or concentration: the most common are acetylene, carbon monoxide, ethylene oxide, hydrogen, and hydrogen sulfide, selenide, and telluride (becoming of greater prominence in solid-state device fabrication research). Others can form sensitive explosives readily, for example, ammonia forms mercury fulminate with mercury (do not use mercury manometers), and acetylene forms copper acetylide (do not use copper or ordinary brass or bronze lines, gauges, or fittings). Some downstream hazards can be minimized by use of hoods, others by monitoring, and some only by use of robots or remote manipulation.

When corrosive gases are used, frequent working of the valve is necessary to prevent freezing. Gauges, valves, and fittings should be flushed with dry nitrogen or air as soon as possible after use. When toxic gases are used, in addition to monitoring, suitable respirators should be either worn or at hand. When radioactive or radiotracer gases are used, if they are an external hazard, then in addition to the precautions used for internal radiotoxicity protection (much like those for toxic gases per se) all the precautions taken to minimize exposure time, maximize distance, and assure best shielding should be taken (cf. above).

To summarize, in using compressed gases both users and response teams should be trained in coping with all the hazards that can occur in the projected uses. Assurance of prompt first aid and physician response is mandatory. The proper type of respirators and other protective equipment should be close at hand in a ready to use condition (proper sizes, personnel fitted and trained). Safety glasses should be worn.

Eyewash fountains and emergency showers should be nearby but not in the area likely to be damaged or contaminated. Recently inspected and approved dry chemical (sometimes Halon) fire extinguishers should be near the area.

Liquefied gases are removed from cylinders using manual (needle) valves, perhaps with a special flow regulator. Removal must avoid pressure-drop flash vaporization, sudden force-out of liquid contents, or cooling that drops flow and pressure (if the cylinder must be heated, do not exceed 125°F.). With carbon dioxide possible dry ice formation can block the exit and make it impossible to determine whether the main valve has really been closed. All liquid transfer lines should have safety relief devices to release sudden pressure buildups.

For nonliquefied gases an automatic pressure regulator may be used, consisting of a spring or gas loaded diaphragm to throttle the orifice down to delivery pressure as set. If the diaphragm is a vulcanized elastomer, it should not be used with oxygen or other oxidizers, particularly ozone, and it should be checked to see if it has been aged by atmospheric ozonolysis. There are single-stage and double-stage regulators, the latter preferable. The former varies slightly, but sometimes objectionably, in delivery pressure as cylinder pressure varies. The two-stage regulator delivers a more consistent pressure. It has a smaller drop in delivery pressure on increase of the flow rate. It also has a lower "lockup" pressure (pressure excess that stops flow to assure delivery only at the set pressure).

While manual flow controls are sometimes advantageous, they provide no way to automatically handle the development of excess pressures in a clogged or plugged up system. In an automatic pressure regulator, the delivery pressure adjusting screw setting the diaphragm and the flow control valve, with indication, respectively, by the cylinder pressure gauge and the delivery pressure gauge, permits delivery pressure to exactly balance the delivery pressure spring to give a relatively constant delivery pressure. If necessary, the system itself can have an arranged relief just a trifle higher in pressure than the delivery pressure setting.

For nonliquefied gases the pressure decreases proportionally to the amount of gas withdrawn. For liquefied gases the pressure remains the vapor pressure at that temperature until all the liquid has been withdrawn and only the gas remains. Because of the cooling effect already remarked, weighing the cylinder is the best way of determining how much gas remains at any time, which is why the tare weight should be determined before accepting delivery. Whatever the fittings, a cylinder should not be opened in use more than a quarter of a turn of the valve. Then, if a downstream problem develops, fast closing is possible. Otherwise there is the spectacle of someone turning and turning and turning while things keep on mishappening.

10.8.4   Oxygen and Oxidant Gases

Pure oxygen gas is five times as concentrated as it is in air, and hence it is a stronger oxidant than air. Since cylinders are prone to leaks, oxygen cylinders should not be stored close to hydrogen, other flammable gas cylinders, or near flammable solvents. The same applies to compressed air cylinders; a compressed gas is many times more concentrated per unit volume than an uncompressed gas, and so its activity is enhanced. As seen previously this will not be as much a concern with compressed air as with liquid oxygen and liquid air, but when under pressure oxygen and air contacting organic materials have blown systems apart. Chlorine can cause amazing fires by abstracting hydrogen from organic compounds, leaving carbon that is so finely divided that sometimes it is pyrophoric.

Fluorine, chlorine dioxide, and ozone are other oxidant gases that can endanger closed systems and they can do so at practically any pressure. System materials should be carefully chosen so that they are not reactive with these gases or with hot or compressed oxygen or air. The problem arises with some of the instruments that have not been tested for use in the systems using these gases; the tests themselves reveal unexpected hazards. For example, copper wire is supposed to form a protective layer of oxide, fluoride, or chloride, as the case may be, with these gases. But, if the reactants introduced into, say, a copper–glass sealed system continuously reduce the oxide (or fluoride, chloride) to the metal powder, the seal can fail and the apparatus leak.

10.8.5   Other Cylinder Gases

In general, other than the specific cylinder precautions noted above, the same precautions should be taken for any gas release from cylinders that would be taken with atmospheric pressure releases of that gas.

Gases not commonly used before in laboratories are now being introduced for use in the processing of silicon chips and related devices. Besides fluorine and hydrogen fluoride, which are not unfamiliar, the solid-state and semiconductor processing gases are diborane, boron trifluoride, silane, phosphine, arsine, and germane. Some of these are pyrophoric. These, and some of the other gases and liquids the electronics industry uses corrode cylinders and their fittings quickly and severely. What they will do to the recently introduced aluminum cylinders remains to be seen, but may pose similar problems.

10.8.6   Acetylene

Acetylene is the maverick gas. Compressed pure, it may polymerize exothermally and explode. Compressed rapidly with undissipated heating, it may decompose to acetylene black and hydrogen, sometimes explosively. The acetylene black is quite graphitelike and is not pyrophoric. To prevent these didoes acetylene is dissolved (compressed) in acetone which is in turn absorbed in a porous filler in a cylinder that is entirely different in form and in safety devices (multiple metal plugs that melt in boiling water). Full cylinder pressure is only 250 psig at 70°F. It is recommended that acetylene never be used outside of the cylinder at a pressure exceeding 15 psig. If a higher pressure is wanted special precautions must be taken against explosion. The cylinder must always be upright at least half an hour before use so that in use acetone will not be expelled along with acetylene.

The recent prominence of acetylene as the fuel in chromatography has vastly increased the number of laboratories that use it. Since acetylene forms explosive copper acetylide, the use of copper lines in such laboratories for other gases will cause problems if acetlyene is introduced into that line. In one Canadian hospital, the entire laboratory was endangered by use of acetlyene in copper lines. In a U.S. research laboratory, when the wall facing was removed near the piped-in acetylene line, the proper noncuprous fittings on both ends of the line that were visible extending from the wall when the covering was in place, were found to lead, inside the wall structure, to copper tube. Polymerization of acetylene is a further problem; the reaction is catalyzed by a variety of semiconductor materials. Polymerization has occurred unexpectedly in systems not stripped of residuals prior to the use of acetylene.

10.9   PIPED-IN GASES

Besides vacuum services that use piping, laboratories have gases piped in at various pressures, for example, fuel gas at low pressure, and compressed air at slightly higher (sometimes much higher; compressors and policies vary) pressures. The variability of pressure can be a trap; pressures drop with the amount of use and laboratory workers sometimes test for air pressure in a hazardous manner. One student opened the compressed air valve fully while his palm was faced toward the outlet, to test the pressure. His hand was severely bruised and, in addition, rust from the line was driven into the bruised area. In another instance helium was manifolded into the laboratory from cylinders outside the laboratory; a student testing the pressure in that line found her hand swelled to a balloon.

10.10   HIGH PRESSURE SYSTEMS

All pressure systems can rupture, and at higher pressures, cold or hot, the pressure release upon rupture could be explosive. If the released contents are flammable, and sources of ignition are present, the first explosive release could be followed by either a fireball or a second explosion. What happens depends on system content. Thus in hydrothermal systems, there is always the threat of release of superheated steam. In other systems toxic or corrosive products might be released.

Hazards that are serious at atmospheric pressures in open systems may become far more serious at high pressures in closed systems. Thus, if a catalyzed or autocatalyzed reaction at atmospheric pressure is overheated because it has an induction period and is slow to start, the penalty for the impatience is a runaway reation that may cause a fire or an explosion. At 3000 to 10,000 atm the starting base pressure is already so high that a catastrophe results; thrown materials go through walls.

10.11 ULTRAPRESSURE SYSTEMS

Amazingly, ultrapressure systems have almost no accident literature. One reason is that they are generally small compared to high-pressure systems or to gas cylinders. A few cubic centimeters may be all that can be brought up to the enormous pressures used. The pressures are attained mechanically, usually in one of three ways. Bridgman used piston pressure with anthophyllite packing. The same packing material is also used in the General Electric tetrahedral anvil (used originally to synthesize both diamonds and borazon, see above). The third way is the Battelle girdle, circumferential transmission. In all three methods there are two main risks. The first is that uneven packing or compression, or failure of a member, can cause a Munro or "shaped-charge" effect jet squirting with tremendous force or velocity. The second is that going to a high temperature, say 8000°C, by putting in heat electrically faster than it flows out can trigger an unexpected reaction that raises the pressure beyond the confinement capability.

The usual way of failing safe is to impose the pressure to secure the temperature confinement first, and then to raise the temperature incrementally, noting the rise curve by the implanted instrumentation. If the rise steepens, then the imposed raises are made smaller. By that time enough data has been gathered so a computer forecasting program can give some indication of future *safe* and *hazardous* time–temperature relations. Therefrom, future raise–rise regimes are tested, with a stop if control is equivocal. The saving grace is pressure quenching. This can be so fast that danger is quickly diverted. Of course, temperature quenching is too slow. The pressure quenching gives samples that tell how to do the next run more safely. Better a failed experiment than a gutted laboratory.

10.11.1 Shock-Wave Systems

Shock-wave systems may be the safest ultrapressure-range systems because they are so small and because the force they use can be so well directed.

Nevertheless they deal with gas jets that move at 90 km/s, and like explosion shock waves they can exercise the Munro or shaped-charge effect to punch holes in what they pass through. Explosion, implosion, spark snap, and other means are used to form the shock-wave jets that transiently compress the target materials. Phases not attainable by other techniques have been observed.

One variation of these systems is the use of lasers to apply enough compression to pellets of thermonuclear fuel to start fusion reaction and power generation. The principle is that of implosion. Both the jet impact and implosion systems are being introduced into chemical laboratories as special chemical compositions amenable to their phenomenology are uncovered.

It is likely that in a decade or so it will be practical to subject hydrogen to pressures high enough to exploit its metallic properties and similarly to make, at high pressures, solid-state current carriers of inner-shell electrons of other elements. These developments will require corresponding new safety protection methods.

10.12   LOW TEMPERATURE AND CRYOGENIC SYSTEMS

Frostbite may be the best-known hazard of working with cryogenic systems, but it is far from being the only one. Temperature affects the phases stable for a given substance or mixture, sometimes with surprising results. Iron, steel, and ferrous alloys in general become brittle at cryogenic temperatures. Tin is a silvery metal above 18°C and a gray semiconductor or semimetal (resembling germanium in part) below 19°C. Rubber stops being elastic and crumbles to a powder.

Unexpected chemical as well as physical changes occur. It is well known that hydrogen embrittlement of metals can occur because the hydrogen can readily enter in between the atoms of metals and weaken the crystal lattice. At low temperatures, where the hydrogen is liquid, there are many more atoms in a given volume than at higher temperatures, when it is a gas. So, despite the slowed molecular motion, liquid hydrogen quickly embrittles metals. This concentration effect has been mentioned before for oxygen and oxidizing gases. It will be revisited in the next section.

10.12.1   Liquefied Gases

Like compressed gases, liquefied gases are extremely concentrated relative to room temperature gases. Frankfurters not burn well in air; they char. Dipped in liquid air they burn well. Dipped in liquid oxygen, they burn vigorously. Dipped in ozone, they may catch fire or explode before they are lit. So, some gases that are considered safe at room temperature experience a significant increase in their hazardous properties on liquefaction. The increased activity of oxidizing gases is understandable. But consider liquefied natural gas, LNG. When spilled at large volume at sea, it freezes itself into an "ice" boat while evaporating, losing mainly methane. And then, very often, it spontaneously explodes; a fire and second explosion usually follow. Probably the first explosion is initiated by discharge of the static potential difference built up during the rapid evaporation of a part of the spilled liquid gas. The heat from the first exposion vaporizes more liquid and ignites it. The same events could occur following a laboratory spill.

When liquid air is used as a cryogen and is replenished as it evaporates away, hazard blossoms. Air is a mixture that has a 4:1 ratio of nitrogen to oxygen by volume. Liquefied and open to the atmosphere, it boils. The lighter, lower-boiling-point nitrogen gas boils off faster than the oxygen. Indeed, its boiling condenses some of the oxygen that boils off, dropping it back into the vessel. It is also condenses some of the heavier atmospheric gases that diffuse into the vessel top, including ozone. Ultimately the liquid air is a high-oxygen mixture laced with some ozone—and much more dangerously oxidizing than liquid air. The safety approach is to adopt a handling and replenishment regime that keeps the danger minimal.

The obvious hazard of liquefied gases is that a large spill can result in vaporization that is dangerous; this has been discussed already. Even if the gas is breathable, and suffocation or asphyxiation is not a problem, the cold gas can sear the nasal and breathing passage mucous membranes by freezing them. Other, less common, hazards are discussed in ref. 28; compare refs. 29 and 30.

10.12.2 Solidified Gases and Dry Ice

Usually solidified gases are formed by freezing of liquid gases and so are colder than the gases they are formed from. On warming, a slurry of flaked solid gas in the liquid gas will reach the melting point temperature of the gas and maintain it until all of the solid has melted. This is an alternative to the use of a liquid cryogen, where the temperature maintained is that of the boiling point of the liquefied gas. Availability of the solidified gases somewhat extends the options for working at low temperatures. There is one solidified gas that is different.

That gas is carbon dioxide; as is well known, at atmospheric pressure the solid sublimes instead of melting. Even in moist air, the solid does not become wet; condensed water vapor on the surface of the solid carbon dioxide is, of course, also a solid. The layer of water ice is so cold that it cannot be melted by a moderate pressure increase as, for example, when a piece of dry ice is held in the bare hand using a soft touch. Tightly held, it will induce frostbite. In picking up dry ice bare handed, even a soft touch works for only a few pieces; the cumulative chilling becomes damaging. A different hazard is associated with the release of liquid carbon dioxide from a cylinder. As it is released it vaporizes and is cooled, and some of the released material forms dry ice that can block outlets and gas lines as discussed in Section 10.8.3. This hazard is too little mentioned in the commercial literature on carbon dioxide.

10.13   THERMOLABILITY HAZARDS

It is well known that heating of thermally unstable or thermolabile systems above room temperature is risky. But it is not generally remembered that a like hazard exists for warmup of very cold systems. This has already been pointed out as one hazard of electrical power outages for refrigerators in chemical laboratories. At lower than refrigerator temperatures a product might have been made, or a system assembled, that would not have been stable at refrigerator temperatures. Hence, when it is warmed up to refrigerator temperature from the lower temperature, it can reach a temperature of catastrophe en route.

Few such catastrophes have been reported. One that did kill the investigator (and probably lost the causative evidence in the accident) might be instructive. For hazardous waste detoxification and disposal, it seemed that if wastes containing explosive nitrogen trichloride, NCl_3, and wastes containing explosive sodium azide, NaN_3, were mixed, the successful formation of salt and nitrogen gas would result:

$$3NaN_3 + NCl_3 \rightarrow 3NaCl + 5N_2 \qquad (2)$$

Since this could be explosive if exothermic, the investigator made a small-scale pure chemical run at cryogenic temperatures. Sodium chloride seemed to form, but no gas was emitted. After discussing this, he went back to watch and see if warmup would liberate the missing gas. The explosion occurred, leaving only broken glass and some salt. Was there a disastrous intermediate formation of solid nitrogen triazide that then blew up?

$$3\,NaN_3 + NCl_3 \rightarrow 3\,NaCl + N(N_3)_3 \qquad (3)$$
$$N(N_3)_3 \rightarrow 5\,N_2 \qquad (4)$$

Or was reaction (2) itself the culprit?

10.14   HIGH TEMPERATURES

Because they cause and are caused by accidents, high temperatures are a large subject area in chemical safety—with scanty documentation outside of fire chemistry. But burns not involving fire or flammable solvents are common laboratory occurrences. The common causes are use of heating equipment and heat, and taking them for granted. Every chemist knows that in general the speed of a reaction is doubled with a rise of 10°C. But this seems to be overlooked in many laboratory accidents.

10.14.1   Common Heat Sources

Gas and electric heating devices, commonly used in open systems, are ever present in laboratories. Unlike the closed systems to be discussed, open systems do not have the pressure gradients that result from rise of temperature. The open heat sources are generally used to elevate temperature to facilitate and speed up chemical reactions. In cases where more than just heating is required—glassblowing, soldering, welding, and so on—the user is in closer proximity to the heat source and for a longer time time at risk for a burn. Also, the special sources used are either not as remote or as tractable as heating elements and the like. Arc welders, soldering irons, welding and glassblowing torches, even jeweler's torches are such sources.

Usually, gas heating simply involves a burner and a heated vessel. Or a heating bath may be used, in which case the choice of heating bath medium affects the hazard (cf. water, oil, sand, and sulfuric acid). Or electric heating may be used; usually the choice is either a hot plate or heating mantle. When gas is used, there is always the possibility of a gas leak. Electricity is not particularly mobile, but at least when gas is turned off the heating stops—unless it was a bath that was heated. Neither hot plates nor mantles cool rapidly. Select that method of heating that provides the best control of potential hazards.

Thus, consider the classic case of heating a catalyzed or autocatalyzed reaction mixture that has an induction period prior to rapid reaction. If such a reactant mixture is overheated, and a gas burner is the simple heat source, shutting off and removing the burner helps limit the extent of the runaway reaction. But if a bath, hot plate, or mantle is used, the heating goes on, the continuing reaction is faster than when a burner is used. The escalation of hazard is apparent.

10.14.2   Other Residual Heat Hazards

Whenever a substance or object has been heated and the heating has stopped, there may remain a potential for damage. The hazard becomes more insidious as the signs of high temperature vanish. As glows vanish, boilings stop, and heat and its transfer become less apparent, hot items look no different than when they are cold. The just-blown glass is touched and dropped. The recently used soldering iron is picked up the wrong way. The cooling but not yet cool beaker is picked up and dropped.

10.14.3   Hot Closed Systems

Some vagaries of these systems have already been discussed, particularly as they relate to pressure gradients that may pose hazards. A hot closed system may be unintended; running an open system so that it plugs up forms a metastable closed system. Closed systems that open may do so explosively, and if they contain flammables there may be a fire and second explosion, as in the case of pressure systems. It is hardly ever safe to expose to air any substance that has been heated above its flash point, fire point, self-ignition, deflagration, decomposition, or detonation temperature. All of these could occur in either an overheated or catalyzed closed system.

10.15   OBVIOUS BUT OVERLOOKED HAZARDS

Each hazard described here is known to have been the direct or remote cause of at least one accident that caused serious personal injury. Note that many are so obvious they can be easily overlooked. Some in the list usually cause only minor injury, but not always. Readers are invited to add their own.

Personal Habit and Manner of Dress. Hair styling that is ignitable, that is, "frizzy"; unkempt beards; long loose hair; hair treatment that leaves ignitable film on the hair; loosely woven and flammable synthetic garments; loose clothing; sleeves, ties, necklaces, bracelets, large-stone rings, shoe strings; open-toe shoes; absorbent articles of clothing that may not be protected by a lab apron or coat, such as shoes, belts, wallets, key cases, wristwatch straps made of cloth, leather, or synthetic polymer; perfume and tobacco odors that impede user's and neighbors' detection of organoleptic warnings of the presence of harmful vapors.

Protective Equipment. Ventilation system not operating at necessary capacity, or operating so as to induce turbulence, thus reducing capacity; leaking gloveboxes; eye and face protection below ANSI or other standards; pinholes in protective gloves; gloves made of material permeable to chemicals being handled; gloves of "proper" material relied on to remain nonpermeable after several hours of use; inside surfaces of glove contaminated with hazardous chemical; reliance on so-called barrier cream instead of gloves; facial beards that prevent proper fitting of eye protection equipment and proper fitting of respirators.

Electrical Power and Distribution. Defective switches; worn or frayed wiring; overloaded socket outlets; no ground connection; sockets improperly wired; lack of ground fault circuit interruption protection; wiring hidden from view is not encased in conduit or equivalent.

Laboratory Environment. Poor housekeeping, general clutter; hoods used both for casual or longer storage and for work needing controlled venting; tripping hazards; inadequate shelving; open shelves in earthquake-prone areas; clogged drains; inadequate lighting; poor inventory control; inadequate means of egress.

Work Practices or Habits. Labels pasted over labels; labels overwritten with different content identification; unjustified personal chemical hoarding; failure to treat the results of a presumed to be minor incident; failure to report incidents deemed to be inconsequential.

Emergency Equipment. Safety shower, eyewash fountain not recently tested; safety shower water flow < 50 gal/min; eyewash fountain water flow < 2.5 gal/min; first aid supplies deteriorated, components missing, inappropriate for hazards encountered in the workplace; emergency rescue equipment not in operating condition, personnel unfamiliar with manner of use; alarm system inoperative, alarm sounds not recognized by workers.

REFERENCES

1. Nuclear Regulatory Commission (NRC); 10 CFR parts 20, 34, 71, and 150.
2. Bureau of Radiological Health (BRH); 21 CFR subchapter J.
3. Occupational Safety and Health Administration (OSHA); 29 CFR suppart 1910.96.
4. Environmental Protection Agency (EPA); 40 CFR parts 260 to 270 and 300.
5. U.S. Department of Transportation (DOT); 49 CFR part 170.
6. U.S. Postal Service (USPS); 39 CFR part 123.
7. National Council on Radiation Protection and Measurement, report Nos. 8, 22 (amended), 30, 38, 39, 50, 57, and 58, Bethesda, MD.
8. H. M. Parker, *The Squares of the Natural Numbers in Radiation Protection*, Lauriston S. Taylor Lectures, Lecture No. 1, published by NCRP, Bethesda, MD, 1977.
9. American Conference of Governmental Industrial Hygienists, *Safe Use of Lasers*, ACGIH Publication No. 4, Cincinnati, OH, 1976.
10. A. Mallow and L. Chabot, *Laser Safety Handbook,* Van Nostrand Reinhold, New York, 1978.
11. D. H. Sliney, "Safety with Biomedical Lasers," in *The Biomedical Laser,* Chapter 3, L. Goldman, Ed., Springer-Verlag, New York, pp. 11–24.
12. C. J. Mackety, *Perioperative Laser Nursing,* Parts I and II, Slack, Thorofare, NJ, 1984, pp. 1–43.
13. E. G. Magison, *Electrical Instruments in Hazardous Locations,* Plenum, New York, 1966.
14. Compressed Gas Association, CGA publication C-7, Arlington, VA.
15. Compressed Gas Association, CGA publications S-1.1, S.1.2, and S.1.3.
16. *Code of Federal Regulations,* Title 49, subpart 173.34(d).
17. Compressed Gas Association, CGA publication C-14, Arlington, VA.
18. Compressed Gas Association, CGA publication C-2, Arlington, VA.
19. Compressed Gas Association, CGA publication AV-1, P-1, and P-4, Arlington, VA.
20. Compressed Gas Association, CGA publications C-6 and C-6.1, Arlington, VA.

21. Compressed Gas Association, *Accident Prevention in Oxygen-Rich and Oxygen-Deficient Atmospheres,* Arlington, VA, 1983.

22. Compressed Gas Association, *Handbook of Compressed Gases,* 2nd ed., Van Nostrand Reinhold, New York, 1980.

23. W. Braker and A. L. Mossman, *The Matheson Gas Data Book,* 6th ed., Mathison Gas Products, Secaucus, NJ, 1981.

24. W. Braker, A. L. Mossman, and D. Siegel, *Effect of Exposure to Toxic Gases—First Aid and Medical Treatment,* Matheson Gas Products Lyndhurst, NJ, 1977.

25. *Working with Compressed Gases,* Matheson Gas Products, East Rutherford, NJ.

26. *Guide to Safe Handling of Compressed Gases in Laboratory and Plants,* Matheson Gas Products, Secaucus, NJ, 1983.

27. The Linde Division of Union Carbide Corporation, South Plainfield, NJ and New York, publication Nos. F-3499-C, 100M, 86-0584, 1976.

28. M. G. Zabetakis, *Safety and Cryogenic Fluids,* Plenum, New York, 1967.

29. Compressed Gas Association, *Safe Handling of Cryogenic Liquids,* P-12, Arlington, VA, 1980.

30. F. D. Eskuty and K. D. Williams, Jr., *Liquid Cryogens,* 2nd ed., CRC Press, Boca Raton, FL, 1983.

CHAPTER 11

Storage of Laboratory Chemicals

J. R. Williamson

Eastern Michigan University, Ypsilanti, Michigan

and

W. K. Kingsley

J. T. Baker Chemical Co., Phillipsburg, New Jersey

11.1 INTRODUCTION

Improper storage of laboratory chemicals is the cause of many laboratory accidents; if a statistically valid database were available, improper storage might well be found to be the major cause, outranking any other category. These incidents will illustrate this fact.

- Diethyl ether was stored in a nonair-conditioned laboratory where temperatures often reached 100°F. While a technician was unscrewing the cap from a can of ether, the pressure of the vapors inside the can forced the cap upward out of the technician's hand, allowing the ether vapor to stream upward at high velocity. The vapor was ignited by the hot surface of a bare incandescent light bulb in the ceiling. The explosion destroyed the laboratory and severely injured the technician and others in the laboratory.

- Methyl bromide was stored for years in a metal can in an out of the way corner of the laboratory. Eventually, pinholes developed in the metal. The two occupants of the laboratory were exposed to the vapors for several days before the cause of their daily headaches was discovered. The older of the two victims suffered permanent mental disfunction.

- The label on an old acid bottle had long ago been destroyed by laboratory fumes. When the bottle fell off the shelf and the contents splashed on the victim, inadequate first aid based on a conjecture about the properties of the splashed liquid resulted in severe scars for the victim.

- Sodium hypochlorite was stored near oxalic acid in a humid storage area; occasionally the floor, for example, was wet with condensed water. Shortly after the two chemicals were accidentally spilled near each other on the wet floor, the resulting fire involved nearby flammables and destroyed the entire storage area.
- When setting up for a project estimated to require about a year and use a few grams of sodium for each run, a laboratory worker acquired 5 lb of the metal still sealed inside the original can, by now rusty with age. Then, the project was canceled and the sodium, still in the corroded can, was stored in the back corner of the laboratory's equipment cupboard—directly under an automatic water sprinkler. A fairly small fire about 1 month later actuated that sprinkler and the original fire was quickly extinguished; the ensuing fire, originating from the water-soaked sodium, destroyed the laboratory.

Of course, few laboratories reach 100°F, nor do many laboratories have bare hot surface lights in the ceiling. Not every laboratory stores methyl bromide in metal or other containers. In most cases of decrepit labels, enough remains of the label to at least identify the contents as corrosive, and proper first aid can be administered. And of course, it is rare that a storage area reaches 100% relative humidity. Even less often is sodium stored out of sight in a deteriorating container under a sprinkler. The fact remains, each of these readily avoidable storage errors has occurred, at least once. Other kinds of improper storage are more common, and all possess accident potential.

11.2   COMMON EXAMPLES OF IMPROPER STORAGE PRACTICES

More than one of the following examples of improper storage practices can be found in far more than half of the laboratories or their associated stockrooms and storerooms:

- Chemicals are stored in alphabetical order by name.
- Chemicals are stored according to poorly chosen categories; thus, all acids are stored together, or all organics are stored in a single allocated area.
- Chemicals are stored in the laboratory hood while the hood is used for other designed purposes.
- Chemicals are stored in an ordinary refrigerator.
- Chemicals are stored in a so-called explosion-proof refrigerator, and so is the laboratory worker's lunch.
- Chemicals are stored on shelves above eye level.

- One bottle is supported by another bottle, underneath the first.
- Storage shelves are so crowded that it would be impossible to put even one more container on the crowded shelf, or so crowded that when a single bottle is to be removed, other nearby containers must be moved, or themselves removed, to get at the bottle of interest.
- Chemicals are stored in out of the way locations instead of in assigned areas.
- Chemicals are put on laboratory bench tops and not returned to their proper shelf location for several days, long after the need to be on the bench top has passed.
- The shelving on which chemicals are regularly stored is not strong enough to support the weight.
- Tiers of shelves are independent, not firmly secured to wall or ceiling, and can topple easily.
- More than three flammable liquid cabinets are located within 100 ft of each other.
- Flammable liquid cabinets are not vented, or are vented by ducting to a hood duct instead of directly, and independently, to the outside at both top and bottom.
- Peninsula and island shelf tiers can be accessed from both sides and the shelves are not fitted with raised-lip edging to prevent containers from falling off the *other side* when a container is put on *this side* of the shelf.
- Inventory control is poor or nonexistent; many containers are not identified with date of receipt or the person responsible for that container of that chemical. Some chemicals are so old that the most senior laboratory worker cannot remember why that chemical was purchased, or when.
- Some containers have no labels, or illegible labels, or labels that describe neither the contents nor the hazards of the contents.
- Bottles or cans of flammable liquids are stored near other materials that are in cardboard boxes.

Combinations of these storage errors are also possible; thus in an ordinary refrigerator there might be volatile compounds, nonvolatile toxic materials, and all in a crowded condition.

11.3   THE PRINCIPLES OF CHEMICAL STORAGE

For safe and proper storage of laboratory chemicals, two major principles apply. Maintain control of the inventory, and segregate mutually incompatible chemicals each from the other(s). The application of these principles is not as straightforward as it might seem.

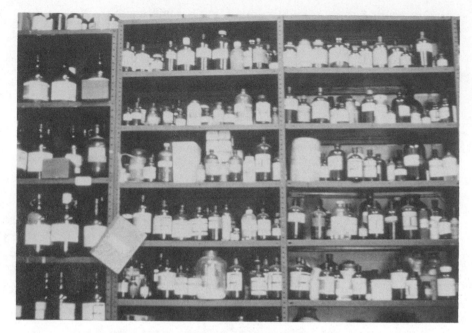

Figure 11.1.   A poor example of inventory control.

11.3.1   Inventory Control

Figure 11.1 illustrates one attempt at inventory control. Every bottle on the shelves is recorded in the notebook seen in the lower left corner of Fig. 11.1 in chronological order of receipt. Proper inventory control consists principally of these elements:

- Purchase limited amounts. In the extreme, never acquire more than a year's supply and preferably limit purchased quantities to amounts that, as nearly as can be foreseen, will be completely used in, say, 3 to 6 months. The lesser unit cost both of large quantity and of annual bid purchases is outweighed by the costs of the additional storage facility that is required and certainly by the costs of rebuilding a laboratory that was destroyed when an old chemical or two, long forgotten and now deteriorating, erupted in a dramatic manner.

- Assign personal responsibility for each different container of chemical that will be on the premises more than, say, one day, whether purchased from an outside source or prepared in house. The identity of the person responsible and the date of acquisition or of preparation should be marked on the container label or on an accessory label that is affixed to the container. These markings and their placement on the container should be standardized.

- At the time of acquisition or preparation, assign expiration dates to each container of each chemical. The expiration date should not be longer than 1 yr from the date of acquisition or preparation and, depending upon the stability of the chemical involved, its foreseen rate of use, storage conditions, and so on, may often be much less than a year. Computer based inventory systems are easily modified to present a list of "this week's expired chemicals" every Friday. An expired chemical is either marked for destruction or disposal (see Chapter 14), or, if warranted, given a new expiration date.
- Personally inspect every storage location at regular intervals, preferably at least semiannually. The purpose of the inspection is to identify unknown storage problems—the expired bottle that somehow escaped the computer flag, water dripping from the ceiling, a fire extinguisher that has lost its charge, a chemical in a deteriorated condition, and so on. For this policy to be effective, it must be known that no chemicals are stored in any other (secret?) locations. In some cases, it would certainly be prudent to also personally inspect likely and unlikely out of the way nooks and crannies; see Fig. 11.2, for example.

11.3.2 Segregation

Numerous publications have warned against nonsegregated, alphabetical storage of chemicals; see Chapter 7 and Pipitone (1), for example. As is obvious, alphabetical storage simplifies inventory control. It also increases the hazards of storage. These four examples of reactively incompatible pairs of chemicals, each pair with potentially alphabetically adjacent members, illustrates this point.

1. Acetic acid and acetaldehyde
2. Copper(II) sulfide and cadmium chlorate
3. Hydrogen peroxide and iron(II) sulfide
4. Sodium nitrite and sodium thiosulfate

In case of a spill, fire, or other commotion in a storage area, the presence of reactively incompatible and unsegregated pairs of chemicals enhances the likelihood of still more, and perhaps dramatic, commotion with consequent injury to persons and damage to property. Clearly, reactive incompatibles ought to be separated by barriers in storage areas.

The problem is how to create the segregated conditions; what barriers are suitable? Again obviously, the best barrier is a combination of sturdy impenetratable physical walls and distance—usually more costly than can be warranted. A related concern deals with the number and variety of segregated classes; the more and the more varied the classes to be segregated, the greater the cost. The point of diminishing return can be exceeded very quickly.

Figure 11.2. Concentrated nitric acid and phenol, stored under a sink behind closed doors.

Thus, segregation into broad classes at first seems reasonable: acids stored here, bases there, some distance away; flammables in another storage area, oxidizers of course not stored near the flammables, all organic compounds not in any of the preceding groups in a different place, and so on. The difficulty is that, for example, some acids are also reducing agents, for example, glacial acetic acid, and some are also oxidizing agents, for example, nitric acid. These two acids should be separated in storage. Or, some oxidizing agents are incompatible, for example, hydrogen peroxide and potassium permanganate, and should be kept separate.

Clearly, segregation by broad classes is unsatisfactory; perhaps further separation into subclasses is acceptable. Table 11.1 is such a solution.

The eight classes, with subclasses as shown in Table 11.1 should each be segregated by physical barrier and/or distance, each from the others. Further, the toxic subclass within each class should be stored, with that class, in

Table 11.1. A Storage Segregation Scheme

Class	Types of Chemical, by Property, in Each Class
1	Flammable or combustible and not highly toxic; flammable or combustible and toxic; peroxidizable; provided that in all cases the chemical is compatible with water.
2	Identical to Class 1, except that all in this class are not compatible with water
3	Oxidizing agents and non flammable/combustible, not highly toxic; oxidizing agents and nonflammable/combustible that are toxic; each compatible with water
4	Identical to Class 3 except none are compatible with water
5	Air sensitive and not highly toxic; air sensitive and toxic
6	Require refrigerated storage and not highly toxic; require such storage and toxic
7	Compressed gas cylinders and other gas containers, divided into four subclasses, oxidizing agents, reducing agents, corrosives, and highly toxic substances; each subdivided into two sub-subclasses, empty and full
8	Unstable chemicals (exlosives, short shelf life, etc.)

an area with strictly limited access. The scheme in Table 11.1 is designed to attend principally to segregations that will prevent fire, explosion, and toxic exposure incidents. Note that it has no provision for radiation hazards and is silent regarding acid–base incompatibility. Thus, both sulfuric acid and sodium hydroxide could be stored near each other, unsegregated, in Class 4. A scheme designed to attend to the prevention of fire, explosion, toxic exposures, acid–base incompatibility, and perhaps even other hazards would be complex.

Probably, an ideally perfect segregated storage scheme would not be practical since many chemicals that are to be stored for laboratory use have more than one hazardous property (if they did not, they would indeed be less useful for laboratory purposes). In order to maximally minimize the segregated storage of reactive incompatibles, the number of classes and subclasses would be unmanageable, leading to still other hazards.

The practical solution seems to be to adopt a manageable number of classes and then to apply sound judgment to solve the inevitable problems associated with that limited number of classes. Refer to the material safety data sheets for the chemicals involved and to Bretherick (2) as aids in making the necessary decisions.

Each chemical could be initially assigned to one of five classes, toxic, flammable, reactive, corrosive, or low degree of hazard. Chemicals that possess more than one hazard, say reactive and corrosive, are assigned to the class that in the judgment of the assigning person (or committee) presents the greater (or greatest) hazard for that laboratory. If that assignment would result in a potential incompatibility with other chemicals in that class, then the multiple hazard chemical is further assigned to a special subclass within the originally assigned class.

An example will illustrate. Hydrochloric acid is corrosive, as is any strong acid; the same applies to any strong base, such as sodium hydroxide. If both strong acids and strong bases are initially assigned to the corrosive class, then because of acid–base incompatibility, either all the strong acids, or all the strong bases, would be further assigned to a corrosive subclass. Then in storage, HCl and the other strong acids would be segregated from NaOH and the other strong bases.

Some suppliers of laboratory chemicals have adopted a color coding storage class and subclass indication on their labels for laboratory chemicals. Thus, toxic labels have a blue colored background or smaller area; flammables are coded red; reactive chemicals, yellow; corrosives, white; and low degree of hazard, orange. Chemicals that are reactively incompatible within each such class have the same color code as the other members of the class, except that the color is diagonally striped to indicate that such a label-coded chemical should be stored separately from the other chemicals in that same class but having a solid color label code. Be aware that different suppliers use different colors to indicate identical hazard classes.

Color code label decisions made by a supplier might not fit the pecularities of all laboratory storage conditions. Prudence suggests reference to the material safety data sheets and to Bretherick (2) as well as to a supplier's color codes. Apply small squares of durable, colored adhesive backed paper to each container when received; the applied color, or striped color, square indicates the segregated storage location for that container. Some of these laboratories also preprint the colored squares with blanks to be filled out at the time of receipt: Date rec'd: ___; Initials _____ (of person responsible for that chemical).

11.3.3   Further Practical Matters

Typically, laboratory chemicals are stored in three different places, in the laboratory itself, in a stockroom that serves one or more laboratories, and in a storeroom that serves the stockroom(s). In addition to the exercise of inventory control and segregation both of which apply to all three, other matters also apply to these three.

See, for example, the errors described in Sections 11.1 and 11.2 Figures 11.1–11.5 illustrate a few of those improper storage practices. Stated positively, these matters also apply to all three storage places: Limit access to those few persons who are directly responsible for inventory control and segregation. Store chemicals at eye level or below; 6 ft shelf height is a

Figure 11.3.   Improper storage of chemicals and equipment in a hood also used for chemical reactions.

suggested maximum, 5 ft is better. Store equipment and apparatus elsewhere, not on the same shelf or tier of shelves with chemicals. An operating vacuum pump on the same shelf with filled reagent bottles is a glaring example of what not to do. Whenever possible, solids should be stored above liquids in the same tier of shelves.

Spill cleanup supplies appropriate for the chemicals being stored should be available near those chemicals. The same applies to fire extinguishers; the type of extinguisher nearby should be consistent with the chemicals stored in that location. Avoid mixed storage of combustible/flammable types. Thus, keep stored flammable liquids (Type B) away from wood and cardboard (Type A) boxed chemicals. The glowing embers from a charred cardboard fragment can reignite the vapors from an extinguished flammable liquid fire. Be certain that laboratory preparations and smaller containers of chemicals that have been filled from a larger volume of bulk supply are fully labeled: Content identity, nature of hazard(s), date of preparation or filling, and identification of person responsible, at a minimum. An atomic absorption laboratory-prepared standard labeled "Cu AA Std in acid" is not acceptable.

Chemicals should not be stored on floors, or on shelves so as to protrude into traffic areas; the same of course applies to equipment and apparatus.

Compressed gas cylinders, large, small, full, empty, or being used should

Figure 11.4. Examples of unacceptable, cryptic labeling.

always be individually secured to a sturdy support by a device designed to prevent toppling; lecture bottles may alternatively be mounted in a heavy large diameter base support. Cylinders should always be capped when not in use and stored in a cool, dry location, segregated from other stored chemicals and segregated from each other according to the local reactive incompatibility segregation scheme. Preferably, the storage area should be under roofed shelter with at least one side partially or fully open, away from ignition sources and excessive heat; above 125°F, the steel walls of gas cylinders can be distempered. Stored cylinders should be examined regularly for signs of leaks and corrosion. Empty stored cylinders should be clearly so marked.

When compressed gases are piped into the laboratory, it is preferable that the bank of cylinders in use be located away from stored, not used, cylinders. The previously mentioned applicable details, however, apply to such cylinder banks.

Cylinders that are full, empty, or in between should be transported only when strapped securely to the transporting cart or hand truck, and should of course always be capped. Storage of full or empty unused cylinders in the laboratory is not considered good practice. When in use in the laboratory, cylinders should be located out of traffic flow, preferably in an area specifically designated for that purpose.

Refer to the CGA (3) and Matheson publications (4) for further information.

Figure 11.5. Overcrowded shelves and overloading storage on cantilevered power supply box.

11.4 Storage in the Laboratory

In addition to all of the concerns addressed thus far in this chapter, two other matters peculiar to laboratory storage should be addressed, the pack rat syndrome and the storage of flammable and combustible liquids.

For understandable and justifiable reasons, laboratory workers often feel all but compelled to retain a few bottles of special chemicals for their own particular use. From the point of view of safe storage, this practice cannot be condoned; the possibility of squirreling away two or more reactively incompatible chemicals is high. The consequences could be disastrous as Fig. 11.2 dramatically illustrates. Laboratory managers need to find ways to promote the safe storage of those chemicals that a laboratory worker legitimately needs to reserve for exclusive use.

Thermodynamic calculations show that if the vapor from one liter of almost any flammable liquid, uniformly mixed with the stoichiometrically balanced amount of air, is ignited, the amount of energy released in the next instant is approximately equal to the energy released when 10 sticks of dynamite are detonated. It is not likely that a liter of flammable liquid would be so dispersed as vapor in the air in a laboratory; but it could happen with, say, 100 mL of flammable liquid. No one wants even one stick of dynamite or its equivalent to be detonated in their vicinity. The policy for the storage of flammable and combustible liquids in laboratories is obvious: Minimize the amounts stored. Whenever purity requirements permit, store the amounts that must be present in a laboratory in properly labeled metal safety cans rather than in glass bottles. As much as possible, keep the safety cans and glass bottles inside closed door properly vented flammable liquid storage cabinets. Less strict recommendations for amounts allowable in laboratories as well as other important details can be found in two National Fire Protection Association Manuals (5).

11.5   STORAGE IN STOCKROOM AND STOREROOM

These additional safe storage considerations apply to stockrooms that serve laboratories and to storerooms that serve stockrooms.

Each should have at least two exits. Each should be protected with automatic sprinkler systems or equivalent, water, foam, carbon dioxide, halon, or other suitable agent. Obviously, if chemicals that are incompatible with the chosen agent are stored in that room there must be provision to prevent contact built into the construction of the room. Each room should have at least one self-contained escape respirator, and all who work in or even occasionally enter the space should know how to use this equipment. Electrical power, lights, switches, sockets, and so on should be explosion proof. Motors, such as fan motors, ventilation equipment motors, and other electrical devices used in the space should be nonsparking. The ventilation vents should be at both floor and ceiling levels, with at least six air turnovers per hour.

The storeroom floor should be diked, and the stockroom floor preferably ought to be diked, so as to confine liquid spills. The diked area should be fitted with a special drain that leads to a separate pond for treatment of waste generated by a spill.

Each stockroom and storeroom should have at least one large sink, one safety shower, and one eyewash fountain. The rooms should be equipped with fire, smoke, flammable vapor, and temperature limit alarms, and personnel should be able to distinguish between and among the alarm sounds.

Shelving should be sturdy, able to carry more than the maximum foreseeable loading, and well braced to prevent toppling. There should be at least 3-ft aisle width between tiers of shelves and no aisles with dead ends. Unless divided in the center with partitions, all shelf edges in island and penninsula shelf tiers should be lipped. If wood shelving is used, all components should be at least 1-in. thick and treated to prevent mold and insect attack. Metal shelving should be corrosion resistant; expanded metal for the shelves themselves has the advantage of allowing better air circulation than solid shelving.

The storeroom, or storerooms if segregation needs so require, should be isolated, as far away from occupied buildings and areas as is practical. Stockrooms should be physically separated from the laboratories they serve and as distantly located from other occupied rooms and areas as practical.

11.5.1 Special Stockroom and Storeroom Concerns

The larger the amounts of these classes of chemicals that are stored, the greater the imperative to attend to each of these related and additional concerns.

1. *Flammables and combustibles.* Follow or exceed the precautions described in the NFPA Manuals (5), also see Chapter 6. A separate storeroom building is preferable, ground level fireproof construction, one floor only. Doors should be self-closing firedoors, provide blow out panels if possible, the building should have at least one outside wall. All exposed metal should be bonded and grounded; a grounding cable should be affixed in position so as to be readily connected to dispensing containers; suitable connecting cables should be available, one end already connected to the grounding cable. Several bonding cables should be available, each with paint and corrosion piercing clamps at the ends. The continuity of the bonding and grounding should be checked regularly.

At least one fire extinguisher of suitable type and capacity should be mounted on the exterior, near the primary entry door and at least one also conveniently available inside. No fire Class A materials should be used in the building construction, nor stored there, except possibly the wood shelves described previously. All containers should be metal or, if necessary, glass; none should be *plastic* because such containers cannot be bonded and, unlike glass, they tend to promote the generation of static charge. Containers should not be stacked on top of each other. Post suitable warning signs in all appropriate locations.

2. *Highly toxic and controlled substances.* Provide a separate, locked room with strictly limited access. The room should be equipped with a decontamination and shower facility; persons entering the room should wear suitable personal protective equipment. There should be no drain in the floor. Floor, wall and door joints, and windows, if any, should be sealed; the ventilation should be independent from other ventilation systems and equipped with filters to prevent dispersion of toxic contaminates. Take frequent inventory; post warning signs.

3. *Compounds that have a short shelf life or are unstable.* Purchase with inhibitors present, for example, a trace of phenol inhibitor in vinyl chloride, at least 10% water in picric acid. Insure consumption before inhibitor is consumed. Protect from heat, light, mechanical shock, rapid temperature change, and high temperature. If explosive, store in an isolated explosive magazine. Post warning signs.

4. *Water and air sensitive chemicals.* Store in isolated location on high ground. Containers should be waterproof and/or sealed against air exchange. Preferably store air sensitive compounds under inert gas. Keep containers well sealed; inspect frequently. The area should not have any water service. If the agent is compatible, other than automatic water extin-

guishers should be considered. Post with warning signs, particularly advising fire fighters of the water sensitivity of the stored materials.

ACKNOWLEDGMENT

Figures 11.1 through 11.5 are reproduced through the courtesy of the J. T. Baker Chemical Company, Inc., hereby gratefully acknowledged.

REFERENCES

1. D. A. Pipitone, Ed., *Safe Storage of Laboratory Chemicals*, Wiley-Interscience, New York, 1984.
2. L. Bretherick, *Handbook of Reactive Chemical Hazards*, 3rd ed., Butterworths, Stoneham, MA, 1985.
3. *Handbook of Compressed Gases*, 2nd ed., Van Nostrand Reinhold, New York, 1980.
4. *Guide to Safe Handling of Compressed Gases*, Matheson Gas Products, Secaucus, New Jersey, 1983.
5. NFPA Standard Code No. 45 *Fire Protection for Laboratories Using Chemicals, 1975* and NFPA Standard Code No. 30 *Flammable and Combustible Liquids Code, 1984*, National Fire Protection Association, Quincy, MA.

APPENDIX 2

REQUIREMENTS

CHAPTER 12

Federal Regulations

Flo Ryer

Formerly U.S. Department of Labor, Occupational Safety and Health
Administration, Washington, D.C.

12.1 INTRODUCTION

Laboratory activities related to chemical health and safety are governed by
more than 100 separate regulations promulgated by two federal agencies,
the Occupational Safety and Health Aministration (OSHA) and the Environmental Protection Agency (EPA). In some states, similar but generally
more restrictive regulations apply; they are not discussed here. Most readers
know that the EPA is responsible for waste disposal regulations (see
Chapter 14) but are not aware of other EPA regulations in Title 40, *Code of
Federal Regulations* that may apply to laboratory work. There are more than
50 OSHA regulations governing health and safety in chemistry laboratories.
The OSHA regulations that deal with permissible exposure limits (PELs)
are generally well known, but other OSHA regulations are also applicable.
More recently, the OSHA "Right to Know" or "Hazard Communication"
regulation, OSHA's currently proposed rule for occupational exposures to
toxic substances in laboratories, and the EPA regulation for reporting and
notification requirements under the Superfund Amendments and Reauthori-
zation Act (SARA) also affect laboratory work.

All of the OSHA regulations that pertain to laboratory safety and health
are found in Title 29, *Code of Federal Regulations,* part 1910 (29 CFR 1910),
known as "General Industry Standards." Part 1910 is subdivided into sub-
parts, and these in turn are further divided into sections. Thus, subpart Z on
toxic and hazardous substances is divided into about 25 sections, 1910.1000
through 1910.1500 (obviously, some numbers are skipped in this OSHA
numbering system). A typical citation to an OSHA regulation is written as
29 CFR 1910.157—which deals with portable fire extinguishers; the citation
notation can be more complex, such as 29 CFR 1910.157(f)(15)(ii)(C)—
which requires that the person who is testing noncompressed gas-type cylinders (used for some fire extinguishers) by means of a hydrostatic test must

be protected by a cage or other barrier that permits that person to see the
cylinder while it is being hydrostatically tested.

12.2. REGULATIONS PERTAINING TO PERSONAL PROTECTION

Five OSHA regulations deal with the protection of persons. These are 29
CFR 1910.132–1910.134, 1910.136, and 1910.212. Paraphrased, an employer
must provide and employees must use protective equipment for eyes, face,
head, and extremeties, and similarly for respirators, shields, and other barriers whenever work hazards could cause harm if the protective equipment
was not used. The equipment must be maintained in a sanitary and useable
condition. If an employee chooses to furnish their own protective equipment, the employer is required to make sure that the equipment is suitable
and is kept in a sanitary and useable condition. Eye protection is described
in detail in 29 CFR 1910.133, respiratory protection in great detail in 29
CFR 1910.134, safety-toe footwear is described in 29 CFR 1910.136 by
reference to another standard, ANSI (American National Standards Institute) Z41.1-1967. Machine guarding, for example, a belt guard for the belt
and pulleys of a vacuum pump, is described in 29 CFR 1910.212.
    Eye protection should be worn at all times in the laboratory, as is well
known. The regulations include the requirements that eye protection should
be comfortable to wear, easily cleaned, durable, and meet the requirements
of ANSI Z87.1-1968.
    Ordinarily, the ventilation system in a laboratory should obviate the need
for respiratory protection. When it is needed, the several requirements of 29
CFR 1910.134 apply. These include written standard operating procedures
governing their selection and use, training of employees in the use of respirators, keeping respirators clean, repairing or replacement of worn or damaged parts, and a periodic review of the medical status of the respirator
users. Further, respirators must provide adequate protection against the
hazard for which they were designed and they must be selected in accordance with ANSI Z88.2-1969.

12.3 REGULATIONS PERTAINING TO FLAMMABLE AND
COMBUSTIBLE LIQUIDS

Flammability and combustibility are the subject matter of Chapter 6. The
safety recommendations in Chapter 6 derive from the same sources upon
which the pertinent OSHA regulations have been based. The OSHA regulations are found in 29 CFR 1910.106; related regulations pertaining to fire
fighting are in 29 CFR 1910.157 through 1910.163, and fire detection and
alarm systems are described in 29 CFR 1910.164 and 1910.165, respectively.
These regulations constitute the major part of the Subpart L regulations;
however, Subpart L has five appendices that also relate to fire protection.

Most of the requirements of 29 CFR 1910.106 pertain to the safe use, handling, and storage of large quantities of flammable or combustible liquids, such as would be required by a manufacturing facility, whereas the discussion in Chapter 6 is concerned with lesser, laboratory used, quantities. It would be prudent, even if only laboratory use is contemplated, to review the OSHA requirements and to ascertain that the laboratory use of these substances does not violate the regulations.

Similarly, laboratory supervisors should make sure that their installations of fire extinguishers are in accord with 29 CFR 1910.157. For example, 29 CFR 1910.157(b)(2) modifies the recommendations of Chapter 6 if the employer has an emergency action plan that designates only certain employees to use portable fire extinguishers; 29 CFR 1910.157(d)(3) permits substituting a sprinkler system for fire extinguishers under certain conditions.

12.4   REGULATIONS PERTAINING TO COMPRESSED GASES

The general requirements for handling compressed gases are summarized in 29 CFR 1910.101 by reference to other publications. Chapter 10 describes these precautions. In addition, however, 29 CFR 1910.103 through 1910.105 each deal with particular compressed gases, 1910.102 for acetylene, 1910.103 for hydrogen, 1910.104 for oxygen, and 1910.105 for nitrous oxide. The discussion of safety precautions in the regulations is particularly useful for hydrogen and oxygen. Inspection of compressed gas cylinders is described in 29 CFR 1910.166; typcially, this service is performed for laboratories by the supplier, but it is useful to know what the supplier can be expected to do, and to check occasionally to make sure that the service has been provided. Similarly, laboratory supervisors may wish to inform themselves about the variety of safety pressure relief devices used on some compressed gas cylinders; for this information, see 29 CFR 1910.167.

12.5   REGULATIONS PERTAINING TO RADIATION

The standards for protection against ionizing radiation hazards are discussed in Chapter 10; the OSHA regulations are in 29 CFR 1910.96. The OSHA regulations for nonionizing radiation hazard protection are in 29 CFR 1910.97; again, see Chapter 10 for a discussion of the appropriate precautions.

12.6   REGULATIONS PERTAINING TO ELECTRICAL HAZARDS

Rubber protective equipment is required for electrical workers by 29 CFR 1910.137. This regulation requires rubber gloves, matting, blankets, sleeves, and so on, as described in the ANSI standard J6 series. Similar protection should be used by laboratory workers if they are exposed to such hazards. Detailed OSHA regulations for protection against electrical hazards have not yet been promulgated. At present, 29 CFR 1910.301 through 1910.308 contain design safety standards for electric equipment and installations that provide power and light to laboratories and other workplaces. Direct application to laboratory safety is limited; for example, 29 CFR 1910.304(b) requires ground-fault circuit protection on construction sites and 29 CFR 1910.305(j)(4)(iii) similarly for certain motors. Clearly, whether or not required by regulation, electrical circuits in laboratories should have ground-fault interruption protection.

X-Ray devices must be equipped with a remotely operable means to be disconnected from the power supply according to 29 CFR 1910.306(f)(1)(i).

The regulation for emergency power systems, 29 CFR 1910.308(b), requires isolation of the emergency circuit wiring and that in a space lit by emergency lights "the burning out of a light bulb cannot leave any space in total darkness."

12.7   REGULATIONS PERTAINING TO MATERIALS HANDLING AND STORAGE, SANITATION, AND HOUSEKEEPING

These regulations, 29 CFR 1910.176 and 1910.141, require adequate clearance in passages and aisles for mechanical handling equipment, hazardless storage of materials, storage areas free from accumulation of materials that would otherwise cause a tripping, fire, explosion, or pest harboring hazard, either dry floors or platforms raised above the water or waterproof footwear, covered containers for putresible refuse, vermin control, adequate toilet and lavatory facilities, and employer prohibition of eating and drinking or storing of food in any areas exposed to toxic materials.

12.8   REGULATIONS PERTAINING TO MEANS OF EGRESS

The applicable regulations are 29 CFR 1910.35–1910.37. Clearly marked exits with unobstructed access sufficient to provide means of emergency egress for all occupants must be provided. If the reasonable safety of occupants of any area would be endangered by the blocking of a single exit from that area, then at least one more exit must be provided, so arranged as to minimize the possibility that all exits could be blocked by an emergency condition. These regulations mandate several other requirements, for example, an exit must discharge directly to an open space or to a clear passage to the open air; any devices installed on an exit door to prevent its improper use cannot impede use as an exit in case of an emergency, even if the device fails; doors that are likely to be mistaken as exit doors must be signed to clearly indicate the contrary.

12.9 REGULATIONS PERTAINING TO AIR CONTAMINANTS

These regulations, 29 CFR 1910.1000 through 1910.1500 limit employee exposure by inhalation of vapors, dusts, or mists of approximately 450 named substances, many of which are commonly used in laboratory work. Tables 12.1–12.3 illustrate this concept.

Each of the tables lists for the substances named therein one or more numbers. In Table 12.1 the numbers in the column headed ppm are the permissible exposure limits (PEL) for that substance in parts per million; the numbers in the column headed mg/m^3 are the PELs expressed in milligrams per cubic meter. The numbers in Table 12.2 have similar significance, described in further detail later. The data in Table 12.3 are expressed as millions of particles per cubic foot of air, Mppcf, or as mg/m^3.

Refer to Table 12.1. An employee's exposure to any substance listed in this table and preceded by the letter C (meaning ceiling limit) shall at no time exceed the value listed for that substance in the table. For all other substances listed in Table 12.1, an employee's exposure, averaged over an 8-h work shift of a 40-h work week shall not exceed the value listed in the table for that substance. For example, if an employee were exposed during an 8-h work shift to substance X for 2 h at 150 ppm, 2 h at 75 ppm, and 4 h at 50 ppm, the average would of course be 81.25 ppm. This calculated result should then be compared with the PEL listed in parts per million for X to determine whether 81.25 ppm is greater than that PEL. If greater, the standard has been violated. The *skin* notation for some substances in the list means that direct contact should be avoided.

In laboratory work, employees are typically exposed to a variety of inhalation exposures. In such cases, an *equivalent exposure* is calculated for the mixture to which the employee has been exposed. The equivalent exposure is the sum of the quotients of exposure to the various substances, that is, the actual exposure divided by the PEL. If the actual exposure to, say, acetone was 500 ppm, averaged over 8 h, the quotient of exposure to acetone would be 500 ppm/1000 ppm, or 0.5. If the same employee was also exposed to, say, toluene, with a quotient of exposure of 0.7, and not exposed to any other named substance during that 8-h period, then that employee's equivalent exposure would be 1.2, the sum of the two quotients of exposure. Whenever the quotient of exposure exceeds unity, the standard has been violated.

Given the difficulties in measuring inhalation exposures (see Chapter 13) and given the likely variety of exposures to different substances in ordinary laboratory work, it is obvious that it is impractical to enforce this OSHA

Table 12.1. Permissible Exposure Limits for Several Compounds and Mixtures[a]

[a]Adapted from *Code of Federal Regulations*, Table Z-1, 29 CFR 1910.1000.

Substance	ppm	mg./m^3
Acetaldehyde	200	360
Acetic acid	10	25
Acetic anhydride	5	20
Acetone	1,000	2,400
Acetonitrile	40	70
Acetylene dichloride, see 1, 2-Dichloroethylene		
Acetylene tetrabromide	1	14
Acrolein	0.1	0.25
Acrylamide—Skin		0.3
Aldrin—Skin		0.25
Allyl alcohol—Skin	2	5
Allyl chloride	1	3
C Allylglycidyl ether (AGE)	10	45
Allyl propyl disulfide	2	12
2-Aminoethanol, see Ethanolamine		
2-Aminopyridine	0.5	2
Ammonia	50	35
Ammonium sulfamate (Ammate)		15
n-Amyl acetate	100	525
sec-Amyl acetate	125	650
Aniline—Skin	5	19
Anisidine (o, p-isomers)—Skin		0.5
Antimony and compounds (as Sb)		0.5
ANTU (alpha naphthyl thiourea)		0.3
Arsenic organic compounds (as As)		0.5
Arsine	0.05	0.2
Azinphos-methyl—Skin		0.2
Barium (soluble compounds)		0.5
p-Benzoquinone, see Quinone		
Benzoyl peroxide		5
Benzyl chloride	1	5
Biphenyl, see Diphenyl		
Boron oxide		15
C Boron trifluoride	1	3
Bromine	0.1	0.7
Bromoform—Skin	0.5	5
Butadiene (1, 3-butadiene)	1,000	2,200
Butanethiol, see Butyl mercaptan		
2-Butanone	200	590
2-Butoxy ethanol (Butyl Cellosolve)—Skin	50	240
Butyl acetate (n-butyl acetate)	150	710
sec-Butyl acetate	200	950
tert-Butyl acetate	200	950
Butyl alcohol	100	300
sec-Butyl alcohol	150	450
tert-Butyl alcohol	100	300
C Butylamine—Skin	5	15
C tert-Butyl chromate (as CrO$_3$)—Skin		0.1
n-Butyl glycidyl ether (BGE)	50	270
Butyl mercaptan	10	35
p-tert-Butyltoluene	10	60
Calcium oxide		5
Camphor		2
Carbaryl (Sevin^R)		5
Carbon black		3.5
Carbon dioxide	5,000	9,000
Carbon monoxide	50	55
Chlordane—Skin		0.5
Chlorinated camphene—Skin		0.5
Chlorinated diphenyl oxide		0.5
C Chlorine	1	3
Chlorine dioxide	0.1	0.3
C Chlorine trifluoride	0.1	0.4
C Chloroacetaldehyde	1	3
a-Chloroacetophenone (phenacylchloride)	0.05	0.3
Chlorobenzene (monochlorobenzene)	75	350
o-Chlorobenzylidene malononitrile (OCBM)	0.05	0.4
Chlorobromomethane	200	1,050
2-Chloro-1,3-butadiene, see Chloroprene		
Chlorodiphenyl (42 percent Chlorine)—Skin		1

Table 12.1 (Continued)

Substance	ppm	mg./m³
Chlorodiphenyl (54 percent Chlorine)—Skin		0.5
1-Chloro, 2,3-epoxypropane, see Epichlorhydrin		
2-Chloroethanol, see Ethylene chlorohydrin		
Chloroethylene, see Vinyl chloride		
C Chloroform (trichloromethane)	50	240
1-Chloro-1-nitropropane	20	100
Chloropicrin	0.1	0.7
Chloroprene (2-chloro-1,3- butadiene)—Skin	25	90
Chromium. sol. chromic, chromous salts as Cr		0.5
Metal and insol. salts		1
Coal tar pitch volatiles (benzene soluble fraction) anthracene, BaP, phenanthrene, acridine, chrysene, pyrene		0.2
Cobalt, metal fume and dust		0.1
Copper fume		0.1
Dusts and Mists		1
Crag^R herbicide		15
Cresol (all isomers)—Skin	5	22
Crotonaldehyde	2	6
Cumene—Skin	50	245
Cyanide (as CN)—Skin		5
Cyclohexane	300	1,050
Cyclohexanol	50	200
Cyclohexanone	50	200
Cyclohexene	300	1,015
Cyclopentadiene	75	200
2,4-D		10
DDT—Skin		1
DDVP—Skin		1
Decaborane—Skin	0.05	0.3
Demeton^R—Skin		0.1
Diacetone alcohol (4-hydroxy-4-methyl-2-pentanone)	50	240
1,2-diaminoethane, see Ethylenediamine		
Diazomethane	0.2	0.4
Diborane	0.1	0.1
Dibutyl phosphate	1	5
Dibutylphthalate		5
C o-Dichlorobenzene	50	300
p-Dichlorobenzene	75	450
Dichlorodifluoromethane	1,000	4,950
1,3-Dichloro-5,5-dimethyl hydantoin		0.2
1,1-Dichloroethane	100	400
1,2-Dichloroethylene	200	790
C Dichloroethyl ether—Skin	15	90
Dichloromethane, see Methylenechloride		
Dichloromonofluoromethane	1,000	4,200
C 1,1-Dichloro-1-nitroethane	10	60
1,2-Dichloropropane, see Propylenedichloride		
Dichlorotetrafluoroethane	1,000	7,000
Dieldrin—Skin		0.25
Diethylamine	25	75
Diethylamino ethanol—Skin	10	50
Diethylether, see Ethyl ether		
Difluorodibromomethane	100	860
C Diglycidyl ether (DGE)	0.5	2.8
Dihydroxybenzene, see Hydroquinone		
Diisobutyl ketone	50	290

Substance	ppm	mg./m³
Diisopropylamine—Skin	5	20
Dimethoxymethane, see Methylal		
Dimethyl acetamide—Skin	10	35
Dimethylamine	10	18
Dimethylaminobenzene, see Xylidene		
Dimethylaniline (N-dimethyl- aniline)—Skin	5	25
Dimethylbenzene, see Xylene		
Dimethyl 1,2-dibromo-2,2-dichloroetnyl phosphate, (Dibrom)		3
Dimethylformamide—Skin	10	30
2,6-Dimethylheptanone, see Diisobutyl ketone		
1,1-Dimethylhydrazine—Skin	0.5	1
Dimethylphthalate		5
Dimethylsulfate—Skin	1	5
Dinitrobenzene (all isomers)—Skin		1
Dinitro-o-cresol—Skin		0.2
Dinitrotoluene—Skin		1.5
Dioxane (Diethylene dioxide)—Skin	100	360
Diphenyl	0.2	1
Diphenylmethane diisocyanate (see Methylene bisphenyl isocyanate (MDI)		
Dipropylene glycol methyl ether—Skin	100	600
Di-sec, octyl phthalate (Di-2- ethylhexylphthalate)		5
Endrin—Skin		0.1
Epichlorhydrin—Skin	5	19
EPN—Skin		0.5
1,2-Epoxypropane, see Propyleneoxide		
2,3-Epoxy-1-propanol, see Glycidol		
Ethanethiol, see Ethylmercaptan		
Ethanolamine	3	6
2-Ethoxyethanol—Skin	200	740
2-Ethoxyethylacetate (Cello-solve acetate)—Skin	100	540
Ethyl acetate	400	1,400
Ethyl acrylate—Skin	25	100
Ethyl alcohol (ethanol)	1,000	1,900
Ethylamine	10	18
Ethyl sec-amyl ketone (5- methyl-3-heptanone)	25	130
Ethyl benzene	100	435
Ethyl bromide	200	890
Ethyl butyl ketone (3- Heptanone)	50	230
Ethyl chloride	1,000	2,600
Ethyl ether	400	1,200
Ethyl formate	100	300
C Ethyl mercaptan	10	25
Ethyl silicate	100	850
Ethylene chlorohydrin—Skin	5	16
Ethylenediamine	10	25
C Ethylene glycol dinitrate and/or Nitroglycerin—Skin	0.2	1
Ethylene glycol monomethyl ether acetate, see Methyl cellosolve acetate		
Ethylene imine—Skin	0.5	1
Ethylidine chloride, see 1,1- Dichloroethane		
N-Ethylmorpholine—Skin	20	94
Ferbam		15

227

Table 12.1 (Continued)

Substance	ppm	mg./m³
Ferrovanadium dust		1
Fluoride (as F)		2.5
Fluorine	0.1	0.2
Fluorotrichloromethane	1,000	5,600
Formic acid	5	9
Furfural—Skin	5	20
Furfuryl alcohol	50	200
Glycidol (2,3-Epoxy-1- propanol)	50	150
Glycol monoethyl ether, see 2-Ethoxyethanol		
Guthion^R, see Azinphosmethyl		
Hafnium		0 5
Heptachlor—Skin		0.5
Heptane (n-heptane)	500	2,000
Hexachloroethane—Skin	1	10
Hexachloronaphthalene—Skin		0.2
Hexane (n-hexane)	500	1,800
2-Hexanone	100	410
Hexone (Methyl isobutyl ketone)	100	410
sec-Hexyl acetate	50	300
Hydrazine—Skin	1	1.3
Hydrogen bromide	3	10
C Hydrogen chloride	5	7
Hydrogen cyanide—Skin	10	11
Hydrogen peroxide (90%)	1	1.4
Hydrogen selenide	0.05	0.2
Hydroquinone		2
C Iodine	0.1	1
Iron oxide fume		10
Isoamyl acetate	100	525
Isoamyl alcohol	100	360
Isobutyl acetate	150	700
Isobutyl alcohol	100	300
Isophorone	25	140
Isopropyl acetate	250	950
Isopropyl alcohol	400	980
Isopropylamine	5	12
Isopropylether	500	2,100
Isopropyl glycidyl ether (IGE)	50	240
Ketene	0.5	0.9
Lindane—Skin		0.5
Lithium hydride		0.025
L.P.G. (liquified petroleum gas)	1,000	1,800
Magnesium oxide fume		15
Malathion—Skin		15
Maleic anhydride	0.25	1
C Manganese		5
Mesityl oxide	25	100
Methanethiol, see Methyl mercaptan		
Methoxychlor		15
2-Methoxyethanol, see Methyl cellosolve		
Methyl acetate	200	610
Methyl acetylene (propyne)	1,000	1,650
Methyl acetylene-propadiene mixture (MAPP)	1,000	1,800
Methyl acrylate—Skin	10	35
Methylal (dimethoxymethane)	1,000	3,100
Methyl alcohol (methanol)	200	260
Methylamine	10	12
Methyl amyl alcohol, see Methyl isobutyl carbinol		
Methyl (n-amyl) ketone (2- Heptanone)	100	465
C Methyl bromide—Skin	20	80
Methyl butyl ketone, see 2- Hexanone		
Methyl cellosolve—Skin	25	80
Methyl cellosolve acetate—Skin	25	120

Substance	ppm	mg./m³
Methyl chloroform	350	1,900
Methylcyclohexane	500	2,000
Methylcyclohexanol	100	470
o-Methylcyclohexanone—Skin	100	460
Methyl ethyl ketone (MEK), see 2-Butanone		
Methyl formate	100	250
Methyl iodide—Skin	5	28
Methyl isobutyl carbinol—Skin	25	100
Methyl isobutyl ketone, see Hexone		
Methyl isocyanate—Skin	0.02	0.05
C Methyl mercaptan	10	20
Methyl methacrylate	100	410
Methyl propyl ketone, see 2- Pentanone		
C# Methyl styrene	100	480
C Methylene bisphenyl isocyanate (MDI)	0.02	0.2
Molybdenum:		
Soluble compounds		5
Insoluble compounds		15
Monomethyl aniline—Skin	2	9
C Monomethyl hydrazine— Skin	0.2	0.35
Morpholine—Skin	20	70
Naphtha (coaltar)	100	400
Naphthalene	10	50
Nickel carbonyl	0.001	0.007
Nickel, metal and soluble cmpds, as Ni		1
Nicotine—Skin		0.5
Nitric acid	2	5
Nitric oxide	25	30
p-Nitroaniline—Skin	1	6
Nitrobenzene—Skin	1	5
p-Nitrochlorobenzene—Skin		1
Nitroethane	100	310
C Nitrogen dioxide	5	9
Nitrogen trifluoride	10	29
C Nitroglycerin—Skin	0.2	2
Nitromethane	100	250
1-Nitropropane	25	90
2-Nitropropane	25	90
Nitrotoluene—Skin	5	30
Nitrotrichloromethane, see Chloropicrin		
Octachloronaphthalene—Skin		0.1
Octane	500	2,350
Oil mist, mineral		5
Osmium tetroxide		0.002
Oxalic acid		1
Oxygen difluoride	0.05	0.1
Ozone	0.1	0.2
Paraquat—Skin		0.5
Parathion—Skin		0.1
Pentaborane	0.005	0.01
Pentachloronaphthalene—Skin		0.5
Pentachlorophenol—Skin		0.5
Pentane	1,000	2,950
2-Pentanone	200	700
Perchloromethyl mercaptan	0.1	0.8
Perchloryl fluoride	3	13.5
Petroleum distillates (naphtha)	500	2,000
Phenol—Skin	5	19
p-Phenylene diamine—Skin		0.1
Phenyl ether (vapor)	1	7
Phenyl ether-biphenyl mixture (vapor)	1	7
Phenylethylene, see Styrene		
Phenyl glycidyl ether (PGE)	10	60

Table 12.1 (Continued)

Substance	ppm	mg./m³
Phenylhydrazine—Skin	5	22
Phosdrin (Mevinphos^R)— Skin		0.1
Phosgene (carbonyl chloride)	0.1	0.4
Phosphine	0.3	0.4
Phosphoric acid		1
Phosphorus (yellow)		0.1
Phosphorus pentachloride		1
Phosphorus pentasulfide		1
Phosphorus trichloride	0.5	3
Phthalic anhydride	2	12
Picric acid—Skin		0.1
Pival^R (2-Pivalyl-1,3- indandione)		0.1
Platinum (Soluble salts) as Pt		0.002
Propane	1,000	1,800
n-Propyl acetate	200	840
Propyl alcohol	200	500
n-Propyl nitrate	25	110
Propylene dichloride	75	350
Propylene imine—Skin	2	5
Propylene oxide	100	240
Propyne, see Methylacetylene		
Pyrethrum		5
Pyridine	5	15
Quinone	0.1	0.4
Rhodium, Metal fume and dusts, as Rh		0.1
Soluble salts		0.001
Ronnel		15
Rotenone (commercial)		5
Selenium compounds (as Se)		0.2
Selenium hexafluoride	0.05	0.4
Silver, metal and soluble compounds		0.01
Sodium fluoroacetate (1080)—Skin		0.05
Sodium hydroxide		2
Stibine	0.1	0.5
Stoddard solvent	500	2,900
Strychnine		0.15
Sulfur dioxide	5	13
Sulfur hexafluoride	1,000	6,000
Sulfuric acid		1
Sulfur monochloride	1	6
Sulfur pentafluoride	0.025	0.25
Sulfuryl fluoride	5	20
Systox, see Demeton^R		
2,4,5T		10
Tantalum		5
TEDP—Skin		0.2
Tellurium		0.1
Tellurium hexafluoride	0.02	0.2
TEPP—Skin		0.05
C Terphenyls	1	9
1,1,1,2-Tetrachloro-2,2-difluoroethane	500	4,170
1,1,2,2-Tetrachloro-1,2-difluoroethane	500	4,170
1,1,2,2-Tetrachloroethane—Skin	5	35
Tetrachloromethane, see Carbon tetrachloride		
Tetrachloronaphthalene—Skin		2
Tetraethyl lead (as Pb)—Skin		0.075
Tetrahydrofuran	200	590

Substance	ppm	mg./m³
Tetramethyl lead (as Pb)— Skin		0.075
Tetramethyl succinonitrile— Skin	0.5	3
Tetranitromethane	1	8
Tetryl (2,4,6-trinitrophenyl- methyl-nitramine)—Skin		1.5
Thallium (soluble compounds)—Skin as T1		0.1
Thiram		5
Tin (inorganic cmpds, except oxides		2
Tin (organic cmpds)		0.1
C Toluene-2,4-diisocyanate	0.02	0.14
o-Toluidine—Skin	5	22
Toxaphene, see Chlorinated camphene		
Tributyl phosphate		5
1,1,1-Trichloroethane, see Methyl chloroform		
1,1,2-Trichloroethane—Skin	10	45
Titaniumdioxide		15
Trichloromethane, see Chloroform		
Trichloronaphthalene—Skin		5
1,2,3-Trichloropropane	50	300
1,1,2-Trichloro 1,2,2-trifluoroethane	1,000	7,600
Triethylamine	25	100
Trifluoromonobromomethane	1,000	6,100
2,4,6-Trinitrophenol, see Picric acid		
2,4,6-Trinitrophenylmethyl- nitramine, see Tetryl		
Trinitrotoluene—Skin		1.5
Triorthocresyl phosphate		0.1
Triphenyl phosphate		3
Turpentine	100	560
Uranium (soluble compounds)		0.05
Uranium (insoluble compounds)		0.25
C Vanadium:		
V₂O₅ dust		0.5
V₂O₅ fume		0.1
Vinyl benzene, see Styrene		
Vinylcyanide, see Acrylonitrile		
Vinyl toluene	100	480
Warfarin		0.1
Xylene (xylol)	100	435
Xylidine—Skin	5	25
Yttrium		1
Zinc chloride fume		1
Zinc oxide fume		5
Zirconium compounds (as Zr)		5

Table 12.2. Permissible Exposure Limits, Ceiling Limits, and Acceptable Peak Excursions[a]

Material	8-h Time Weighted Average	Acceptable Ceiling Concentration	Acceptable Maximum Peak Above the Acceptable Ceiling Concentration for an 8-h Shift	
			Concentration	Maximum Duration
Benzene	10 ppm	25 ppm	50 ppm	10 min
Beryllium and Be compounds	2 µg/m³	5 µg/m³	25 µg/m³	30 min
Cadmium dust	0.2 mg/m³	0.6 mg/m³		
Carbon disulfide	20 ppm	30 ppm	100 ppm	30 min
Carbon tetrachloride	10 ppm	25 ppm	200 ppm	5 min in any 4 h
Chromic acid and chromates		1 mg/m³		
Ethylene dibromide	20 ppm	30 ppm	50 ppm	5 min
Ethylene dichloride	50 ppm	100 ppm	200 ppm	5 min in any 3 h
Formaldehyde	3 ppm	5 ppm	10 ppm	30 min
Hydrogen fluoride	3 ppm			
Hydrogen sulfide		20 ppm	50 ppm	10 min, once only if no other exposures
Fluorides, as dust	2.5 mg/m³			
Mercury		1 mg/10 m³		
Methyl chloride	100 ppm	200 ppm	300 ppm	5 min in any 3 h
Methylene chloride	500 ppm	1000 ppm	2000 ppm	5 min in any 2 h
Organic (alkyl) mercury	10 µg/m³	40 µg/m³		
Styrene	100 ppm	200 ppm	600 ppm	5 min in any 3 h
Tetrachloroethylene	100 ppm	200 ppm	300 ppm	5 min in any 3 h
Toluene	200 ppm	300 ppm	500 ppm	10 min
Trichloroethylene	100 ppm	200 ppm	300 ppm	5 min in any 2 h

[a]Adapted from *Code of Federal Regulations*, Table Z-2, 29 CFR 1910.1000.

standard for laboratory work. Accordingly, OSHA has proposed a new rule, to be known as 29 CFR 1910.1450, as a practical solution; see Section 12.12.

The five columns in Table 12.2 contain (from left to right) the name of the substance, the PEL, the ceiling concentration, the maximum peak concentration, and the maximum (peak value) duration. The PEL is as stated previously, the maximum allowable value of a time weighted average exposure during an 8-h work shift. If, at any time during the 8 h the exposure exceeds the PEL value, that higher value and its time duration is of course included in the calculation of the average. However, if that excessive exposure is greater than the ceiling value, it may not be longer than the time duration specified, and it may not exceed the maximum peak concentration. That is, if the time weighted average exceeds the PEL, there has been a violation; or if the exposure exceeds the ceiling limit for a time longer than the maximum duration, there has been a violation; and if the exposure ever exceeds the maximum peak concentration, there has been a violation. Table 12.3 lists the PELs for a few mineral dusts.

The remaining air contaminant regulations, 29 CFR 1910.1001 and higher, are substance specific and delineate exposure limits, methods of handling, required warning signs, and so on, for asbestos, coal tar pitch volatiles, 4-nitrobiphenyl, α-naphthylamine, methyl chloromethyl ether, 3,3'-di-chlorobenzidine and its salts, bis(chloromethyl)ether, β-naphthylamine, benzidine, 4-aminodiphenyl, ethyleneimine, β-propiolactone, 2-acetylamino-azobenzene, N-nitrosodimethylamine, vinyl chloride, inorganic arsenic compounds, lead, coke oven emissions, cotton dust, 1,2-dibromo-3-chloropropane, acrylonitrile, and ethylene oxide.

It should be emphasized here that until 29 CFR 1910.1450 or a similar regulation is promulgated, the provisions of 29 CFR 1910.1000 and the

Table 12.3. Permissible Exposure Limits for Mineral Dusts[a]

Substance	Mppcf	mg/m^3
Crystalline silica (respirable)	250	10
Amorphous silica	20	80
Silicates		
Mica	20	
Soapstone	20	
Talc (nonasbestiform)	20	
Portland cement	50	
Graphite	15	
Inert or nuisance dust		
Respirable	15	5
Total	50	15

[a]Adapted from *Code of Federal Regulations*, Table Z-3, 29 CFR 1910.1000.

following are effective in all workplaces governed by the OSHA regulations, laboratories included.

12.10   REGULATIONS PERTAINING TO ACCESS TO MEDICAL RECORDS

This regulation, 29 CFR 1910.20, provides that if an employer has medical and exposure records of employees, then employees or their designated representative may have access to these records. This standard defines the scope and manner of access in detail; for example, if knowledge of the information in the records would be likely to be inimical to the employee's health, then only the employee's designate may have access even if it is known that the designate will relay the information to the employee.

12.11   REGULATIONS PERTAINING TO EMPLOYEES' "RIGHT-TO-KNOW"

This regulation, 29 CFR 1910.1200, is the most recently promulgated OSHA regulation that is related to work in chemical laboratories. Among other things, the regulation requires, for example, that employees be informed of the hazardous properties and of the appropriate precautions related to the chemicals to which they are exposed in their workplace. The regulation mandates a detailed, written training program based on information such as is found on chemical labels and in the material safety data sheets. Other provisions in the regulation require that suppliers of hazardous chemicals provide proper warning labels on containers and full information in the material safety data sheets they provide to their customers. Employers, in turn, must insure that workers have access to those material safety data sheets during their working hours.

With respect to laboratory workers, however, the regulation is less demanding with only three requirements. Specifically, labels on incoming containers may not be removed or defaced as long as the hazardous chemical remains in that container. (The labeling of other containers used in the laboratory is not covered in this regulation.) Material safety data sheets that are received for laboratory chemicals must be readily accessible to laboratory workers. Employers must be sure that laboratory workers are apprised of the hazards of the chemicals in the laboratory in accordance with the details of the training program prescribed by the regulation.

12.12 THE PROPOSED RULE FOR LABORATORIES

This rule, which will probably be promulgated as 29 CFR 1910.1450, is intended as a practical response to the need for laboratory worker protection. In essence, it retains the requirement for compliance with PELs but eliminates the requirement that only measurement of the concentration of an air contaminant suffices to demonstrate compliance (or noncompliance).

Note that all other OSHA regulations, such as those for personal protection, for means of egress, for materials handling and storage, and so on, will still apply to all laboratories. At present, as noted previously, laboratories are covered under 29 CFR 1910.1200. But, as presently proposed, all laboratories except those associated with group health practice, or dental or veterinary facilities must comply with 29 CFR 1910.1450.

The proposed rule contains two major provisions: compliance with PEL requirements and compliance with a locally produced chemical hygiene plan (CHP). The CHP, in turn, is comprised of several required parts, discussed in the following paragraphs.

Compliance with the PEL requirements is determined by an *exposure evaluation* conducted by the chemical hygiene officer (CHO), see below. An exposure evaluation is an assessment of the conditions that are present in a laboratory for the purpose of determining whether an employee has been overexposed (e.g., the PEL exceeded) to a hazardous chemical. The assessment may be by measurement of the concentration of the contaminant in a sample of laboratory air, by a more elaborate measurement of contaminant in the breathing air, by a calculation based on the vapor pressure of a volatile hazardous substance, or by less formal means. If the exposure evaluation indicates that an employee may have been overexposed, the employee is entitled to a medical consultation with a licensed physician. The consultation may be as simple as a telephone conversation with the physician, or quite elaborate with several different tests involved. The consultation shall be at no cost to the employee.

The employer must develop a CHP to protect laboratory employees from health hazards if any toxic substances are used in the laboratory. The CHP must be implemented by a CHO, who must be qualified by training and experience to provide guidance in the development and implementation of the CHP.

The CHP must include an indication of specific measures to insure the protection of the laboratory workers and include each of these elements:

Standard operating procedures to be used when working with toxic substances.

Criteria to determine and to implement control measures that reduce exposure to toxic substances. These include engineering controls, personal protective equipment, and personal hygiene practices.

A requirement that fume hoods and similar equipment function properly and detailing specific measures to be taken to insure proper functioning.

Laboratory employee training to inform employees about potential health hazards and the precautions to be taken, including

The existence of the CHP.

A description of 29 CFR 1910.1450.

The use of appropriate protective equipment.

The PELs and other similar exposure limits such as TLV®s.

The availability of reference works that deal with the handling of toxic substances including but not limited to material safety data sheets.

A description of the circumstances that require prior approval before proceeding with laboratory work.

Provisions for employee exposure evaluations when a laboratory worker reasonably suspects or believes that they have sustained an overexposure to a substance in the laboratory.

Provision for employee medical consultation and, if so determined by that consultation, provisions for further medical evaluation.

The identity of the CHO and other persons responsible for implementing the CHP.

Provisions for additional employee protection when working with carcinogens or potential carcinogens, to include:

Establishment of a regulated area, that is, a designated laboratory or laboratory space to which access is limited to those persons who are aware of the hazards of the substances handled in that area and of the precautions to be taken.

A requirement that work in a regulated area is conducted in a hood, closed system, or other protective device.

Specification of procedures for safe waste disposal.

Specification of appropriate personal hygiene practices.

Specification of procedures to prevent contamination of vacuum lines and pumps.

Requirements that appropriate protective clothing and equipment be used in the regulated area.

The CHP must be reviewed at least annually and updated whenever necessary. The training provided to laboratory employees in the manufacturing industry under this proposed rule is in place of that required under 29 CFR 1910.1200. Clearly, many employers in the manufacturing industry with both industrial workplace and laboratory employees will choose to combine the 29 CFR 1910.1200 training with their CHP training; the two are in no sense contradictory.

Other requirements of the proposed rule include a requirement for keeping records of exposure evaluations, medical consultations and examinations; the records to be maintained and made accessible as required in 29 CFR 1910.20.

12.12.1   The Final Rule

The Occupational Safety and Health Administration asked for comments and scheduled a hearing on the proposed rule. More than 100 comment letters were received; the hearing required three days. Often, a final rule is different from the proposed rule; the changes reflect the content of the suggestions in comment letters and hearings. This summary of those suggestions indicates possible changes that may be incorporated into the final rule.

Except for the expected few letters that averred the proposed rule to be either unnecessary because everybody was already complying in spirit or not nearly strong enough because no one was presently even remotely in compliance, all of the comments directly or indirectly supported the requirements for a detailed CHP, administered by a designated CHO; for training laboratory workers; for the opportunity to obtain medical consultation; and for a standard operating procedure when working with toxic substances. Several comments suggested that only potent carcinogens needed to be restricted to a regulated area and that less potent carcinogenic substances could be handled safely with less restrictive precautions. Some letters asked for a clarification of who should be trained so as to include (other letters: exclude) janitors, secretaries, and others who do not regularly work full time in or may only briefly visit a laboratory; for clarification in the definition of a laboratory to exclude laboratory bench work located in a manufacturing or production area. Other letters urged an expansion of the definition of toxic substance to include well recognized toxic chemicals that are not listed in the OSHA regulations. Still other comments urged expansion of the rule to include the other three hazards, flammability, corrosivity, reactivity, that are ordinarily attributed to chemicals.

Laboratory supervisors who wish to be prepared in advance for the effect of the final regulation could, it would seem, begin now to implement the provisions of the proposed rule that were almost unanimously favored in the comment letters.

12.13   RELATED EPA REQUIREMENTS

Manufacturers of new chemicals, that is, chemicals that are not currently listed in the EPA inventory, must notify the EPA of their intention to manufacture a new chemical. In principle, this requirement would apply even if the new chemical were at first to be *manufactured* in a laboratory for laboratory study. Under 40 CFR 720.36 and 720.78, the EPA has issued regulations, see 51 FR 15096–15103 April 22, 1986, that modify this in principle requirement for new chemicals that are used in a laboratory. The effect of these EPA regulations closely resembles the proposed requirements of 29 CFR 1910.1450 in some respects.

First, note that any new chemical that is synthesized solely for noncommercial research and development purposes by, for example, a nonprofit institution, need not be reported. However, a new chemical synthesized for commercial purposes must be reported unless the EPA regulations 40 CFR 720.36 and 720.78 apply so as to waive the reporting requirement. No report is necessary if the following conditions are met:

Only a small quantity is synthesized, sufficient for and used solely for research and development.

All employees in the laboratory who would potentially be exposed to the new chemical are informed of any risks that may be associated with the new chemical.

The laboratory work is supervised by a technically qualified person.

A technically qualified person is a person who:

By education, training, or experience can understand the health and environmental risks associated with the chemical.

Is responsible for enforcing appropriate methods of experimentation to minimize such risks.

Is responsible for safety assessments and clearances in the procurement, storage, use, and disposal of chemicals as these are related to the conduct of research and development.

To determine any health risks employers must review all information in their possession and control that relate to health hazards the chemical might possess, other available published or private information, and any other reasonable sources as well. There are other requirements; see the text of the regulation for these details.

If portions of a new research chemical are given to another laboratory, that laboratory must be given written notice of the hazards and advised to restrict their use to research and development.

However, a laboratory that can certify by documentation that research and development are conducted using prudent laboratory practices and can prove that portions of a new research chemical are given only to similar laboratories is exempt from the requirement to ascertain if exposure to the new chemical would entail any risks to health.

Prudent laboratory practices mean those that are "accepted by health and safety experts as effective in minimizing the potential for employee exposure to toxic substances." The regulation cites *Prudent Practices for Handling Hazardous Chemicals in Laboratories* (1) and *NIH Guidelines for the Laboratory Use of Chemical Carcinogens* (2) as describing acceptable prudent laboratory practices. It seems reasonable that a laboratory conforming in good spirit with the OSHA proposed rule would also be exercising prudent laboratory practices, although this should be determined by inquiry to local EPA offices, in advance.

12.14   THE CHEMICAL EMERGENCY PREPAREDNESS PROGRAM

To encourage and support chemical emergency planning efforts at local and state levels, EPA has promulgated new regulations as 40 CFR 355.10, 355.20, 355.30, 355.40, and 355.50, with Appendices A and B. The purpose of the regulation is to provide the public with information regarding potential chemical hazards present in the community and to enable local and state authorities to establish contingency plans in advance of chemical accidents that could, without such plans, potentially inflict health and environmental damage and perhaps cause community disruption.

Briefly described, the regulation establishes "Threshold Planning Quantities" (TPQ) for approximately 400 "extremely hazardous substances" and requires that the owner, operator, or controller of a facility possessing more than the TPQ of any chemical as listed in Appendix A (alphabetical order) or Appendix B (ordered by CAS number) both notify their State Emergency Response Commission that the facility is subject to 40 CFR 355 and designate a "Facility Representative" who will participate in the local emergency planning process.

Most of the approximately 400 chemicals listed have been assigned TPQ amounts that are much larger than would be stored or used in laboratory operations. But for a few the TPQ's are within laboratory use amounts. Thus, the TPQ for phosgene is 10 lb, for chromium(III) chloride the TPQ is 1 lb if it is in solution or, as a solid, is a powder with a particle size less than 100 μm; the TPQ is 10,000 lbs, otherwise, unless it is molten; if molten, the TPQ is 3.33 lbs. If chromium(III) chloride fell within the National Fire Protection Association reactivity categories of 2, 3, or 4 (which it does not) the TPQ would be 1 lb without regard to any other qualifications.

Refer to 52 FR 13378–13410 (April 22, 1987) for a complete description of these and other requirements.

REFERENCES

1. *Prudent Practices for Handling Hazardous Chemicals in Laboratories,* National Academy of Sciences Press, Washington, DC, 1981.
2. *NIH Guidelines for the Laboratory Use of Chemical Carcenogins,* Government Printing Office, Washington, DC, 1981.

CHAPTER 13

Air Sampling in the Chemical Laboratory

Stephen K. Hall

Medical College of Ohio, Toledo, Ohio

13.1 INTRODUCTION

As discussed in Chapter 8, exposures to toxic air contaminants can be harmful, especially when the levels exceed threshold or permissible limits. Laboratory work, as distinguished from work in the chemical process industry, typically involves potential exposures to a variety of air contaminants for short time periods under conditions such that the level of exposure could be excessive. For this reason, an introductory discussion of air sampling, one aspect of the art and science of industrial hygiene, is in order.

Air sampling is done for a variety of reasons. The primary reasons are to identify and to quantify specific chemical contaminants that are present in the work environment. In the chemical laboratory, air sampling is simplified by two factors. First, the laboratory worker usually knows which contaminants are present in the laboratory from the nature of the reaction plus a knowledge of the raw chemicals, end products, and wastes. Therefore, identification of laboratory contaminants is rarely necessary and, as a rule, only quantification is required. Second, usually only a single contaminant of importance is present in the laboratory atmosphere and the absence of obvious interfering substances often permits great simplification of procedures. Nonetheless, one must be on guard continually to detect the presence of subtle and unsuspected interferences.

Other important reasons for air sampling in the chemical laboratory include routine surveillance and evaluating compliance status with respect to various occupational health standards. In addition, air sampling is conducted to determine exposures of laboratory workers in response to complaints, to evaluate the effectiveness of engineering controls, such as ventilation systems installed to minimize workers' exposure. Epidemiology of diseases of occupational origin and many other areas of research associated with chemical health are dependent on accurate evaluations of working and nonworking exposure to toxic chemical substances.

13.2 AIR CONTAMINANTS

13.2.1 Nature of Air Contaminants in the Laboratory

The type of air contaminants that occur in the chemical laboratory may be divided into a few broad groups depending on physical characteristics.

Gases are fluids that occupy the entire space of their enclosure and can be liquified only by the combined effects of increased pressure and decreased temperature, for example, carbon monoxide and hydrogen sulfide. Vapors are the evaporation product of substances that are also liquid at normal temperatures, for example, methanol and water. Although thermodynamically gases and vapors behave similarly, the reason for making the distinction is because in many instances they are collected by different devices.

Particulate matter may be broadly divided into solids and liquids. In the solid group, there are three categories based on particle size and method of sampling. Dusts are generated from solid inorganic or organic materials reduced in size by mechanical processes such as crushing, drilling, grinding, and pulverizing. The concern is for particles with a diameter < 10 μm because these remain suspended in the atmosphere for a significant period of time and are respirable. Fumes are formed from solid materials by evaporation and condensation. Metals such as lead, when heated, produce a vapor that condenses in the atmosphere. Smokes are products of incomplete combustion of organic materials and are characterized by optical density.

Liquid particles are produced by atomization or condensation from the gaseous state. Some of the terms used to describe liquid particles include mists and fogs but the distinction between these two terms has not been fully defined.

13.2.2 Environmental Factors Affecting Sampling Performance

Evaluating the impact of airborne comtaminants necessitates the accurate determination of the amount of the contaminant present in a unit volume of air. This value defines the concentrations of the contaminant and is determined either directly from the airstream or following collection on a suitable medium. Calibrations are performed to establish the relationship between the instrument, between the instrument's response, and the airborne chemical contaminant concentration being measured. The reference standards used must be accurate and precise to produce well-characterized and reproducible calibrations.

Water vapor competes effectively with organic vapors for activated charcoal sites. High humidity, therefore, can significantly reduce the adsorption capacity of the charcoal that can be reduced by as much as 75% at relative humidities above 90%. The amount of reduction in adsorption capacity depends on the test compound and must be determined individually. Changes in barometeric pressure and sampling temperature will also affect

the collection efficiency and consequently the reported results. If the barometric pressure, elevation, or temperature conditions at the sampling site
are substantially different from the calibration site, it would be necessary to
recalibrate the sampling instruments at the sampling site where the same
conditions are present. Generally, if the barometric pressure difference is
> 3 kPa, an elevation difference > 300m, and a temperature difference >
10°C, sampling site calibration of instruments is deemed necessary.

13.2.3   General Sampling Criteria

An air sampling program for any purpose should be designed to yield the
specific information desired. In devising an air sampling program, it is essential to consider the following basic requirements. Any manipulation of sampling equipment in the field should be kept to a minimum and the sampled
air must follow the shortest possible route to reach the collection medium.
The sampling instrument should provide an acceptable efficiency of collection for the contaminants involved and this efficiency must be maintained at
a rate of air flow that can provide sufficient sample for the intended analytical procedure for subsequent sample analysis. The collected air sample
should be obtained in a chemical form that is stable during transport to the
analytical laboratory. Consequently, any use of unstable or otherwise hazardous sampling media should be avoided.

13.3   SAMPLING CONSIDERATIONS

The first step in conducting air sampling in any occupational environment is
to become as familiar as possible with the particular operation. The person
evaluating the chemical laboratory should be aware of the reactions being
run, the reagent chemicals used, and the end products, by-products, and
chemical wastes encountered. He should also know what personal protective
measures are provided, how engineering controls are being used, and how
many workers are exposed.

An experienced professional investigator such as a trained industrial hygienist often can evaluate quite accurately the magnitude of chemical and
physical stresses associated with an operation on a qualitative walk through.
However, in order to document the actual airborne concentration of contaminants, it would be necessary for the investigator to use air sampling devices
and instrumentation. Regardless of the objectives of the air sampling program, the investigator must take into consideration the following parameters
in order to be able to implement the correct air sampling strategy.

13.3.1   Area versus Personal Sampling

The most common method of evaluating occupational exposure is to measure workroom contamination in the area of the workers in question. This

practice introduces a certain degree of uncertainty when evaluating the precise personal and specific exposure of the worker. It is, however, likely that many of these sampling systems will continue in operation to monitor the effectiveness of engineering controls as well as administrative controls. Furthermore, area sampling can generate valuable information on background exposure levels.

Ideally, one wishes to characterize the environment in the breathing zone of workers to evaluate their specific exposure. Breathing zone is defined by the OSHA to be a hemisphere forward of the shoulders with a radius of approximately 15 to 23 cm. Personal sampling devices are especially useful for monitoring those who move from place to place and engage in a variety of operations that involve different amounts of air contaminants of a diverse nature. For contaminants that are primary irritants or have permissible exposure limits that include ceiling concentration, short period personal samples are especially useful to define short period maxima.

13.3.2 Sample Volume

The volume of air sample to be collected is based upon the sensitivity of the analytical procedure, the estimated contaminant concentration, and the permissible exposure limit of the particular contaminant. Thus, the volume of air sampled may vary from a few liters where the estimated contaminant concentration is high, to several cubic meters where low contaminant concentrations are expected. Generally the maximum amount of air to be sampled should be a balance between increased sensitivity of the analytical procedure and the economy of time.

13.3.3 Sample Duration

Brief period samples are often referred to as *instantaneous* or *grab* samples, whereas longer period samples are termed *continuous* or *integrated* samples. Although there is no sharp dividing line between the two categories, grab samples are generally obtained over a period of <2 min and are best for determining peak concentrations of contaminant, whereas integrated samples are taken for longer periods, ranging anywhere from a few minutes up to a full shift of 8 h.

The usual objective in airborne chemical contaminant monitoring is to maintain an environment below the permissible exposure limit, which is called threshold limit value (TLV®) by the ACGIH. The TLV®s refer to airborne concentrations of substances and represent conditions under which it is believed that nearly all workers may be repeatedly exposed day after day without adverse effect. Brief period samples include the threshold limit value–ceiling (TLV®–C), the concentration that should not be exceeded even instantaneously, and the threshold limit value–short term exposure limit (TLV®–STEL), the maximal concentration to which workers can be exposed for a period up to 15 min continuously. The longest period sample is the threshold limit value–time-weighted average (TLV®–TWA), the time-weighted concentration for a normal 8-hr workday for a 40-hr workweek. Closely similar short-term and long-term sampling periods are specified in the *Code of Federal Regulations* 29 CFR 1910.1000, Tables Z-1 through Z-3, and are incorporated into current legal standards for evaluating the exposure of workers to airborne contaminants in the workplace (see also Chapter 12).

13.3.4 Sampling Rate

Gas mixtures resist separation into components under the influence of centrifugal or inertial forces no matter how strong they may be. Consequently, gas sampling presents no special problems with respect to sampling rate or velocity of entry into the sampling device. This is not the case for particulate matter sampling for particles > 5 μm in aerodynamic equivalent diameter, which is the most widely used definition of particle size in aerosol science. This definition is based on the way the particle behaves when airborne in a field of force. The formal definition is "the diameter of a unit density sphere which has the same settling velocity as the particle in question." This definition is based on behavior in a force field rather than on particle appearance.

In addition to the aforementioned factors, other important factors to be considered in sampling rate are the total sampling time required, the dynamic characteristic of the sampling device, and the increase in sampling media resistance such as filters used in organic particulate matter collection.

13.3.5 Sampling Efficiency

One of the most important factors in the collection of atmospheric contaminants is the efficiency of the sampling device for the particular contaminant in question. In many cases, the efficiency need not be 100% as long as it is known and is constant over the range of concentrations being evaluated. It should, however, be above 90%. For many types of sampling devices, the collection efficiency of the concentrating device must be measured.

Several methods for determining the sampling efficiency of collecting devices are available: (a) by series testing where enough samples are arranged in series so that the last sample does not recover any of the test compound; (b) by sampling from a gas tight chamber containing a known gas or vapor concentration; (c) by comparing results obtained with a device known to be accurate; and (d) by introducing a known amount of gas or vapor into a sampling train containing the absorber being tested. The chief advantage of grab sampling methods is that collection efficiency is essentially 100%, provided correction is made for completeness of evacuation of the container and assuming no losses as a result of leakage.

13.3.6   Number of Samples

The number of samples to be collected depends to a great extent on the purpose of sampling and the type of sampling device used. There are no set rules regarding the number of samples to be collected.

Evaluation of a worker's TWA exposure is best accomplished, when analytical methods will permit, by allowing the workers to work a full shift with a personal breathing zone sampler attached to their body. Such a full shift single sample can demonstrate the worker's true average exposure level and the status of compliance or noncompliance with the exposure standard of the chemical contaminant. The major disadvantage of a full shift single sample is that all environmental fluctuations of the chemical contaminant are averaged out during the sampling period.

Ideally, sampling results should reflect the true exposure levels relating to the worker's activities. In other words, the sampling period should represent some identifiable period of time of the worker's exposure, such as a complete cycle of activity. This is particularly important in studying nonroutine or batch-type activities, which are characteristic of many laboratory operations. For this reason a continuous series of exposure measurements, either equal or unequal time duration, should be obtained for the full duration of the desired time-averaging period. This type of sample is known as a full shift consecutive sample.

Other sampling schemes include (a) partial shift single sample, (b) partial shift consecutive samples, and (c) random grab samples.

13.4   SAMPLING PROCEDURES AND DEVICES

There are two basic methods for the collection of airborne contaminants. The first method involves the use of an air moving device to obtain a definite volume of air at a known temperature and pressure. This method is known as active sampling and, depending on the concentration and the analytical method used, the contaminant may be analyzed either with or without further concentration. The second method does not involve any air moving device. Instead, the air sampling device depends entirely on the phenomenon of diffusion of airborne contaminants to achieve trapping in a collection medium. This method is known as passive sampling. Whereas active sampling can be applied to gases, and vapors, as well as particulate matter, passive sampling is currently used for gases and vapors only. Each of these sampling methods is discussed in greater detail in the following sections.

13.4.1   Active Sampling

Active sampling can involve either grab sampling or integrated sampling. In grab sampling, an actual sample of air is taken in a suitable container and

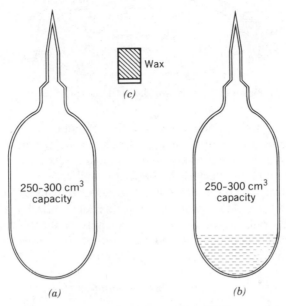

Figure 13.1. Evacuated flasks. (*a*) Empty evacuated flask. (*b*) Evacuated flask containing only liquid absorbent and its vapor. (*c*) Wax plug, used to seal flask after sampling.

the collected air sample is representative of the atmospheric conditions at the sampling site at the time of sampling. Numerous types of grab sampling devices are available. These include evacuated flasks (Fig. 13.1), gas or liquid displacement containers (Fig. 13.2), flexible plastic containers, and hypodermic syringes.

Integrated sampling is used when the concentration of airborne contaminant is low, or when the contaminant concentration fluctuates with time. It is also used when only the TWA exposure value is desired. The contaminant in these cases is extracted from the air and concentrated by an absorbing solution or collected on an adsorbent. Absorbers vary in characteristic depending on the gas or vapor to be collected. Four basic types of absorbers are available: simple bubbling or gas washing bottles, spiral and helical absorbers, fritted bubblers, and glass bead columns (Fig. 13.3). The function of these absorbers is to provide sufficient contact between the contaminant in the air and the absorbing solution.

For nonreactive and insoluble vapors, adsorption is the method of choice. Commonly used absorbents include activated charcoal, silica gel, and fluorisil. Adsorption tubes, which are useful for personal integrated sampling of most organic vapors, contain two interconnected chambers in series filled with the absorbent. In an activated charcoal tube, for example, the first chamber contains 100 mg of charcoal; this chamber is separated from the backup section, containing 50 mg of charcoal, by a plastic foam plug (Fig. 13.4). The contents of the two chambers are analyzed separately to deter-

Figure 13.2.   Gas or liquid displacement collector.

mine whether or not the charcoal in the first chamber has become saturated and lost an excessive amount of the sample to the second chamber. Sampling results are not considered valid when the second chamber contains >10% of the amount collected on the first chamber.

An active sampling train may consist of an air moving device or source of suction, a flow regulator such as an orifice or a nozzle, a flowmeter to indicate the flowrate, a collection device such as an absorber, adsorber, or a filter, a probe or sampling line, and a prefilter to remove any particulate matter that may interfere with sample collection or laboratory analysis.

13.4.2   Passive Sampling

One of the more noteworthy developments in air sampling technology has been the availability of passive dosimeters for a broad list of vapors and gases. These monitoring devices are small, lightweight, and inexpensive. They have no moving parts to break down and can be conveniently used unattended. They depend solely on permeation or diffusion of gaseous contaminants to achieve trapping in a collection medium. Commercially available units usually require no calibration since this is generally provided by the manufacturer.

Permeation devices utilize a polymeric membrane of silicone rubber as a barrier to ambient atmosphere. Gaseous contaminants dissolve in the membrane and are transported through the membrane to a collection medium such as activated charcoal, silica gel, or ion exchange granules. Permeation across the membrane is controlled by the solubility of the vapor or gas in the membrane material and by diffusion of the dissolved molecules across the membrane under a concentration gradient.

Diffusion devices are provided with a porous barrier to minimize atmospheric turbulence. Molecules diffuse through the barrier to a stagnant air layer and then are collected on the adsorbent material. Specific information on sampling rates of the airborne contaminants is supplied by the manufacturers.

All the organic passive monitoring dosimeters use activated charcoal as the collection medium. Both permeation and diffusion devices are commercially available. They can be used to monitor any organic compound that can be actively sampled by the charcoal tube method. It is important to note that the sampling and analysis of these organic passive dosimeters are affected by environmental factors such as atmospheric pressure, ambient temperature, relative humidity, and face velocity. The overall accuracy of pas-

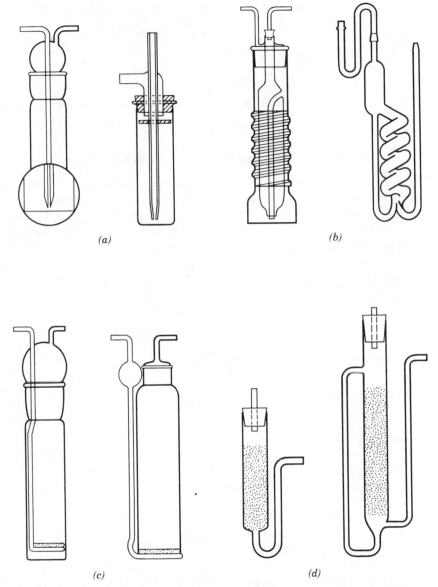

Figure 13.3. Absorbers. (*a*) Simple bubblers. (*b*) Helical absorbers. (*c*) Fritted bubblers. (*d*) Glass bead columns.

sive dosimeters is ±25% in the range of 0.5 to 2.0 times the environmental standard. It should be stressed that the manufacturer's recommendations on how and under what conditions the passive monitors may be used should be strictly followed.

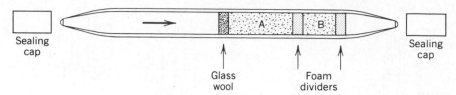

Figure 13.4. Activated charcoal tube. A, Sample portion (100 mg). B, Back-up portion (50 mg).

13.4.3 Direct Reading Colorimetric Indicator Tubes

Colorimetric indicator tubes provide compact direct reading devices that are convenient to use for the detection and semiquantitative estimation of gases and vapors in atmospheric environments. Presently there are tubes for about 300 atmospheric contaminants on the market. Indicator tubes have been widely advertised as being capable of use by unskilled personnel. While it is true that the operating procedures are simple, rapid, and convenient, many limitations and potential errors are inherent in this method. The results may be dangerously misleading unless the sampling procedure is supervised and the findings interpreted by a trained industrial hygienist.

Colorimetric indicator tubes are glass tubes filled with solid granular material, such as silica gel or aluminum oxide, which has been impregnated with an appropriate chemical reagent. The ends of the glass tubes are sealed during manufacture. The use of indicator tubes is extremely simple. After its two sealed ends are broken open, the glass tube is placed in the manufacturers holder, which is fitted with a calibrated squeeze bulb or piston pump. The recommended volume of air is then drawn through the tube by the operator. Adequate time must be allowed for each stroke. The observer then reads the concentration in the air by examining the exposed tube.

The earlier types of indicator tubes are provided with charts of color tints to be matched by the solid chemical in the indicating portion of the tube. Recent types of indicator tubes are based upon producing a variable length of stain on the indicator gel. A scale is usually printed directly on the tube and the result of sampling can be obtained instantly. The range in the reading of results by different operators may be large, since in many cases the end of a stain front is not sharp.

A trained operator would take care to see that the pump valves and connectors are free of leaks. At periodic intervals, the flowrate of the apparatus should be checked and maintained within specifications. With most types of squeeze bulbs and piston pumps the sample air flowrate is high initially and low toward the end when the bulb or pump is almost filled. This has been claimed to be an advantage because the initially high rate gives a long stain and the final low rate sharpens the stain front. The general certification requirement for the accuracy of indicator tubes is ±25% of the true value when tested at one to five times the TLV®. This accuracy requirement is modified to ±35% at one half the TLV®. At best, indicator tubes may be regarded as only range finding and approximate in nature. Furthermore, many of the indicator tubes are far from specific. An accurate knowledge of the possible interfering gases that may also be present is very important.

13.5 SAMPLE ANALYSIS

The choice of which analytical method to use is frequently influenced by the sensitivities of the approaches available. In the field of industrial hygiene, heavy reliance has been placed on two techniques: atomic absorption spectrophotometry and gas chromatography. This does not constitute a claim that these two methods dwarf other analytical methods in importance. It does indicate the analytical versatility of these two popular techniques. Other analytical methods used in the Standards Completion Program of NIOSH include UV–VIS spectrophotometry, gravimetric analysis, fluorescence spectrophotometry, and high pressure liquid chromatography, as well as others. Emerging analytical techniques that are beginning to find their way into the field of industrial hygiene include inductively coupled plasma (ICP) optical emission spectrophotometry, electron spectroscopy for chemical analysis, X-ray fluorescence, ion chromatography, derivative spectroscopy, and Fourier transform infrared spectroscopy (FTIR).

13.6 INTERPRETATION OF RESULTS

Since the implementation of the Occupational Safety and Health Act of 1970, statistical tests for noncompliance or compliance are being applied more consistently when environmental data are used to make a decision concerning a worker's exposure to chemical contaminants. The employer's responsibility is the protection of the employees. The employer's goal is to keep the probability of declaring compliance when the true state is noncompliance as small as possible. In statistical decision terms, therefore, the employer wants to assume for a given worker that he or she is in a state of noncompliance (null hypothesis). Then data are collected to show that the hypothesis can be rejected, with a goal of keeping as small as possible the probability of wrongly rejecting the null hypothesis.

The following procedures should be used by the industrial hygienist when comparing sampling results with the applicable occupational health standard. The statistics have been oriented toward determining whether noncompliance exists.

13.6.1 Full Shift Single Sample

For a time-weighted average standard, the sample must have been taken for the entire period for which the standard is defined. The variability of the sampling, expressed as a standard deviation, and the analytical methods used to collect and analyze the sample, must be well known from previous measurements. The statistical test is the one-sided comparison-of-means test using the normal distribution at the 95% confidence level. Only if the lower confidence limit (LCL) of the sample exceeds the standard is there 95% confidence that the true average concentration exceeds the standard. Thus a condition of noncompliance exists. The LCL can be expressed as

$$LCL = \chi - 1.645\sigma$$

where χ is the measurement being tested and σ is the standard deviation of the samping and analytical method.

13.6.2   Full Shift Conservative Samples

This procedure involves several consecutive samples taken for the entire time period for which the standard is defined. The standard deviation of the sampling and analytical methods must be well known from previous measurements. The statistical test is the one-sided comparison-of-means test using the normal distribution at the 95% confidence level. Only if the LCL of the mean of the consecutive samples exceeds the standard is there 95% confidence that the true average concentration exceeds the standard and that a condition of noncompliance exists.

$$LCL = \bar{\chi} - 1.645\sigma_{\bar{\chi}}$$

where $\bar{\chi}$ is the TWA of all consecutive samples in the shift and $\sigma_{\bar{\chi}}$ is calculated from the known standard deviation of the sampling and analytical method by the following equation:

$$\sigma_{\bar{\chi}} = \frac{\sigma}{T} (T_1^2 + T_2^2 + T_n^2)^{1/2}$$

where T is the total duration for the n consecutive samples and T_1, T_2, and so on, are the respective durations for the individual samples.

Statistical procedures for partial shift single sample, partial shift consecutive samples, and random grab samples are quite involved and beyond the scope of this chapter. They are not discussed here.

13.7   CONCLUSION

Evaluating the impact of airborne chemical contaminants necessitates the accurate determination of the amount of the contaminant present in a unit volume of air by a trained industrial hygienist. This can be accomplished either directly from the airstream or following collection upon a suitable medium. A successful operation involves four areas that are equally important in the accurate monitoring of airborne contaminants: (1) sampling, (2) calibration, (3) analysis, and (4) data handling. The accuracy and conversely the error in the determination of the concentration of a chemical contaminant is a function of all four of these aspects of air monitoring, and to consider one without the other three often results in the incorrect or even improper determination of airborne concentrations.

CHAPTER 14

Safe Disposal of Hazardous Waste

Robert E. Varnerin

Boston University, Boston, Massachusetts

14-1 INTRODUCTION

14.1.1 Prologue

Do you have a list of the chemicals and chemical products in your laboratory, plant, or other work area? Do you have records of the approximate age and condition of these materials? Can you estimate quantities of chemicals consumed in your processes in the past year? Can you justify the quantity of waste you generate and your method of hazardous waste disposal?

If you answer affirmatively to some or all these questions, you probably already are in substantial compliance with waste management regulation.

If not, you might first read Section 14.2.1 Managing Chemical Inventories. Inventory control programs will be the foundation of your compliance initiative for hazardous waste as well as hazard communication. This same effort could strengthen sound business leadership. It could result in new productivity and economies.

In addition, if you have not previously been subject to or aware of disposal statutes, you might then read Sections 14.2.2 Identifying Hazardous Waste, 14.2.9 Selecting a Waste Disposal Contractor, and 14.2.8 Initial Compliance Strategies. These sections will assist you in the initial stages of your compliance effort. As your program progresses, other sections, for example, on Waste Minimization and Recycling of Waste will suggest procedures for further development. Many operations can realize substantial economies in their waste management. These savings often exceed the costs of implementation.

14.1.2 Who Should Read This Chapter?

The potential audience for a survey of hazardous waste management is vast:

Generators and regulators.

Litigants or their lawyers in a trial.

Administrators.

Waste disposal managers.

Laboratory and production line workers.

The target audience for this review is primarily one who manages, hands-on, disposal programs or handles, transports, or treats hazardous waste in any quantity. Those who are exempt from legislation but comply as socially responsive citizens will also benefit from these pages.

14.1.3   Background

In 1976 the EPA received a mandate. Congress enacted the Resource Conservation and Recovery Act (RCRA) to protect human life and the environment from improper waste management practices. In issuing regulations under RCRA, the EPA first regulated large generators who produce the major portion of hazardous waste. Regulations that the EPA published on May 19, 1980, exempted small quantity generators (SQG) (i.e., operations producing 100 to 1000 kg (1) of hazardous waste in a calendar month) from most of the hazardous waste requirements.

The Hazardous and Solid Waste Amendments (HSWA) were signed into law on November 8, 1984. These amendments to RCRA directed the EPA to establish new requirements for SQGs. On March 24, 1986, the EPA published the final regulations which went into effect on September 22, 1986. In general, the new regulations are based upon the hazardous waste regulatory program already in force for large quantity generators (LQG) (1).

The EPA mandated states to enforce regulations. Many states enacted their own statutes, which exceed the federal requirements. Consequently, readers should presume that federal regulations represent the minimum requirements. They should consult their state hazardous waste offices to determine their full obligation.

14.1.4   Scope of Problem

Hazardous waste disposal is not a single end-of-the-line event. It does not terminate with the final delivery of wastes to secure chemical landfills. It begins with designing processes, plants, and laboratories long before one orders materials. After disposal, the generator retains long-term liability as long as the material retains its identity and may cause harm to health or the environment.

Waste management is a comprehensive cradle-to-grave process. It involves:

- Initial design of processes and procedures that use chemical feedstocks and supplies.
- Review of purchasing procedures and policies.
- Adequate inventory turnover.
- Safe and legal storage.
- Fine-tuning of online processes.
- Inventory control and safe storage facilities.
- Recycling and reuse of *spent* materials.
- Conservation or resale (rather than disposal as waste) of surplus and previously used chemicals.
- Competent, safe, and legal disposal.

In brief, it requires the cooperation and commitment of many departments, of many individuals and, especially, of top management within an institution. It crosses the jurisdictional barriers, politics, and budget lines of administrative units.

Many university and other laboratories, for example, have serious ventilation, storage, and waste disposal problems. The laboratories and storage areas most often are located in buildings of mixed occupancy. As a result, odors permeate classrooms and office areas. Storage areas are often inadequate and not up to code. Hood facilities, heating, ventilation, and air conditioning systems in the work areas are often inadequate for traditional procedures using chemicals. Costs to repair or modify existing facilities exceed available budget allocations. Elimination of the laboratory activity would terminate certain curricula or essential business activities. No single administrative unit has jurisdiction to solve all these problem areas. Faculty, quite correctly, cannot eliminate laboratory science. Plant managers cannot stop production lines. Construction budgets cannot fund new facilities or renovation costs.

The chemistry faculty at Bowdoin College faced and resolved such a dilemma. Mayo et al. (2) describe an entirely new way of conducting the undergraduate organic laboratory: "everything on a very small scale." They note in the Foreword that their book

> . . . represents a striking new departure from lab manuals. . . . Many colleges, held hostage by their macro glassware, will be unable to change over to microscale immediately . . . chemists in academia are profoundly conservative in terms of didactic approaches to their subject matter. New insights into chemistry are eagerly incorporated into the substance of what is taught; in contrast, the structure and methodology or courses are altered slowly and reluctantly. . . . The success of organic chemistry in the United States, both at the academic and industrial levels, provides a cogent argument that what is being done in teaching is effective and productive. At the same time, it may blind us to the opportunity for making a quantum leap in the teaching of

organic chemistry...(the book) describes an entirely new way of doing the undergraduate lab in organic chemistry: everything on a very small scale.

The results exceeded expectations. The new approach to laboratory science developed students' reliance, increased cost-effectiveness and safety, virtually eliminated toxic chemical wastes, and provided better laboratory experience for students. "In short," the authors conclude, "the teaching process has been demonstrated to be successful at the only level that really counts. Theoretical arguments faithfully translate into practical benefits." And this was accomplished without costly renovation or construction costs.

Many businesses and other laboratories confront similar dilemmas. Processes designed in an era of inexpensive waste disposal options become less profitable as disposal costs escalate. Plant managers and executives resist modification of processes and production strategies that had been profitable and "efficient" for years. Administrators and stockholders, to be sure, are concerned with diminishing returns of inefficient procedures and processes but are unable to fund expensive renovation or construction. Often modification to state-of-art processes can reduce hazardous waste generation and make processes more efficient and profitable. The conversion usually can be accomplished in the same facility, without relocation of personnel and with a minimum of shut-down time.

14.1.5 Objective

Many guides are available for hazardous waste generators, for example, see refs. 3a–3c. The objective of this chapter is to assist and provide some initial approaches for those who initiate new or upgrade existing programs. It will also benefit those with experience but who are insufficiently familiar with technical and management details of the new regulations and waste disposal procedures. The chapter reproduces some key references and tables for the convenience of the reader who does not have ready access to reference materials. This service should provide some foundation for readers to consult other reference works. As the need arises, they should consult legal and technical specialists in waste management. This chapter does not attempt to be a single, self-sufficient compendium or authoritative source of all regulation and practice.

Individual generators retain total responsibility to ensure compliance with all federal, state, and local statutes. A generator cannot legally transfer this obligation to the transporter when the latter removes the materials from the generator's premises. Transporters, to be sure, assume their own set of legal responsibilities which supplement those of the generator. To become more knowledgeable with their responsibilities, generators should consult EPA (see Table 14.2), local, and state waste control offices (see Table 14.1) and, if necessary, their own attorneys. The author cannot assure that the information contained in this chapter fulfills all local statutes or includes changes made in regulation subsequent to publication.

14.2 COMPLIANCE PROCEDURES

14.2.1 Managing Chemical Inventories

If there is any single message in this chapter, it is to urge generators of waste and users of chemical substances to develop total management systems for chemical inventories. No other management option is more effective in reducing quantities of hazardous waste generated. Excess stock is often orphaned when investigations and processes terminate. Surplus chemicals in storage create unnecessary safety risks. Handling and disposal costs of excess chemicals is high.

Company wide policy should limit new purchases of chemicals to materials required only for short-term needs. A central receiving area and stockroom could coordinate procedures and needs of end users. Each container of chemical substances should bear a label with the date of receipt, the name of the requisitioner, and the party responsible for use and storage. Shelf life should be curtailed to ensure rapid inventory turnover through consumption or reassignment of stock to another user. No materials should remain in stock unless listed on an active inventory list. Thereby, surplus materials of one user could be made available to or transferred or sold to other users. Finally, dead-stored materials should be weeded out of inventories, if necessary by disposal, without delay. Inventories should remain lean but sufficient for immediate needs. This measure will ensure a significant decrease in waste generated. Purchase contracts and frequent deliveries from suppliers will ensure availability of materials when needed.

Companies without active inventory control should set up systems to maintain records of all purchases, use, and disposal of chemical substances. In this manner, a reasonable material balance of chemical substances is quite possible for most institutions.

A library cannot run without a card (or computer) catalog and a record of withdrawals and returns. Yet, many laboratories and plants do not have accurate records of hazardous materials and wastes within their domain. Such inadequate management motivates hazardous waste and right-to-know legislation.

14.2.2 Identifying Hazardous Waste

The RCRA stipulated *cradle-to-grave* waste management system. Yet, the EPA required many years to evolve a definition of *waste*. The definition is still too broad and has led to much controversy and confusion. In practice, chemical plant and laboratory directors should presume that waste products and surplus chemicals are hazardous waste until proven otherwise. The EPA provides some guidelines and lists.

The EPA (4, 40 CFR 261.3) classifies waste as any material that usually is discarded or has no economic value. Waste can be either previously used (i.e., *spent* or partially used) or unused surplus chemicals. *Characteristic waste* (4, 40 CFR 261.10) is any material that is

- Ignitable (i.e., capable of fire hazard under normal conditions or operation) (4, 40 CFR 261.21).
- Corrosive (i.e., capable of eating through normal containers) (4, 40 CFR 261.22).
- Reactive (i.e., tendency to explode under normal use conditions, to react violently with water, or to generate toxic gases under specified conditions) (4, 40 CFR 261.23).
- Extraction procedure (EP) toxic (i.e., contain certain toxic materials that can be released in acidic water) (4, 40 CFR 261.24).

Listed waste is anything contained in the EPA "F" list (4, 40 CFR 261.31) of hazardous wastes, "P" list [4, 40 CFR 261.33 (e)] of toxic wastes, and "U" list [4, 40 CFR 261.33 (f)] of acutely hazardous wastes. *Statutory waste* is anything that can cause harm to human health and the environment. It even contains substances not currently federally regulated under characteristic or listed wastes.

Hazardous waste generators must also identify individual chemical components of waste products. Transporters and processing facilities cannot accept *unknown* chemicals for disposal. The generator also must identify the contents of containers whose labels have deteriorated. Labels that do not accurately identify the contents must be corrected. Consequently, company and institutional waste managers must insist that directors of various units and departments specify accurately and completely the chemical contents of every hazardous waste their operation generates.

14.2.3   Liability and Rising Costs

Robert W. Franz, (5) Counsel—Environmental Pollution, General Electric Company, highlights the potential high costs associated with environmental liabilities vis-à-vis hazardous waste:

> The most significant federal statute imposing liability associated with hazardous substances is the Comprehensive Environmental Response Compensation and Liability Act (CERCLA) . . .

> CERCLA allows the government to spend money from the Superfund . . . to conduct a cleanup action wherever there has been a release or is a substantial threat of a release, and then collect the costs expended from responsible parties.

Note that the responsible parties are jointly and severally responsible; that is, any one party, or some of them, or all parties collectively can be held liable for total damages. The responsible parties include generators (e.g., a laboratory), transporters, and owners or operators of a building, structure, or real estate, at any of which a hazardous substance has come to be located. Continuing from ref. 5:

> RCRA . . . authorizes EPA to bring action against any person contributing to an imminent and substantial endangerment.

One of the most pressing issues at present is liability. Insurance coverage, if available, is becoming more prohibitively expensive for lesser and lesser levels of coverage. The problem is so severe that it forces many disposal companies out of business and requires others to increase cost for service unpredictably. Large segments of industry reconsider insurance coverage and, in fact, some adopt the self-insurance option. It is forcing many waste generators to accept lower levels of protection for their own operation as well as for that of their waste contractors.

The insurance crisis alone could lead to an abrupt termination of waste disposal for long or short periods. In November, 1985, it precipitated a vast reduction of landfills. It contributed to a crisis, for example, with contaminated waste disposal in Boston in 1985. Waste haulers refused to pick up waste at many hospitals for more than a week. The resolution doubled the cost of contaminated waste disposal for most hospitals.

For the long term, insurance and liability issues will force closure of more landfills and mandate waste disposal by newer and more expensive destructive disposal techniques.

14.2.4 Escalating Regulation

The regulatory community continually increases the pressure on generators and waste disposal contractors. The long-term prognosis is clear. Regulation provides strong economic incentive for generators to reduce quantities of waste they produce. The ultimate goal of regulators is to allow disposal only of treated and already stabilized waste. For the short term, federal and state agencies lack staff and funding for complete surveillance. Spot checks and unannounced inspections to ensure compliance efforts will continue. Pressure from the general public will increase until confidence is restored.

The end result need not be pessimistic. Technologies do not now exist for total elimination of waste. Regulation will catalyze ingenuity and motivate creative solutions from the scientific community. It is time now for proactive measures to reduce the production of waste. Otherwise, further, possibly more onerous, regulation might be promulgated.

14.2.5 Those Regulated and Rationale

Insufficient attention to managing risk foreshadows many emergencies and will induce future crises. Thus, the SQGs might argue that LQGs are the primary target of regulation. They might contend that these large generators are the primary potential polluters; that LQGs compliance efforts are sufficient to protect health and the environment.

Joan B. Berkowitz, a Vice President of Arthur D. Little, recently may have reinforced such a conclusion and cast doubt on the extent of the hazardous waste disposal problem. Kaufman (6) reports she

> . . . described the problem as being 30 percent technical, 30 percent legal, and 40 percent political. It is complicated and serious and we don't need emotionalism. She suggested that the 'hazardous waste clean-up is a preoccupation of the affluent.' And, 'the $2,000,000,000 may not be justified by the risk.'

Perhaps her analysis of the problem vis-à-vis the technical, legal, and political sectors is reasonable; but, if emotionalism solves problems, this waste impasse would have gone away long ago! Further, one could argue quite reasonably that the immediate benefit of cleanup programs might not always justify the expenditure. The process of hazardous waste risk reduction certainly does justify the expense. Neglect of hazardous waste will erode margins of safety that even now are too tenuous.

14.2.6   Uncertainties and Accommodations

Block and Scarpitti (7, p. 43) discuss this oversight, often inadvert, of the risk reduction process:

> . . . at the beginning of the decade (1970s), hazardous waste was not even considered a health and environmental problem. In fact, the 1970 Report of the Council on Environmental Quality contains no mention of hazardous waste, nor did EPA say anything about it in the early years of the Agency's existence. It appears that both official and private concern for the environment centered upon air and water pollution, but not on the potentially dangerous by-products of industrial production. That, of course, would change over the next ten years.

Disposal into Love Canal and other dump sites that cause so much concern today was legal at the time. It probably was the best available technology. Escalating qualities and diversity of waste has been a by-product of the lifestyle of the industrial revolution. Waste generation has exceeded the ability of nature to clean itself up in many areas. In addition, scientists do not agree on the harmful impact of even trace quantities of chemicals in the environment. These considerations justify controlling waste generated in any quantity.

Until the late 1970s, corporate and legal policy seemed to be to pursue activities that could withstand legal challenge in courts. Even when technical evidence provided strong indication of potential dangers, products and processes were considered legally "safe" and allowed to continue until proven otherwise. In the early 1980s, the perspective changed drastically. Voluntary initiatives often exceeded requirement of law. Courts rendered decisions based on what knowledgeable persons should do rather than on what the law prescribed. After Bophal, the chemical industry intensified its efforts further to provide public information on the hazards of chemicals (8). The forum was changed from the courtroom to the public domain. How much should I do to protect health and the environment? and safety initiatives started to replace paper compliance. A good-faith effort to reduce risk often exceeded requirements of the law. Regulators, in such cases, are inclined to allow variances when strict legal compliance imposes a serious burden.

The EPA, because of insufficient staff and budget, restricted the original enforcement of RCRA to LQGs. Effective September 22, 1986, the regulations will apply to SQGs because "hazardous waste from generators of 100 to 1000 kg/month could pose a substantial risk if improperly managed (4, 51 FR, 10146–10178, March 24, 1986)."

14.2.7 Violations

Block and Scarpitti (7, p. 60) recount the tactics and violations of some operators during the early stages of federal enforcement on large generators:

> . . . Generators who once took care of their own waste now found it cumbersome and possibly prohibitively expensive to do so. Increasingly, they turned to waste haulers who provide the service they once provided for themselves . . .

> . . . Faced with seemingly prohibitive costs for disposing of their wastes in a non-hazardous manner, some generators contracted the services of illegal haulers and site operators, whose fees, although high, were more acceptable . . .

> . . .'Midnight dumpers' have used every means imaginable to get rid of waste they collect . . .

They continue,

> . . . An Associated Press report described . . . : One operation that allegedly runs out of Hartford, Connecticut only works in foul weather. A driver watches the forecast for rain or snow, then picks up a tanker of chemicals. With the discharge valve open he drives on an interstate until 6,800 gallons of hot cargo has dribbled out. 'About 60 miles is all it takes to get rid of a load,' boasted the driver, 'and the only way I can get caught is if the windshield wipers or the tires of the car behind me start melting.'

The problem of expense does not impact only large generators. Kaufman (6) reports:

> Dr. Michael Strem, President of Strem Chemicals, Inc. of Newburyport, MA, . . . discussed the plight of the small businessman trying to deal with difficult disposal procedures, increasing disposal and insurance costs, liability issues, and record keeping. He emphasized the need for methods other than direct regulation and for better information transfer.

Voluntary measures are insufficient to prevent environmental damage and violations from small quantities of waste.

Recently, a director of safety discovered 218 lb of old, reactive, and potentially explosive surplus chemicals in various laboratories in his institution. Ironically, even after the new federal regulations become effective on September 22, 1986, this amount will still be unregulated if these chemicals were the total waste generated in that month and if no additional local statute applied. The cost of disposal was $15,000, a considerable percentage of the budgets for chemicals of the laboratories involved! A director of another laboratory abandoned one third of an ounce of methyl ethyl ketone peroxide when he resigned from a company. Years later, this dangerous, autooxidizing compound, long forgotten in the inventory, broke the seal of its container and started fuming. It caused evacuation of the building and emergency fire department procedures. Even that small quantity could have initiated fires or explosions within the building. Disposal into regular trash, which probably would have been legal when the laboratory director resigned years before, could have caused a fire in a dumpster or trash truck or to a front-loader operator or scavenger in a dump site! Yet, illegal disposal, to be sure, continues as an option for the uninformed or unscrupulous.

How does a SQG, especially those new to waste disposal regulation, comply with seemingly conflicting statutes? The LQGs, who have been regulated for a number of years have found compliance options. State-of-art compliance is expensive; to a certain degree it is a good-faith effort to reduce risk. The objectives of social concern most often exceed those of legislation. No one, including this author, has all the answers to or solved all the problems. Consequently, the lure of illegal dumping remains strong.

14.2.8   Initial Compliance Strategies

The first effort for those who have no functioning waste disposal program should be a call to their state waste disposal agency. Ream (3b) in this information pamphlet lists these contacts. They are reproduced in Table 14.1 for the convenience of the reader.

The next step is to hire a reliable hazardous waste disposal contractor and work closely with them. You may obtain a list of such contractors from the local EPA office listed in Table 14.2. Competent contractors will advise and assist institutions to establish legal, ethical, and socially conscious waste practices. Many institutions follow this approach quite effectively.

TABLE 14.1. State Solid and Hazardous Waste Agencies

State	Telephone[a]	Code[b]	State	Telephone[a]	Code[b]
Alabama	(205)277–3630	A	Montana	(406)449–2821	A
Alaska	(907)465–2666	A	Nebraska	(402)471–2186	A
Arizona	(602)225–1162	A	Nevada	(702)885–4670	A
Arkansas	(501)562–7444	A	New Hampshire	(603)271–4608	B
California	(916)322–2337	C	New Jersey	(609)292–8341	B
Colorado	(303)320–8333	A	New Mexico	(505)984–0020	A
Connecticut	(203)566–5712	B	New York	(518)457–3274	B
Delaware	(302)736–4781	A	North Carolina	(919)733–2178	A
Florida	(904)488–0300	A	North Dakota	(710)224–2366	A
Georgia	(404)656–2833	A	Ohio	(614)466–7220	A
Hawaii	(808)548–6767	A	Oklahoma	(405)271–5338	A
Idaho	(208)334–4064	A	Oregon	(503)229–5913	B
Illinois	(217)782–6760	B	Pennsylvania	(717)787–7381	A
Indiana	(317)633–0176	B	Rhode Island	(401)277–2797	C
Iowa	(515)281–8853	B	South Carolina	(803)758–5681	A
Kansas	(913)862–9360	B	South Dakota	(605)773–3329	A
Kentucky	(502)564–6716	A	Tennessee	(615)741–3424	B
Louisiana	(505)342–1227	B	Texas	(512)458–7271	A
Maine	(207)289–2651	A	Utah	(801)533–4145	A
Maryland	(301)383–5734	A	Vermont	(802)828–3395	B
Massachusetts	(617)292–5500	B	Virginia	(804)786–5271	A
Michigan	(517)373–2730	B	Washington	(206)459–6301	B
Minnesota	(612)297–2735	C	West Virginia	(304)348–5935	B
Mississippi	(601)961–5171	A	Wisconsin	(608)266–1327	B
Missouri	(314)751–3241	B	Wyoming	(307)777–7752	A

[a]This is a listing (from reference 3b) of information numbers at state agencies responsible for the proper disposal of waste materials.
[b]This code gives an indication of whether the state regulations for small quantity generators are (A) similar to federal regulations, (B) more stringent than federal rules, or (C) do not recognize the federal exemption for small quantity generators.

Also, for example, contact professional societies (9a–9c), trade associations (10), law offices and associations (11), and waste contractors all of which offer a number of training sessions and seminars. The cost is usually modest as a service to their members or clients.

14.2.9 Selecting a Waste Disposal Contractor

The local office of the EPA (Table 14.2) will provide a list of contractors in your area. This list does not fulfill all your obligations or guarantee the reliability of a contractor. You should consult with colleagues in other companies who use the same contractor. You might visit the contractor's treatment and disposal sites or seek reports from colleagues in other institutions who have made such visits. Most of your professional peers will share details

Table 14.2. Environmental Protection Agency Regional Offices(1)

EPA Region 1 JFK Federal Building Boston, MA 02203 (617)223–7210 (CT, MA, ME, NH, RI, VT)	EPA Region 6 1201 Elm St. Dallas, TX 75270 (214)767–2600 (AK, LA, NM, OK, TX)
EPA Region 2 26 Federal Plaza New York, NY 10278 (212)264–2525 (NJ, NY, PR, Virgin Is)	EPA Region 7 726 Minnesota Ave. Kansas City, KS 66101 (913)236–2800 (IO, KS, MO, NE)
EPA Region 3 841 Chestnut St. Philadelphia, PA 19107 (215)597–9800 (DE, MD, PA, VA, WV, DC)	EPA Region 8 One Denver Place 999 18th St.—Suite 1300 Denver, CO 80202 (303)293–1603 (CO, MT, ND, SD, UT, WY)
EPA Region 4 345 Courtland St. Atlanta, GA 30365 (404)881–4727 (AL, FL, GA, KY, MS, NC, SC, TN)	EPA Region 9 215 Fremont St. San Francisco, CA 94105 (415)974–8071 (AZ, CA, HI, NV, American Samoa, Guam, Trust Ter- ritories of the Pacific)
EPA Region 5 230 South Dearborn St. Chicago, IL 60604 (312)353–2000 (IL, IN, MI, MN, OH, WI)	EPA Region 10 1200 Sixth Ave. Seattle, WA 98101 (206)442–5810 (AK, ID, OR, WA)

and recommendations with you. In addition, ask the contractor to provide documentation of credentials, certificate of insurance, and references.

Other factors also contribute to the decision. Can a particular contractor handle all your waste? Or, must you enlist other contractors for special classes of waste? Does the contractor dispose in a landfill or by incineration? Incineration, though somewhat more expensive, is a preferred method. It destroys waste and, thereby, removes it from the environment and from potential future liability. Have colleagues in other institutions found that the contractor is reliable? Lives up to agreements? Completes work on time? Returns required paperwork promptly?

14.2.10 Small Quantity Generators

The EPA gives examples of SQGs that are now federally regulated:

> Vehicle maintenance establishments (such as garages, paint and body shops, and car dealerships), metal manufacturers, printers, laundries and dry cleaners,

chemical manufacturers and formulators, laboratories, equipment repair shops, construction firms, textile manufacturers, pesticide applicators and schools are likely to produce small quantities of hazardous waste. EPA estimates that there are approximately 100,000 businesses which would be affected by the new regulations (1).

If anyone generates 100 kg (220 lb) in any month, the federal law applies. The law, however, does not exempt totally one who averages less than 100 kg/month but who generates more than the 100 kg/month minimum in some months. The regulations give guidelines for the intermittent generator:

The Agency has always taken the position that a generator may be subjected to different standards at different times, depending upon his generation rate in a given calendar month. . . . Thus, a generator of less than 100 kg in one calendar month would be deemed conditionally exempt in that month . . . ; however, if he generates more than 100 kg and less than 1000 kg of any regulated hazardous waste, he is subject to all of the standards promulgated today, as his generator status has changed. Furthermore, if he generates more than 1000 kg in any calendar month, he is deemed to be a large quantity generator, subject to all applicable standards. . . . If such fully regulated waste is mixed or combined with waste exempt or excluded from regulation or waste that is subject to reduced regulation under today's rule, then all the waste is subject to full regulation until the total mixture is removed from the generator's site. If, on the other hand, the generator stores separately the waste generated during a month in which less than 1000 kg (but more than 100 kg) of hazardous waste is generated, from waste generated during a month in which more than 1000 kg is generated, the former is subject to today's reduced requirements, while the latter is subject to full regulation. . . . Therefore, generators who expect to periodically exceed their 1000 kg/mo cutoff for the reduced requirements . . . should be prepared to ship their waste off-site if they wish to avoid being subject to full regulation (4, 51 FR, 10153).

If a generator can restrict waste products to less than 100 kg of hazardous waste in any calendar month, that generator

. . . will remain conditionally exempt from most of the hazardous waste program, as provided in 261.5(g). For example, generators of the less than 100 kg are not required to comply with any manifesting provisions. No additional requirements apply to this class of hazardous waste generator under the existing rules unless the quantity limitations contained in 261.5(g) are exceeded . . . (4, 51 FR, 10147).

14.2.11   Storage of Waste

If SQGs can maintain waste production to below 1000 kg in a calendar month, more liberal storage concessions apply:

Section 3001(d)(6) directs EPA, in developing regulations for 100 to 1000 kg/mo generators, to allow storage of hazardous waste on-site without the need for interim status or a RCRA permit for up to 180 days. In addition, EPA is directed to allow these generators to store up to 6000 kg of hazardous waste for a period of 270 days without the need for interim status or a permit if the generator must ship or haul his waste greater than 200 miles (4, 51 FR, 10161).

In practice, plant and laboratory managers of waste must maintain reliable inventory records (including date of receipt, use, and classification as a waste) of chemicals, spent chemicals, and waste products. In addition, they might suggest modifying processes, production or delivery schedules to minimize waste generation to stay below the 100 or 1000-kg/month limits.

14.2.12   Waste Minimization

Under Section 3002(b) of HSWA:

> . . . all generators must certify on the manifest required under subsection (a)(5) that they have in place a program to reduce the volume or quantity and toxicity of the waste they generate to a degree determined by the generator to be economically practicable. Generators must also certify that their current method of management is the most practicable method available to minimize present and future threat to human health and the environment (4, 51 FR, 10158).

Just as energy was once inexpensive, so chemical feedstocks were relatively cheap when many of today's production lines and laboratory procedures were designed. When the cost of waste disposal was "free" or modest, management had little incentive to reduce waste production. It was more economical to discard surplus or partially spent chemicals than to reuse or recycle them. Economics, rather than environmental factors, favored inefficient, lower-cost processes that produced considerable waste.

Yet, many of these "waste" chemicals might remain sufficiently uncontaminated to be reused or sold without treatment. They might be feedstock, not a waste, and not subject to full hazardous waste regulation. Further, in-house or off-site processes might consume some of these "spent" chemicals. If supplies of these spent chemicals cannot turn over in a relatively short period, disposal is preferred to long-term storage.

14.2.13   Recycling of Waste

Recycling implies processing of spent chemicals before reuse. If the spent chemical is reprocessed in a continuous process, without prior storage, it is not first classified as a waste. If, however, it is stored prior to reprocessing, it must be counted in the total of accumulated waste generated.

14.2.14   Waste Treatment

Waste treatment differs from recycling because reuse is not the objective. In treatment, waste materials are converted to less or nontoxic materials or to lesser quantities of waste. If such treatment is a final stage of a continuous process, a waste treatment license under RCRA probably is not required. If it is a separate process undertaken to reduce waste disposal costs, the waste treatment facility must be licensed under RCRA.

14.2.15   Transportation Issues

If the generator transports waste off-site for storage or recycling,

> The existing standards for transporters of hazardous waste are contained in 40 CFR Part 263, and are applicable to any form of hazardous waste transportation that requires the use of a hazardous waste manifest . . . The standards pertain to compliance with the manifest system, record keeping and actions to be taken in response to spills or discharges of hazardous waste. Taken in conjunction with U.S. Department of Transportation (DOT) requirements under the Hazardous Materials Transportation Act regarding labeling, marking, packaging and placarding (incorporated in 40 CFR Part 262, Subpart C), such standards are deemed by the Agency to be those necessary to protect human health and the environment during transportation of hazardous waste (4, 51 FR, 10167).

If transportation involves only surplus chemicals moved to a supply area for future use without processing, DOT regulations apply.

14.2.16   Requirements

What does the HSWA require of the SQG and LQG? The first inquiry should be directed to local EPA (Table 14.2) or state waste control offices (Table 14.1).

You should note carefully that federal, state, and local regulation often do not agree. It is the generator's primary responsibility to comply. Normally, your licensed waste disposal contractor will advise you but not assume your obligation.

The EPA brochure (1) summarizes requirements and procedures:

1. Generators must obtain an EPA Identification Number. This registration helps states and EPA maintain a database on hazardous waste activities. You may obtain EPA Form 8700-12 from your regional EPA office (Table 14.2) or state hazardous waste management agency. You will receive instructions for completing the form and information on where you can get additional assistance.

2. Use the full Uniform Hazardous Waste Manifest system when shipping hazardous waste. The Uniform Hazardous Waste Manifest–EPA Form 8700-22 is available from some commercial printers. Many state agencies issue their own manifests. Since you must use only licensed hazardous waste contractors that have EPA identification numbers, they will supply proper manifest forms. If not, you should contact your state hazardous waste agency and consider contracting with a different disposal firm.

The multiple copies of the manifest form track hazardous waste shipments from the point of generation to their final destination. The manifest system is obligatory whether you ship waste off-site to another company or via public roads to another location of your company.

3. A 100 to 1000-kg/month generator may accumulate waste for no more than 180 days, or 270 days if the waste is to be shipped more than 200 miles. If the duration of storage exceeds 180/270 days for any quantity or exceeds 6000 kg within the allowed period, the generator must obtain a hazardous waste storage permit. Regulations in some states are stricter.

4. Generators may engage only contractors with an EPA identification number. Your state hazardous waste agency or regional EPA office (Table 14.2) will supply information on hazardous waste transporters and management facilities in your area.

5. The EPA encourages recycling hazardous waste. Waste shipped off-site for recycling requires a manifest. Consult your state hazardous waste agency, EPA regional office (Table 14.2), or trade association for more information on recycling.

6. You may obtain more information from
   EPA RCRA Hotline: (800)424–9346 (In Washington, DC: 382–3000)
   EPA Small Business Hotline: (800)368–5888 (In Virginia: (703)357–1938)

For new SQGs who lack the EPA identification mentioned in item No. 1 the first priority of business should be to obtain their identification number. The next action should be to enlist the services of a hazardous waste contractor (see item No. 4 and Section 14.2.9). The contractor may manage, at least initially, the manifests mentioned in item No. 2. At a later stage, you should plan to take a more active role in packaging waste materials for shipment and preparing the manifest. The contractor also will advise you on item No. 3 and assist you in identifying waste materials. As your program develops, you will start identifying procedures to

Recycle (item No. 5) and minimize waste.
Identify certain wastes for which less expensive disposal options are both legal and safe.

14.2.17 Total Management Strategies

This chapter has outlined initial procedures and information for those who are relatively new to the field of waste disposal. A single individual or administrative unit can initiate many of these measures. In fact, these approaches form the waste management program for many companies. Some companies "get by" with less; others do more. The rewards can be more efficient processes, safer working environment, less waste for disposal, and substantial economies.

Strategies to reduce waste generation on a company-wide basis, however, require cooperation of many departments and individuals. Any change on production lines, in laboratories, or shops has to change decades of "successful" experience, established policy, and prejudice. Waste minimization requires changes, often drastic, in company and department policies and procedures.

The objective, for example, of a safety office is to reduce risk by identifying and controlling chemical substances and hazardous wastes. The research scientist or production manager, on the other hand, is pressured by delivery deadlines. On-time availability of chemical intermediates is their major concern. Stockpiling provides an easy sinecure! The purchasing department delivers supplies at lowest reasonable unit cost. The short-term enticements of quantity discounts and of reduced paperwork for larger, less frequent ordering usually prevail. The administration must avoid budget deficits. Deferred safety—like deferred maintenance—delivers fast short-term revenues to balance the books. Finally, anyone welcomes a gift or a bargain. Unfortunately, such gifts or "fast kills" might make the beneficiary the waste disposal contractor who serves that inadvertent gift recipient.

A few examples of leftover chemicals that "might be used someday" illustrate the safety risks involved. A gift to a university of about 100 lb of a water reactive and potentially explosive chemical almost led to disastrous consequences. Only an alert faculty and responsive fire department avoided a serious crisis. Surplus and very old potassium perchlorate stored (forgotten?) in a stockroom detonated and killed a workman assigned to remove it. Floodwaters from a torrential rain inundated a basement storage area containing dead-stored chemicals. The value of the chemicals did not warrant the risk of environmental impact and potential cleanup costs the flood could have induced. When the only option considered by managers for surplus chemicals is dead storage or disposal, hoarding is a frequent result. When caches of dead-stored chemicals become too large, disposal is the only option to resolve, at exorbitant costs, those overstocked chemical morgues.

And this is the predicament in which many institutions find themselves today. Once these old and dead inventories are depleted, active management of the inventory could guarantee on-time delivery of all required chemicals and asssure rapid turnover of surplus or spent chemicals.

Hazardous waste legislation provides impetus, or the goad, to coordinate seemingly competing management priorities. Efficient management systems make good business as well as safety sense. The savings usually exceed the dollar investment of safety programs.

Ream's (3c) strategy outlines a management system that applies to laboratories as well as chemical production lines:

Step I. Collect and analyze data about the manufacturing process and the waste generated. Explore the available management options.

Step II. Ask: Can hazardous waste generation be prevented by changing the product line, substituting safer materials, or increasing the efficiency of the process?

Step III. Ask: Can hazardous waste generation be minimized by extending the product lifetime, modifying a process, or acquiring more efficient technology?

Step IV: Ask: Can *hazardous wastes* be recovered for reuse, either on-site or off-site? Can the materials be a resource of chemicals rather than become hazardous waste?

Step V. Ask: Can the hazardous waste be detoxified or otherwise rendered nonhazardous through a chemical, physical, thermal, or microbial treatment process?

Step VI. Ask: Is burial or deep-well injection necessary? Can waste be placed in retrievable storage to be used as a material for a new product or to be detoxified when the appropriate technology becomes available in the future? Can the waste be placed in a monitored, long-term storage area?

And, for the residual waste, are methods other than landfill disposal available? Destructive processes, though more expensive, are preferable from an environmental and liability perspective.

14.3   CONCLUSION

At this time, regulation exceeds the ability of waste generators to comply fully and of agencies to enforce globally. The general public is aroused and serious in its demands. Minimally, generators of hazardous waste have an acute public relations and technological problem. The general public demands continued availability of the products, services, and comforts of technology. Yet, it is adamant in its desire to stop pollution of the environment. Industry and others might protest that the risks of chemical wastes do not justify massive cleanup costs; that risks are not proven scientifically; that the provisions of the law are not clear.—No longer do such arguments pacify the general public or sway courts of law.

I end where I began this chapter. If generators of hazardous waste cannot account for their inventory, use, storage, and disposal of chemicals, statutes in the future might deprive them the right to use these materials. Yet, I predict that that option will not be necessary. Instead, regulation will increase; new technologies will be developed for waste elimination; the enforcing agencies will continue to accept good-faith efforts. Incremental progress, providing waste generators use the best available technology, will continue.

ACKNOWLEDGMENTS

The author thanks his colleagues, Henry Littleboy, CSP, PE (Harvard University), Ronald Slade (Boston University), and Jeffrey Dill (Advanced Environmental Technology Corporation) for their review and comments on this manuscript. I especially thank Barbara, my wife, for her patience and invaluable help in composing, writing, editing, and proofreading the manuscript.

REFERENCES

1. Environmental Protection Agency, *Hazardous Waste Requirements: For Small Quantity Generators of 100 to 1000 kg/mo*, EPA 530 SW 86 003, Washington, DC, March, 1986.
2. D. W. Mayo, R. M. Pike, and S. S. Butcher, *Microscale Organic Laboratory*, Wiley, New York, 1986.
3a. *Less Is Better*, American Chemical Society, Washington, DC, 1985.
3b. K. A. Ream, *RCRA and Laboratories*, American Chemical Society, Washington, D.C., 1983.
3c. K. A. Ream, *Hazardous Waste Management*, American Chemical Society, Washington, DC, 1984. Call (202) 872–8725 for a free copy.
4. Environmental Protection Agency publications.
5. R. W. Franz, *Minimizing Liability for Hazardous Waste Management*, American Legal Institute–American Bar Association Course of Study Materials No. C-139, Philadelphia, PA, April, 1986 p. 97s.
6. J. A. Kaufman, "Public Perception and Chemical Realities: Summary of the Hazardous Waste Symposium," *The Nucleus*, Publication of the Northeastern Section of the American Chemical Society, LVII, 6 (1986).
7. A. A. Block, and F. R. Scarpitti, *Poisoning for Profit*. Morrow New York, 1985.
8. Chemtrec (24-hour *Hot Line*, established in 1971, for chemical emergencies involving transportation of chemicals): 1(800)424–9300. Chemical Referral Center (Daytime *Hot Line*, established in 1985, for safety and health information on chemicals): 1(800)262–8200 (Chemical Manufacturers Association, 2501 M St., NW, Washington, DC 20037 provides these services).
9a. "Forum on Hazardous Waste Management at Academic Institutions," Office of Federal Regulatory Programs, Department of Public Affairs, American Chemical Society, Washington, DC.

9b. American Society of Safety Engineers, 1800 E. Oakton St., Des Plaines, IL 60018-2187.

9c. Safety Council Institute, National Safety Council, 444 North Michigan Ave, Chicago, IL 60611.

10. Chemical Manufacturers Association, 2501 M St., NW, Washington, DC 20037.

11. Council on Continuing Professional Education, American Legal Institute–American Bar Association, 4025 Chestnut St., Philadelphia, PA 19104.

APPENDIX 3

PLANS AND DESIGNS

CHAPTER 15

Designing Safety into the Laboratory

Janet Baum

Payette Associates, Boston, Massachusetts

and

Louis DiBerardinis

DiBerardenis Associates, Inc., Wellesley, Massachusetts

15.1 INTRODUCTION

Hazard control in laboratory work, as with any other operation, is the responsibility of management. When the laboratory or other workplace is not built to facilitate safe operation, it is obviously more difficult to achieve safety goals. This chapter is intended to assist management both in planning new laboratory construction and in remodeling already existing laboratory spaces. The principles of safe design, the functions of each participant— owner, architect, building contractor, and others—are essentially the same, whether remodeling or building a new structure.

This chapter treats the principles of laboratory design, the function of the architect, and the responsibilities of the owner as these are related to laboratory safety, concluding with a brief practical discussion on working with design professionals. Laboratory design is a broad topic; it includes matters discussed in Chapter 16 and in the chapters that comprise Appendix II. To avoid repetition, the pertinent topics from those chapters are either briefly addressed or omitted here. Readers should refer to these other chapters for the details.

15.2 LABORATORY DESIGN, SAFETY ASPECTS

Laboratory design involves a variety of components; of these, ventilation, egress, hazard zoning, and emergency facilities are essential.

15.2.1 Laboratory Ventilation Systems

Ventilation of laboratory buildings is needed to provide an environment that is safe and comfortable. This is accomplished by providing measured amounts of supply and exhaust air plus provisions for temperature and humidity control. Good laboratory exhaust ventilation contains or captures toxic contaminants and transports them out of the building. This must be done in a manner that will not contaminate other areas of the building by recirculation from discharge points to clean air inlets, or by creating sufficient negative pressure inside the building to subject hoods to down drafts. The same supply ventilation system may serve to provide makeup air for exhaust air systems plus a comfortable and safe work environment, or there may be a separate supply system for each function.

15.2.1.1 Air Balance.   Laboratory safety requires a careful balance between exhaust and supply air volumes, as well as concern for the location, design, and quantity of exhaust and supply air. Even within the same type of laboratory, requirements may vary depending on the hazard rating of the materials being used, the quantity of hazardous materials that will be handled, and the nature of the laboratory operations.

Ventilations systems for laboratories can be divided into three main categories based on function:

1. *Comfort ventilation* is provided to the laboratory by a combination of supply and return air flows through ceiling and wall grilles and diffusers. The main purpose is to provide a work environment within a specific temperature, air exchange, and humidity range. Part or all of the comfort ventilation may be provided by special systems installed for health and safety purposes, such as chemical fume hoods.

2. *Supply air systems* are required to make up air removed by the health and safety exhaust ventilation systems. Comfort ventilation air may be supplied by one system and makeup air for health and safety exhaust systems by another, or the two supply systems may be combined.

3. *Return air systems* are health and safety exhaust ventilation systems that remove contaminants from the work environment through specially designed hoods and duct openings.

Components of a ventilation system include fans, ducts, air cleaners, inlet and outlet grilles, sensors, and controllers. Automatic fire dampers should be installed when air ducts pass through fire rated barriers.

15.2.1.2 Pressure Relationships.   To avoid intrusion of laboratory air into other areas of the building, it is advisable in all cases to maintain constant differential pressure relationships between and among laboratory rooms (greatest negative pressure), anterooms, corridors, and offices (least negative pressure). For laboratories working with very hazardous materials and biological agents, pressure gradients that decrease incrementally from areas of high hazard to areas open to public access are an essential part of the building's health and safety protective system. Areas such as animal holding rooms, animal laboratories, autopsy rooms, and similar facilities generate unpleasant odors. For these rooms, graded air pressure relationships are usually relied upon to prevent release of foul smelling air. The reverse pressure relationship is required for germ-free and dust-free facilities such as operating room and white room (clean room) laboratories.

15.2.1.3 Supply Air, Exhaust Air.   The location and construction of room air outlets and the temperature of the air supplied are critical. High velocity air outlets create excessive turbulence that can disrupt exhaust system performance at a hood face. Therefore, the supply air grilles should be designed and located so that the air velocity at the occupant's level does not exceed 50 ft/min. See Chapter 16, Section 16.5.2. There is no single preferred method for the delivery of makeup air. Each building or laboratory must be analyzed separately.

The manner in which general ventilation air is exhausted from each laboratory room depends on its size and the nature of the activities and equipment present. In some cases, adequate exhaust of general room ventilation air will be provided by a laboratory fume hood or some other local exhaust air system. In other cases, a combination of general room return air facilities, chemical fume hoods, and additional local exhaust air facilities may be used. The supply and exhaust locations within the laboratory should be situated so as to provide a full sweep of the laboratory leaving no dead spots. To determine these locations, it is very helpful to set up a mock laboratory.

All occupied and unoccupied laboratories require a minimum mechanical exhaust ventilation rate of 0.5 cubic foot per minute (cfm) per square foot of floor area when the fume hood is not operational. When a fume hood is operational, removal of a minimum of 1 cfm/ft^2 or the equivalent of six complete laboratory air changes per hour will be needed. Exhaust grilles for general room ventilation should be sized so that the inflow face velocity is between 500 and 750 fpm. Wall mounted grilles should be placed to provide an airflow direction within the laboratory from the entrance door towards the rear of the laboratory to minimize the escape of fumes to the corridor. These figures represent the minimum exhaust air rates for laboratories; in most cases, larger air exhaust rates will be needed to handle all the air discharged to the atmosphere through laboratory hoods and other local exhaust air facilities inside the laboratory.

15.2.1.4 Spot Exhaust Facilities.   Laboratory hoods are discussed in Chapter 16; other forms of control are possible. For example, if the equipment is large it may not fit into a hood, or it may fit but occupy too much hood space and affect hood performance accordingly. Or, it may be necessary, or more convenient, to work in an area not served by a hood. These cases call for special or supplementary ventilation arrangements that often take the form of high-velocity–low-volume exhaust points consisting of open ended exhaust hoses or ducts. Flexible ducts of 4 to 6 in. diameter, often referred to as "sucker hoses" or "elephant trunks," are useful for this service because they can be moved to locations where they are needed. An advantage of local exhaust hoses is their ability to reduce the total volume of air removed from the laboratory by capturing contaminants at their source with a high-velocity but less total-air-volume air flow than would be required by a hood.

Spot exhaust facilities require a high static pressure exhaust system that must be provided to the laboratory from a system separate from that serving the hoods. This is because hoods are low static pressure devices, whereas spot exhaust points require from 2- to 5-in. water gage, depending on design factors such as the quantity of air and air velocity at the opening. Designing for adequate local exhaust quantities in the planning stages is desirable; it is almost impossible to upgrade the system after installation except by total replacement.

15.2.1.5 Air Intakes. Outside air intakes must be located so as to avoid bringing contaminated air into the building. Identify existing and future likely contaminant sources such as air exhaust stacks, street traffic, parking lots, receiving and loading docks, and their relation to wind patterns before outside air intakes are selected. A minimum distance of 30 ft upwind from air discharge openings and stack outlets is recommended to reduce fume reentry problems but this separation distance is not always sufficient. It is good practice to design for the maximum feasible separation. Air intakes at roof level also are subject to contamination from other laboratory exhaust stacks or high stacks serving other facilities in the vicinity. Outside air intakes at ground level are subject to contamination from automobile and truck exhaust fumes. When buildings contain more than 10 stories, it is often advisable to locate the air intakes at the midpoint of the building. Difficult sites require wind tunnel tests to investigate the fume reentry problem under simulated conditions.

15.2.1.6 Air Discharges. For roof-mounted laboratory hood exhaust installations, the stack on the positive side of the fan should extend at least 8 to 10 ft above the roof parapet and other prominent roof structures in order to discharge the exhaust fumes above the layer of air that clings to the roof surface and prevents contaminants from displacing upward. This arrangement will help to avoid reentry through nearby air intake points. To further assist the exhaust air to escape the roof boundary layer, the exhaust velocity should be at least 2500 fpm. There should be no weather-cap, turns, or other obstructions to prevent the exhaust discharge from rising straight upward.

All exhaust fans should be installed on the building roof to maintain the exhaust ducts inside the building under negative pressure as a health protection measure. This arrangement makes it certain that should duct leakage occur, it will be inward. In cases where exhaust air duct work has to be at positive pressure relative to the building interior, use pressure tests before the building is put to use to ensure that the duct work is airtight. Repeat the testing at regular intervals thereafter. Insure that architectural features and finishes of the building allow access to the entire duct work.

15.2.2 Egress

Safe egress is a critical aspect of laboratory planning. The National Fire Codes 45 and 101 prepared by the National Fire Protection Association (NFPA) and adopted as code in many localities, define all aspects of egress that apply to laboratories (Code 45) and laboratory buildings (Code 101). It is good design practice to adhere to the requirements of these codes. Two means of egress are required from most laboratories. Exceptions to this rule are based on strict limitations of area in the lab and limits on the types and quantities of chemicals contained within. The primary exit door should be no greater than 75 ft in travel distance from the farthest occupiable space of the laboratory. The recommended minimum width of exit doors is 36 in. clear. Exit doors must swing in the direction of egress. Doors opening into corridors should not reduce the required width of the egress passage. Design of building egress outside of individual laboratories is governed by NFPA 101 provisions.

Arrangement of laboratory benches should facilitate egress. Benches arrayed in parallel rows form regularly spaced working aisles. The recommended minimum width of working aisles is 5 ft, maximum 7 ft. Working aisles should join directly to at least one egress aisle of equal or greater width. The egress aisle should lead directly to a fire protected exit.

Working aisles are flanked by equipment and or benches with counter tops. Benches are commonly 24 to 36 in. front to back. Standard floor mounted equipment is between 24 and 42 in., front to back dimension. Plan the floor area so that the minimum aisle clearance is maintained under all conditions. Typical bench heights are 35 to 37 in. for standing workers and 28 to 31 in. for seated work.

15.2.3 Hazard Zoning

In addition to coordinating egress requirements with bench and equipment layout, a renovated or new laboratory can be organized according to the principles of hazard zoning. A hazard may be defined as a potential source of fire, smoke, explosion, intense heat, electrocution, exposure to laser beams, harmful radiation or to corrosive, toxic, or infectious materials, fumes, or aerosols. The source(s) of greatest potential hazard should be located farthest from the primary exit. For many chemical laboratories then, the fume hood will be at that location, but see Chapter 16, Section 16.3 for related information. Secondary exits used for emergencies may be near fume hoods. Less hazardous activities should be next to the exit. Generally this is the paperwork area. Laboratory users absorbed in writing and calculating, and not actively moving about the laboratory are generally at least risk seated near the exit. The desks should be arranged so the chairs will not be pushed out into the egress pathway.

The laboratory designer should question each laboratory manager about the potential hazards in each laboratory. Hazard zoning principles and proper egress should be applied to each laboratory. See Section 15.3 for gathering the information required to design safe laboratories.

15.2.4   Emergency Facilities

Emergency facilities include emergency deluge showers, eyewash fountains, spill kits, fire extinguishers, fire blankets, telephones and communications devices, control panels, and building fire protection systems.

Each laboratory room should be equipped with at least one safety shower and one eyewash fountain. Both services must be within 30-s walking distance from the farthest occupied space in the laboratory. Large open laboratories may require multiple facilities. The shower and fountain should be positioned at least 5 ft apart. If the safety shower and eyewash are closer than 5 ft, they cannot be used simultaneously; usually, that is, a victim requires the help of two others; six persons do not fit easily when both services are close together. Both services should be equipped with tempered water; the necessary 15-min minimum flushing period is difficult to achieve when the flushing water is cold. Emergency deluge showers should have approximately 50-gal/min flow; eyewash fountains, 2.5 gal/min. An adequately sized floor drain under the shower is useful to speed cleanup.

If possible, deluge showers and eyewash fountains should be in standardized locations throughout a laboratory building. Laboratory personnel can quickly and reliably locate them under emergency conditions. Bright distinctive floor markings and signage aid in safety shower and eyewash fountain identification.

For additional information see the American National Standards Institute standard on emergency deluge showers (1) and ref. 2.

Spill kits should be located conveniently near the primary exit. Spill kits should be appropriate for the chemical hazards in each laboratory and checked regularly.

Fire extinguishers should be of the type and size appropriate to the work carried out in the laboratory. They should be mounted on a wall at a comfortable height in a convenient location. It should be in a well marked and standardized location from lab to lab, and placed where coats or equipment cannot impede its use.

The use of fire blankets in laboratories is controversial. While a tightly wrapped blanket may extinguish burning clothing, it may also press molten, burning polymeric fabric deeply into burned skin and tissue. If mounted in a laboratory, they should be within 30-s walking distance from any part of the lab and well marked.

A telephone for emergency calls should be convenient to all laboratory workers and within a 1-min walk from any place in the laboratory. Emergency intercom devices may augment telephones for emergency communications.

The laboratory utility control access panel is usually located just outside the laboratory primary exit door with proper identification signage. Master shut-off controls for each utility serving the laboratory are recommended. They are mounted inside the panel with clear marking as to each function. The access panel door may or may not be locked, as determined locally. If locked, consider installing a key within a break-the-glass cabinet adjacent.

Typically, laboratory fire supression systems are water sprinklers. Water systems are hazardous in high voltage locations and in locations where chemicals that are incompatible with water are used or stored. Alternate fire suppression systems are available for these applications. Fire detection devices should be installed in all areas not protected by fire suppression systems.

15.2.5   Other Issues

Many other design details which pertain to laboratory safety could be mentioned here. The list would be extensive. Those identified here are examples intended to stimulate thorough, thoughtful consideration in the identification of similar concerns. More detail on laboratory design can be found in ref. 3.

15.2.5.1   Laboratory Occupancy.   Laboratories should be designed to accommodate at least two persons. In most jurisdictions, the minimum occupied room must be larger than 70 ft^2 with a minimum dimension of 7 ft. Maximum occupancy of a laboratory not used for instruction is not limited by safety considerations, although 25 persons is a typical maximum.

Very careful consideration should be paid to the maximum number of persons allowed in a laboratory used for instruction. Even if there is a high ratio of trained instructors to students, the number of persons in any one laboratory should be limited. Laboratories used for teaching may be occupied by persons inexperienced in laboratory safety procedures. Special precautions should be taken to contain hazards and promote easy access of instructors to students. As in standard laboratory design, working aisles should lead directly to an exitway. Aisles should be a minimum of 6 ft. This extra width promotes easy access of instructors to any student who may be in trouble. If benches are used for sit-down activities, aisle width should increase to 7 ft, so chairs do not obstruct clear passage. A minimum of 3 linear feet of workbench per student should be provided. A minimum of 35 ft^2 of floor area (measured wall to wall; that is, *not* 35 ft^2 minimum of working aisle) per person is necessary, 50 ft^2 is recommended.

15.2.5.2 Piped Services. Each laboratory should be equipped with at least one sink with hot and cold water supply for hand washing and large enough to accommodate the needs of the work in the laboratory. All utility valve controls should be marked so the on or off condition can be determined at a glance. The outlets for steam, compressed air, other high pressure gases, if any, should point down at an angle of about 45° from horizontal.

15.2.5.3 Laboratory Furnishings. Shelving and cabinets should not be higher than 6 ft from the finished floor. The tops of wall hung cabinets should have sloped tops to prevents storage on top. Hinged doors on cabinets should not project beyond the edge of the benchtop below. Lips or earthquake bars on the edges of open shelving is discussed in Chapter 11.

From a safety point of view, construction of shelves and casework from wood or metal is largely a matter of economy or aesthetics. Wood rots under wet conditions and can burn. Wood can be treated to minimize these disadvantages but not eliminate them. Metal corrodes, especially in typical laboratory atmospheres. Although it can burn, this is not likely in promptly extinguished conflagrations. Metal can be coated to resist corrosion, but cannot be totally protected. The owner should look for good quality finish and construction in any casework selected for laboratory installation.

All wall hung storage units, shelves, and equipment should be securely attached with fittings of sufficient strength to support units under maximum loading conditions. Walls that support storage units should be reinforced to bear the maximum load. In normal construction of gypsum wallboard on metal stud partitions, lateral bracing and blocking is not included. Even lightweight concrete masonry unit wall construction is not structured to hold the weight of typical laboratory storage. Reinforcing for interior partitions must be directly called for in the specifications and drawings.

15.2.5.4 Laboratory Features. Lighting should be designed and installed so there are no shadows and low glare at the laboratory bench height. Normal lighting levels range between 50 and 100 fc (500 and 1000 lux) at countertop heights. The intensity of light depends on the nature of the work. Provide all emergency lighting and exit signs required by code. Large open laboratories and laboratories used for instruction should have supplementary emergency lighting and lit exit signs to facilitate egress. Laboratories with special hazards should have distinct warning lights or lit signs, such as laser or high power magnet in use or UV lights on.

If compressed gas cylinders are used in the laboratory, provision must be made for sturdy supports to which cylinders in use, as well as those awaiting use or empty, can be firmly secured.

A closed or open alcove out of the flow of traffic should be provided for storing personal gear such as coats, hats, raingear, and briefcases. Similar recessed locations are desirable for waste containers. These include receptacles not only for ordinary trash but also for temporary storage of different chemical wastes, which are separated by their hazardous qualities or special method of disposal.

Walls between adjacent laboratories and between laboratories and corridors should have a ¾-hr fire resistant construction rating or greater according to local building codes.

15.2.5.5 Access for Handicapped Persons. Laboratory design for use by handicapped persons is regulated by state and local codes. The regulations apply to all workplaces for provision of access to persons in wheelchairs or otherwise impaired. The prime concern is safe egress. The regulations specify minimum door widths, minimum clearances on the latch side of in-swinging doors, the directions of egress, elimination of threshold conditions, ramp access and allowable slopes, handrail heights, and appropriately designed personal hygiene facilities. In general, slightly more floor space is required to comply with most handicapped regulations. These regulations enhance the safety aspects of good laboratory design.

15.3 LABORATORY PLANNING PROCESS

Laboratories are complex building types that must be carefully planned from the initial inception to build or to renovate. The owner needs some method to inform the architects and engineers of its size, primary use, occupancy, the functions and processes it must contain, the budget, and the schedule. The document in which this information is collected is called a program. The program can be written by a team of owner representatives, often named "the building committee," by a laboratory planning consultant or the architects and engineers. Ideally, a program will list each type of space that is required, the quantity and the preferred size of each type of space. Each room type should be described in as much detail as possible. Room data should include the following:

1. The number and work status (full-time, part-time, etc.) of room occupants.
2. The primary function(s), secondary function(s), and any future uses.
3. List of chemicals to be used or stored, probable quantities, and descriptions of their hazardous characteristics.
4. List of all potentially hazardous processes.
5. Description of the environmental standards to be maintained:
   Ambient temperature range in summer and winter.
   Humidity range in summer and winter.

      Air quality, particle count, air changes, percentage of outside air.
      Differential pressure requirements from adjacent spaces.
      Vibration sensitivity.
      Air turbulence characteristics.
      Radio frequency or magnetic field interference.

6. The list of equipment, instruments, and scientific apparatus within:
      Manufacturer and model number.
      Size, including clearances for loading, servicing, or operation.
      Benchtop or floor mounting.
      Weight and load distribution, if greater than 50 psf.
      Electrical load, voltage, phase, amperage, horsepower or kVA rate.
      Emergency disconnects, grounding, shielding requirements.
      Local or remote alarm systems.
      The estimated period of the connected electrical load.
      Heat load in Btu/h to be dissipated by room ventilation.
      Plumbing connections, rate of consumption, pressure, drain.
      Inlet water temperature, outlet rise.
      Steam or other gaseous utility connection, pressure.
      Valves, gauges, safety monitoring devices.
      Mechanical connection, ducted exhaust, fume canopy.

7. List of work surfaces, area, mounting height, materials to resist chemical stains, heat or cold resistance, loading, and utility outlets.

8. List of storage requirements for each laboratoory:
      Flammable liquid storage cabinets.
      Temperature controlled storage.
      Chemical storage, acid, base, corrosive chemicals.
      Sterile and other clean storage.
      First aid kits and emergency supplies.
      Waste including hazardous or contaminated materials.
      Soiled materials including glassware, laboratory coats.
      Tools, instruments, computer accessories.
      Other nonhazardous storage, books, manuals, coats, glassware.
      A similar list should be developed for each nonlaboratory space in which chemicals, equipment, or related materials will be stored.

9. Description of building finishes required:
      Flooring material.
      Ceiling construction.
      Wall finish and fire rating.
      Door fire rating and special hardware requirements.

10. Criteria for the adjacencies of each space and if possible a description of the desired arrangements of spaces. This may be done effectively with a series of simple diagrams showing critical horizontal and vertical relationships in the building or on the site.

The building program provides the performance criteria of the building or renovations, it is not the design solution. The program is the standard against which the owner can evaluate progress toward the design solution.

15.4 OWNER RESPONSIBILITIES

Owners have responsibilities that pertain to aspects of laboratory building not directly related to safety. The owner must obtain the permits or releases from all relevant local, regional, or national authorities to construct, renovate, or operate a laboratory building on a designated site. Some of these permits require at least minimum standards of public health and safety. Examples of the normal regulatory agencies are the local building inspector, the state plan review board, the fire marshall, board of health, department of environmental quality engineering, and others with similar concerns for public health and safety.

In the United States three basic documents are required for new construction and for most renovations. They are zoning permit, building construction permit, and occupancy permit. Without these, local authorities can prevent an owner from building or occupying any building.

At all times, from the earliest stages of planning to building occupancy, it is the owner's responsibility to ensure that all safety aspects have in fact been included in the plans and specifications, and fully implemented in the building itself. Architects, engineers, and other consultants specializing in laboratory design can assist the owner in fulfilling these obligations. In order to select the design team, the owner should know the key individuals within the architectural and engineering firms who will be assigned to the project. The management of the design firms must make a firm commitment that the selected individuals will follow through the project to completion, that is, to occupancy. The owner should check the references of the firm and of the design team with current or former laboratory clients. The owner should investigate that the experience of the architectural, engineering, and construction firms includes completion of one or more projects of comparable laboratory type, size, budget range, and general site condition. Determine in advance that the owner, or owner's representative and the design team can form a good working relationship. Communication between the owner's representative and the other team members is the most critical factor to ensure satisfactory completion of the project.

15.5 THE FUNCTION OF ARCHITECTS AND ENGINEERS

Architects and engineers are licensed by a state for practice in that state. Each state requires the stamp or seal of a registered architect or engineer on the construction documents prior to approval for a building permit. The stamp or seal on the documents certify that the design complies with all state code requirements. The drawings and specifications become part of the contract between the owner and the general contractor or construction manager.

The architects and engineers advise the owner's representative about those design issues that require the owner's decisions or further information. Architects and engineers provide data and often make presentations to regulatory agencies on behalf of the owner, prior to and during construction. The design team produces complete and accurate drawings and specifications appropriate to the three phases of the design process: schematic design, design development, and construction documents. The design team coordinates and reviews documentation from all the consultants for compliance to codes and the design intent. The design team project manager manages the disbursement of fees to consultants and accounts for the fee budget to the owner.

Copies of completed construction document sets are distributed to one or more general building contractors for either bids or negotiated proposals. The owner selects the contractor. The architects and engineers may offer recommendations based on their experience, and participate in contract negotiations on behalf of the owner.

During the construction phase, the design team ensures that the requirements of the construction documents are met. These activities include checking shop drawings from vendors for suitability to actual construction conditions and compliance to design intent, attending job site meetings to monitor building progress, inspecting construction to evaluate the quality of the work, materials, and methods. The architect or engineer approves the contractor's applications for payment and submits them to the owner.

Other services that qualified architects and engineers can provide for the laboratory building owner are programming, cost feasibility studies, alternate design studies, life-cycle cost analyses, energy conservation studies, interior or landscape design, preparation of reports, and presentations to regulatory and funding agencies.

REFERENCES

1. American National Standards Institute; ANSI Z358.1, New York, 1985.
2. A. Weaver and K. Britt, "Criteria for Effective Eyewashes and Safety Showers," *Prof. Safety,* 22, 38–54 (1977).
3. L. DiBerardinis et al., *A Guide to Laboratory Design: Health and Safety Considerations,* Wiley-Interscience, New York, 1987.

CHAPTER 16

Laboratory Hoods

G. Thomas Saunders

Geneva Research, Durham, North Carolina

16.1 INTRODUCTION

In 1984, I was to present a paper at the spring meeting of the American Chemical Society in St. Louis. The preceding presentation was made by Dr. Gerald Knutson, of Industrial Health Engineering, and a competent and highly respected name in regard to fume hood testing and evaluation. Dr. Knutson made a statement that brought down the house, "there is only one person worse than an IRS Agent, and that is a fume hood manufacturer." When my time arrived, and I am a fume hood manufacturer, what do I say. As strange as it may seem, I had to agree, but with some qualifications.

Without manufacturers trying to capture a larger part of a pretty good sized market, the hoods made today would be exact copies of those made 50 years ago. However, not all things that are presented as advancements are in effect better ideas—some are—but some are merely ways to try and convince a customer that your product is better, best, or whatever.

This chapter is going to dedicate itself to giving lab people and their planning counterparts some respectability as knowledgeable fume hood users and planners. They may not become experts but they can decide that the expertise some people claim may be more fancy than fact.

Sometimes there are no simple answers to the problems encountered in fume hood selection and use. Other times the problems are self inflicted. The poet's admonition still applies: A little knowledge is a dangerous thing.

As you proceed through this chapter, do so with as few preconceived ideas as possible. Start considering fume hoods as part of a total system and demand of the hood manufacturer that he or she minimize fancy and maximize fact. Convince yourself and your architect/engineer that there is no substitute for knowledge.

At the end of the chapter is a list of reference articles. These are to help broaden your fume hood perspective. Please secure them and read them, as this chapter is not all inclusive. You will find that I do not always agree with these references but that is to be expected.

Let us establish, once and forever, the basic reason for having such an unsightly item in the laboratory as the fume hood. It exists to protect you, your fellow workers, your work, and your environmental neighbors (birds, bees, flowers, trees, and people).

I may not repeat these exact words again in this chapter—but the text amplifies this thought and your thought process must do the same. Do not have a "close call" accident without knowing it. Have sufficient knowledge to be able to avoid this exposure. Be a live and contributing member of society—LIFE IS FUN!

16.2   RATING EXISTING FUME HOODS

16.2.1   Exhaust Volume Determination

There is a very simple rule that applies to all laboratory fume hoods: To operate properly they must exhaust air. Therefore, the first step in hood evaluation is to determine this exhaust volume. The simplest method is to utilize a velometer and do a face velocity traverse of the hood face opening with the exhaust system operating. If this hood has its own exhaust blower and is located in a room with additional hoods, each with its own exhauster, then all hoods should be on at the same time. Multiply the face velocity in feet per minute (fpm) times the full face opening of the hood (ft^2) and you have the exhaust volume in cubic feet per minute (cfm).

There are a variety of velometers, from propellers to vanes to hot wires to thermistors, some more expensive than others and some better than others. My personal preference leans towards the Alnor Model 8500 hot-wire thermoanemometer. With its analog readout, it has some built-in (minor) dampening to minimize wild gyrations and this allows you to visually recognize the highs and lows and to establish your own averages. Digital units with time retention memories are difficult for me to use since I do not get a chance to make my own value calls in determining the average readings; I like to see the high and low spikes and be able to self-eliminate. There are excellent instruments made by Kurz and others. An inexpensive hot-wire unit that travels very well (the Alnor 8500 was not made for abusive treatment) is the Dwyer 470-1, which has been factory modified to include 0.6 units of delay (dampening action). If you want to do a respectable evalution, buy a good instrument. It should be recalibrated by the manufacturer on a periodic basis (at least yearly) and maintained as a precision instrument.

To establish the average face velocity for a typical bench hood, use the following procedure; refer to Fig. 16.1.

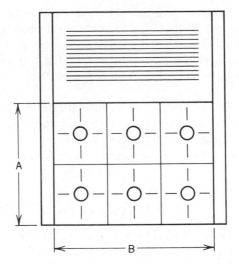

Figure 16.1. Grid for measuring hood face velocity.

1. Measure the full open dimensions of the face area, that is, A and B.
2. Establish an equally spaced grid to represent the average for the face area, with a maximum grid size of 12 × 12 in.
3. Take a velocity reading at the center of each grid coordinate (boxed area above).
4. Add the individual readings and divide by the total number of readings taken. Use this as the average face velocity in feet per minute.
5. Multiply the velocity (fpm) times the open area in square feet to obtain the flow rate in cubic feet of air per minute.

This example applies to a single style of fume hood. For other types and styles, refer to the SAMA Standard (1). This procedure using a 12 × 12 in. grid size yields an average flow rate with an accuracy of about ±20%. If you want to improve the accuracy to about ±10%, then increase the number of grid (reading) points by 50%. The additional tedium involved may or may not be worth the trouble.

For reasons I will explain later, it would be to your advantage to be able to check the velocity/volume calculation against the calculated volume of air going up the exhaust duct. If you have a straight section of exhaust duct 10 duct diameters long—no bends, dampers, transitions up or down—and that section is reasonably accessible, with a duct velocity about 1500 fpm, then you can do this cross check. At a point 7.5 duct diameters (minimum) downstream from a major air disturbance of this straight piece of duct you can make provisions for taking either velocity or velocity pressure readings, and then equate these to volumes. To take either type of reading, refer to the Industrial Ventilation Manual (2). Use your velometer (on the high range) and average the 20 readings. If you have a pitot tube assembly, take a traverse with this and then extrapolate the velocity pressure to duct velocity (2, p. 6–33). Referring to the manual again (2, p. 6–18) and multiply the duct velocity (fpm) by the duct area (ft^2) and you have duct volume in cubic feet per minute.

Now, compare the flow rate (volume) calculated by face velocity alone to the duct flow rate (volume) you have just measured. The safe bet is that the flow rate (volume) measured in the duct was at least 10% greater than that determined by the face velocity readings. Save the results and we shall use them later.

16.2.2. Face Reality

Now we have a reading with which we are reasonably comfortable, as to accuracy, for the face velocity of the hood. Let us ask ourselves what the face velocity actually should be.

A few OSHA regulations (3) flatly say 150 fpm. The SAMA Standard (1) grades hoods from Type A to Type C and suggests 125 to 150 fpm for certain critical operations. I was at a meeting with a Fortune 500 company that was building a new laboratory, and the corporate hygienist said: We've got some real nasty chemicals and are using a 150-fpm face velocity for all hoods. Some professional societies (4) come forth with 150-fpm velocities. So after reviewing all of this prestigious material one would be inclined to believe that 150 fpm is obviously safer than any lesser number.

This may surprise you—but this is just *so much bunk!* There is absolutely no data to support this "more is better" philosophy. As a matter of fact, the most recent data tends to indicate a maximum face velocity of 100 fpm with face velocities of 60 to 80 fpm most adequate given proper attention to other factors that affect hood performance, that is, hood safety.

Now that we have made such a statement—how do we prove it. Let us do it bit by bit.

OSHA has informed the American Chemical Society that the pertinent regulation (3) is not enforced as it applies to laboratories. On July 24, 1986, OSHA published a proposed rule (see Chapter 12), 29 CFR 1910.1450. The overall meaning, if promulgated as proposed, would return to current data and accept 60- to 100-fpm face velocities as being safe if all other factors, such as the information in the ACGIH publication (2), Fuller and Etchells (5), and Caplan and Knutson (6), are considered. To realize the significance of such a turnaround, while not going through the mathematical gyrations here, it could result in an energy savings worth some $350 million per year nationwide. This move would in no way decrease laboratory safety; it would actually improve safety.

The SAMA Standard (1) was published in 1980 but was actually in final form in 1979. It was in 1978 that Caplan and Knutson (6) published ASHRAE

Research Project 70 that upheld the range of from 60 to 100 fpm as acceptable. However, SAMA, like many (voluntary) standards groups approve by consensus of membership and whether the membership is well informed or not they can say no and the so-called standard is rejected or modified; change is hard to come by and the 125-to-150 fpm category was retained.

I mentioned Caplan and Knutson, ASHRAE RP 70 and safe face velocities in the range of 60 to 100 fpm. Concurrently, with RP 70, Fuller and Etchells (5) with E.I. DuPont were performing similar and sustaining work to Caplan and Knutson which they published in 1979; they too were very comfortable with the range of 60 to 100 fpm of face velocities.

We now have in place some quantitative data from two separate and highly respected research groups. In 1982, the ACGIH (2) concurred with ASHRAE RP 70 and included this quantitative testing procedure and the demonstrative data contained therein in their most prestigious manual, thus giving the blessing to reason and away from the "more is better" society. Perhaps too few hygienists have read the manual (2, p. 5-23) or Caplan and Knutson (6) or Fuller and Etchells (5). There are bumper stickers reminding you to hug your kids—how about having hygienists hugging, or at least reading the manual.

Enough of picking away at numbers—suffice it to say that you must pick one, as either a personal or a corporate choice. If you choose 150 fpm, after reviewing current data—fine; if you choose to go with 80 fpm—fine. The value you select is yours to defend.

The SAMA Standard (1) advises that no single grid reading in a face velocity profile should vary more than 10% from the average value. I just do not believe this to be all that important. I do not know the percentage of allowable variation and no one, to my knowledge, has done any acceptable research. I suspect that there are at least two main influences:

1. That the variations could be less (percentagewise) at the lower face velocities.
2. The variation could be more (percentagewise) for an auxiliary air fume hood.

16.2.3   Makeup or Supply Air Systems

"You can't take it out 'til you put it in." Exhaust air has to come from somewhere; supply grills, under doors, through open windows—but it must come into a room before it goes out. So back to a statement made earlier about taking a single hood face velocity (exhaust volume) with all the hoods in the same room operating simultaneously. Equally as important as the exhaust system is the supply system and this is pretty easy to check.

Turn on the hood(s); close the door(s) and window(s); try to open the door (it should open outward); is it hard to open? Open the door and measure the air velocity going through the door opening into the room—is it pretty high (100 fpm or more)? Measure one hood's face velocity with door open and then door closed—much difference? In rooms with multiple hoods, close the door and measure a particular hood's face velocity with no other hoods operating. Then do it again as each additional hood is energized until they are all on simultaneously. If there is a noticable difference, you have problems. If a minor difference, or it takes 30 s or so for face velocities to equalize (due to makeup air volume control lag) you most likely are in pretty good shape.

Now you either do, or do not, have a supply air problem. You can quantify the problem, if you have one, by measuring the volume coming in through the open door. With some luck a rebalance of the supply system can make up the required volume. If this is not possible, then the HVAC mechanical group should evaluate the overall problem. They may or may not like the solution but it is a problem you must solve. If you do nothing, then you are tossing safety out the window, under the door, or wherever.

16.3   LOCATING A FUME HOOD

Over the years, fume hood location was stressed to be

1. Away from doorways since a fire or explosion in the hood could block an exit from the room.
2. Away from traffic patterns to minimize cross traffic and therefore cross drafts. This could be true of doorways especially with the door open.
3. Away from corners.
4. Away from disruptive air discharge (into the room) grills and diffusers.

Now, walk into your lab—or some other person's lab—particularly if it is 10–20 years old—and tell me where the hood is located. I will guess:

1. At least 30% are adjacent to the main doorway.
2. Another 20% are in a corner.
3. Thirty percent have the air supply diffusers blowing into or towards the hood face.

That leaves only 20% of the hoods that have been located for safety and not for the convenience of the architect, engineer, or the HVAC contractor.

I am not going to take offense at a doorway location from a fire exit viewpoint, since most newer facilities require a minimum of two doors, however, traffic caused cross drafts can be a significant problem. Architects have a beast of a time efficiently filling corners (so they say). So what do they do? They cram the hood into the corner. What happens is that rooms set up characteristic air swirl patterns and a hood in a corner can catch more effect from such turbulence than a hood even 3 to 5 ft away from the corner proper. See chapter 15, Section 15.2.1.3. Also note that the air makeup system is not always well defined. The mechanical drawings may be loose; what is well planned may be changed in the construction because of structural obstructions such as plumbing or electrical services, or there may even be a lack of concern, or knowledge, of how important deviations from well-planned entrance air locations can be to safety.

Referring back to Caplin and Knutson and the ACGIH—the location and throw velocities of supply air diffusers is vital to hood safety. The location is easy to see—the throw velocity is easy to measure with your velometer.

The recommended discharge or terminal velocities taken at the exit vanes of the diffuser are 0.5 to 0.7 of the hood face velocity. Where do your parameters fall? Are you safe or a victim of someone's lack of expertise?

So go back to the obvious question. Where do I locate my hood(s)? Simple answer: For maximum safety and for the users workplace convenience. It may cause the architect/engineer to work harder, but remember, once the building phase is over, the architect/engineer can leave and the user is working in the laboratory day in and day out. It is amazing how much an architect/engineer can accomplish if they are challenged and how little if they are not. Your problem then is to learn enough basic parameters to make the architect/engineer perform up to his or her maximum abilities. Not only can it make you more productive, it might just save your life.

16.4 FUME HOOD TESTING

We earlier referred to the SAMA Standard (1) which is dedicated to qualitative hood performance tests: (a) face velocities and (b) smoke patterns.

Nothing is wrong with this initial testing approach. We have spent considerable time on hood face velocities and hood location and the quickest way to visually see if we are headed in the right direction is by judging these two aspects of the actual "hood in the lab."

So, from a qualitative point of view, you take the hood face velocity profile(s). Do it with doors open, doors closed, diffusers covered and uncovered, or masked away from the hood. After all of this—and it really is not too time consuming—put a 30-s smoke bomb (7) in the hood and repeat the door and diffuser sequences. If the smoke flows from the hood chamber into the lab (see Fig. 16.2) then do not go any further with more defined quantitative tests. You have a problem. We shall cover corrective measures at the conclusion of this portion of the chapter.

If, on the other hand, you do not see any smoke flowing from the hood, if your cross drafts are 20 fpm (1) or less and if your face velocity has a swing of no more than ± 10 to 15%, then let us consider really putting the hood(s) through quantitative testing.

It is backtrack time. We made a special point to emphasize Caplan and Knutson's ASHRAE RP 70; this is now ASHRAE Standard 110-1985 (8). The current levels of acceptance are as outlined in the ACGIH ventilation manual (2).

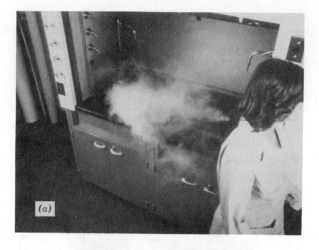

Figure 16.2.   Two examples of smoke patterns indicating poor performance.

I feel very strongly that the fume hood manufacturer should supply the purchasers of new hoods with documents that certify that a hood of a similar design and size has been tested in a fume hood test facility according to ASHRAE 110-1985 and has a rating not to exceed 4AM.05. If the hood under consideration is of a special design, I would have the hood manufacturer test it. This would be a most worthwhile additional cost.

A quick overview of ASHRAE 110-1985: A test gas (Freon 12 or sulfurhexafloride) is discharged from an engineered diffuser, at a known rate and in a predetermined geometrical pattern inside the hood chamber. A department store manikin (simulating the average American laboratory technician) is placed in the front plane of the hood and a continuous sample is taken in the breathing zone of the manikin. This sample is then processed through an instrument capable of a quantitative evaluation of the test gas in the range of 0.01 ppm. The ventilation manual "suggests" that a test average should not exceed 0.1 ppm. I would offer, from doing a large number of these ASHRAE Standard 110-1985 tests, that a 5-min test period is most ample and that you should expect a safe hood to perform at a control level of 0.05 ppm (or less) in a fume hood test facility and at a 0.1-ppm level in actual or AU (as used) configuration. These values are not carved in granite—and may not be for a few years. But if you are going to have your hoods tested by this procedure—or you are going to do it yourself—establish a number and strive to achieve it. A major pharmaceutical research laboratory has instituted an evaluation of all of their fume hoods as they are used and will accept a control level of 0.1 ppm or slightly higher if conditions, equipment, and usage might warrant an increase. I should like to point out that when you performed the smoke test, your visible range was in the area of 200 ppm, now with ASHRAE Standard 110-1985 you are in the 0.1-ppm range. Apply these two numbers against the STEL (9) of the materials you are currently using (and those you could very well use) and determine your level of fume hood safety. This is a judgment call. If you exceed the permitted PPM exposure level of your research or process chemicals—STOP—call for engineering help—it could save a call to the hospital.

I should like to give my opinions and a very brief overview of a fume hood test that was included in the U.S. EPA Fume Hood Standards (10) dated January 15, 1978, and updated in June, 1982. This test procedure is primarily applicable to auxiliary air fume hoods.

This very controversial procedure is usually referred to as the *urinine dye test*. In brief: An aerosol of urinine dye is discharged into the auxiliary air supply duct. Sampling points are established in the auxiliary air system and the exhaust air system. The samples, taken via suction through a filter paper, are prepared for analysis by a fluorimeter. The test is to determine what percentage of auxiliary air is actually captured by the hood over the 5-min test period. A second part of this test is to disperse the urinine dye in the hood chamber, set the auxiliary air at 100% of the exhaust air, and then take samples to determine how much air escapes from the hood into the room.

There are many questions raised about the complexity of performing this test without cross contamination, particle size complications such as flocculation, and so on—factors which compromise the validity of the data and vitiate any conclusions drawn therefrom. But most importantly, even if the test could be conducted to yield reliable, reproducible data, the test simply does not address the safety performance of the hood; it only addresses exhaust and supply air balance; it only addresses one of the factors upon which hood safety depends. Even worse, in a recent survey (11) taken by the National Safety Council, 22% of those responding used this EPA test to evaluate their hood safety performance.

16.5   CORRECTING HOOD PROBLEMS

16.5.1   Exhaust Air Volumes

Earlier we explored hood face velocities, how to measure them, and then how to decide on a face velocity that was consistent with a viable safety program.

If the measured face velocities exceeded the upper limits set, then cutting down this air is a matter of reducing fan speeds, closing dampers, or whatever to reduce the exhaust air volumes and therefore hood face velocities. This is primarily a maintenance function with in-house personnel or by an outside mechanical-ventilation contractor. Remember, when the exhaust volumes are adjusted, you must do the same with the makeup air volumes. This is normally referred to as rebalancing the hood, room, or laboratory building proper.

If the measured face velocities (exhaust volumes) are too low, then there are several items to investigate. First—how much too low are they? If it is in the area of 10% and they are vertical sash hoods, then you can most likely correct this at the hood proper. Let us review how hoods are made.

Where a vertical sash goes up past the hood chamber, there is a gap to allow this up and down travel. It ranges from $\frac{3}{4}$ in. up. See Figure 16.3.

When the sash is up, air from the room goes through this gap and not through the open sash. It does not contribute to face velocity and only wastes expensive room quality air (72°F). This gap is easy to close (see Fig. 16.4); closing can provide an increase in the face velocity of ~ 10%. We took note of this possibility with the hood face velocity traverse and the pitot traverse of the duct proper showing two different volumes. This sash leakage is the reason. The problem is easily corrected and should not cost more than $100/hood to rectify.

If the volume is more than 10% low, then you have two or three possible reasons for this lack of air. First, check the exhaust blower (system). Is the exhaust fan operating to exhaust the proper amount of air? Check the blower belts, the fan rotational speed, and the direction of rotation. Stop the fan, and somehow get a look at the blower wheel (impeller) proper. Is it still whole or has corrosion eaten away on the blower blades? Each of these check points has an obvious answer should they be the culprit of reduced volumes. But what is the course to follow if everything checks out OK?

Now you need to analyze your systems. Early in this chapter we explored makeup air—as I said—You can't take it out 'til you put it in. How did your

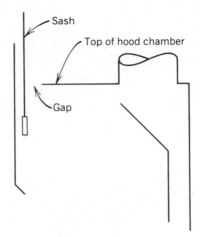

Figure 16.3. Gap at the top of the vertical sash (16). (Copyright by the *Journal of Chemical Education,* reproduced with permission.)

checks come out with doors open and doors closed? If you found that you had insufficient air, do the same fan checks on the supply system as you did on the exhaust system. Again, correct the deficient parts of the system. Now—what to do if the exhaust and supply systems are up to design capacities and there is still insufficient air.

It would appear that the system volume is too small for the fume hood population. Three solutions: eliminate some hoods from the system; increase the system volume (if possible); modify the hoods themselves so that they use less air while maintaining the face velocity you have established as being adequate for safe operation. The first two solutions require staff plus engineering concurrence—also adequate monies to achieve the end result. Hood modification also requires monies plus staff and engineering concurrence but most importantly it requires discipline. That is, the most effective program for the least amount of capital outlay is to restrict the sash opening area when using the fume hood and to enforce such restriction.

If you have a hood (or hoods) with vertical sash then the simplest thing is to add spring loaded stops so that you can only (easily) open the sash to some predetermined percentage of the full open area. If you only have half enough exhaust and supply capacity, put the stops at half open; if two-thirds capacity then two-thirds open, and so on. When the hood is being used for active experimentation, the sash must be lowered to the stops, otherwise your face velocity is well below the value you have set for safe operation.

If this modification is acceptable, then you can enforce it a bit more easily by adding a switch that is activated when the sash is above the stop and this in turn activates a light (or alarm). A flashing yellow light beats an alarm any day because the alarm noise will annoy everyone and sooner or later it

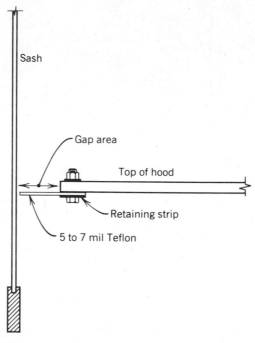

Figure 16.4. Closure of the top sash gap (16). (Copyright by the *Journal of Chemical Education,* reproduced with permission.)

will be disconnected. The light should be mounted on the hood face at about eye level.

The alternative to sash stops is to change the vertical sash to a horizontal sash. This is more expensive than the sash stops but, if made so that the sash panels are not easily removable by the laboratory personnel, it is much easier to enforce. Over and over I hear the remark "horizontal sash hoods are too hard to work in." For some this is true but if the hood is properly designed (sash panels no wider than 15 in.) you would be surprised how easy they are to use for probably 80% of all general hood applications.

Note that a vertical sash in a half-open position provides a horizontally polarized opening. Thus, to reach up into the hood to adjust, open, close, or whatever, some piece of equipment, you either open the sash full or grow 5-ft long arms. On the other hand, a horizontal sash gives you a vertically polarized opening so that you can reach from top to bottom with no problem or obstructions. Two things you must insist on: (a) maximum sash panel width of 15 in. and (b) at least four sash panels in two tracks; a two sash horizontal style hood is a disaster because you cannot reach anything in the center of the work surface.

16.5.2   Hood Location

If your hood(s) are located adjacent to doorways or other high traffic areas, you almost have to face up to the situation as it is very expensive, if not impossible, to change. If, by some stroke of luck, you can divert traffic or cut in another door—great. Most of the time this is pretty tough. If the problem is the relationship between the laboratory supply air and the hood proper, however, there are some things you can do to improve your hood performance.

First, qualify your possible problem. Secure a 30-s smoke bomb,(7) light it, and place it inside of the hood chamber. If it is a real bad problem, you will see it in a hurry. If you cannot see the smoke pour from the hood, it does not mean you do not have a problem—but only by submitting to the ASHRAE 110-1985 standard can you determine the quantitative hood performance.

Start by looking at how the air enters the laboratory. A fairly standard variety of air diffuser injects air into the room at a fairly high velocity. These pieces of equipment can be round, square, or rectangular; Fig. 16.5 is typical. The problem is exit velocity. It should be from one half to three quarters of the hood face velocity. The chances are that when you measure this diffuser's exit velocity, it will be two to five times the hood face velocity.

Where is the diffuser located? Is it discharging part of the air directly at or across the hood face?

Chances are that it would be difficult to add more diffuser(s) or diffuser area or to change the diffuser(s) location, although with drop grid ceilings it can be economically feasible. So you get a bit ingenious and impose a barrier of perforated metal (use small chain or wire) and slow the air down or divert it partially from its previous paths. Each situation is different, but the approach to solving the problem is the same. Expensive it is not, except in time, and it can be most rewarding in improved fume hood performance.

16.6   HOOD DESIGN

At one time a fume hood was just a box with a door or window on the front and an exhaust hole in the top. As time went by a back baffle system was added to give a more even front velocity profile from top to bottom—especially at the bottom (work surface). The streamlined air entrance shape was developed in the 1940s to 1950s and was perhaps the single most important advance to that date. The bottom front air foil was paramount with this front configuration as it greatly reduces front edge turbulance. Concurrently, with the aerodynamic design of the bottom front area, double side walls were included to further reduce turbulence and to contain the mechanical service piping and electrical wiring.

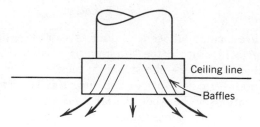

Figure 16.5. Air supply diffuser.

As designed and promoted by various hood manufacturers, there are variations in the back baffle design and the size and shape of the airfoil area and side wall configuration.

16.6.1 Baffle Design

Let us tackle baffle design (refer to Fig. 16.6); the three slot back baffle with top, middle, and bottom locations is almost standard; one manufacturer also has a gap between the baffle and the hood side walls. The physical dimensions of these slots are reasonably identical among the various manufacturers. Size is quite important, as is the ability to adjust them properly. It is very important to have a bottom slot (C) that is more than difficult to block off with the usual and smaller lab equipment. It (C) should be at least 2 in., 3 in. is even better. The center slot (B) is of a fixed size in the range of 1 in. to $1\frac{1}{2}$ in. and the top slot (A) is adjustable from closed to 2 in.

Many specifications for hoods state: "The top and bottom baffle slots shall be adjustable to accommodate lighter than air and heavier than air gases." I think this is mostly sales promotion and cannot be substantiated for actual practice. Experimentation (12, 13) using the ASHRAE 110-1985 procedure would strongly indicate that there is an optimum slot sizing scheme: At the top, slot A is adjustable $\frac{1}{2}$ to $\frac{3}{4}$ in. wide (and to $1\frac{1}{4}$ in. for high heat loading); the middle slot, B, is fixed at 1 to $1\frac{1}{2}$ in.; the bottom slot, C, is fixed to 2 to 3 in. The center slot has little influence on hood performance; it is useful because the bottom slot might get blocked by equipment (see Chapter 11 for an illustrated example). The $1\frac{1}{4}$ in. opening of the top slot is used for a heat load with a 5- to 10-W power dispersion.

Although the vapors and fumes generated in a fume hood usually have densities greater than air, the dilution effect of the inflowing air reduces the practical densities to only slightly greater than air. This general statement does not apply when the rate of vapor generation is large and the vapors are very dense, as, for example, in the case of a bromine spill on the hood floor. In such cases, the hood sash must be closed to the smallest opening that will achieve maximum sweep velocity and rapidly cleanup or neutralize the material that was spilled. Conversely, lighter than air situations are extremely rare except as caused by heat.

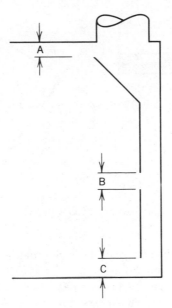

Figure 16.6. Back baffle slot adjustments (16). (Copyright by the *Journal of Chemical Education,* reproduced with permission.)

The bottom slot provides the hood floor sweeping action. It is essential that it not be impeded. Always insure that large apparatus, furnaces, mixers, square cans, and so on, have legs or are on platforms that keep them about $1\frac{1}{2}$ to 2 in. off the work surface. I cannot emphasize this point too many times. Give your hood a break so it is a true safety device.

16.6.2 Aerodynamically Shaped Entrance (Face)

Air does not go around square corners with any degree of grace. As shown in Fig. 16.7 air initially takes a 90° vector in relationship to the corner being turned in. Then, due to the velocity imparted by the 90° opposed air stream, it tries to conform to this new direction but ends up doing a complete 360° pattern and the area becomes turbulent. The turbulence is also caused by some air pressure differentials at the front edge of the work surface—all in all not the ideal situation. The angle shaped entrance of the side walls greatly diminishes the turbulance at the sides and the same is basically true of the top front area above the sash. The bottom front air foil is the main contributor to reduced turbulence. It gives the air a vector in the direction towards the back of the hood as the turning takes place starting outside of the hood and in a confined plenum area. See Fig. 16.7. So instead of turbulent air you have a reasonably laminar sweep of the work surface area of the hood. I have said many a time: If you have a hood that does not have

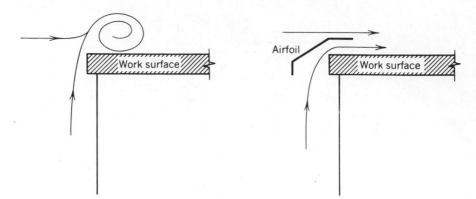

Figure 16.7. Air patterns entering hood with and without bottom air foil.

a bottom front air foil—add one. It is a great investment in safety and should not cost over $100/hood.

A word of caution. The airfoil shapes have to be efficiently designed. Refer to Fig. 16.8. The initial shapes had angle A of from 20 to 30° and dimension B of 6 in. Tests of variations in the lateral dimension showed that B could be reduced to a minimum of $3\frac{1}{2}$ in. and that an increase over 6 in. did not add to the performance. A side wall and or front bottom foil design that gets too small or too angular presents problems. The dimension of the angled section of the foil proper should be a straight section no smaller than $3\frac{1}{2}$ in. and preferably 4 in. to achieve maximum results, and the angle A at from 20 to 30°. The dimension of C should be such that it extends to the back side of the sash configuration.

At one time the airfoil shaped hoods were more expensive than the old square faced conventional hoods. This is not necessarily true any longer. Hood manufacturers, overall, produce more foil shaped hoods than square fronts and the prices are very close. If you consider safety, the foil shaped hood is a real bargain. People who purchase fume hoods should not buy the old styling; hood manufacturers should refuse to make them.

And, old style or air foil fitted, front face turbulence is increased by objects inside the hood that are within 6 in. of the face of the chamber (5). Always work well within the hood chamber, more than 6 in. from the face.

16.7 ENERGY CONSERVATION

Properly designed fume hood systems can contribute a great deal to saving energy. When hoods are not disciplined they use and waste air (energy). Let us set the stage with some simple calculations. A 6-ft vertical sash hood (full open), operating at a face velocity of 100 fpm and being used 10 h/day, 220 days/yr; exhausts:

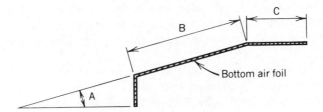

Figure 16.8. Bottom air foil design and dimensions.

13.5 (ft^2 sash opening) × 100 (fpm) × 60 (min) × 10 (h) × 220 (days/yr)
= 178,299,000 ft^3 of air/yr

If the hood operates 24h/day and 365 days/yr, the figure becomes:

= 708,100,000 ft^3 of air/yr

And guess what? This is room quality air, 70 to 72°F in the winter and about the same in the summer (if air conditioned). Depending on the part of the world in which we live, this can be and is a significant amount of money, in some places greater than others.

First, you have to provide ample heating and cooling equipment—so this is an initial construction cost: capital dollars—and then you have to operate and maintain the equipmment: operating dollars. The operating expenses never stop as long as the hoods are operating.

The simplistic answer—and it is true—use less air.

How do you do this without sacrificing safety. It has been awhile but let us remember that this book and this chapter are dedicated to safety.

You need to cut down on expensive, fully treated air volumes, and you can do this in one of several ways:

1. Reduce hood face velocities.
2. Reduce hood face opening area.
3. Add auxiliary air.
4. Add monitoring controller systems.
5. All or combinations of the above.

Earlier we discussed determining a hood face velocity that you considered safe for your particular laboratory. If your number was 100 fpm or more, maybe you would like to review it. If you would consider all factors and say that 80 fpm is acceptable, then you could save from 17 to 70 million ft^3 of air/yr/6-ft-wide hood. The choice is yours, but a lot of good data would support such a change if you weigh the various factors as outlined by Caplan and Knutson (6), Fuller and Etchells (5).

Now, let us be more aggressive and explore cutting down face opening. We touched on this subject as we tried to achieve proper exhaust volumes. We talked about sash stops at a half-open position. This is a potential for a 50% savings.

We mentioned horizontal sash and again you can instantly achieve a substantial reduction in exhaust volume. The problem here is selling the concept of the horizontal sash. I have a suggestion. Have a hood manfacturer come to your facility and modify one hood with a horizontal sash and let the technical staff work or simulate work conditions just to realize how they can be used. Chances are you will get some yeas, some nays, and some "I don't know" votes; I will suggest that if the technical staff is truly objective that the yeas and "I don't know's" will out number the nays.

In this same vein of reduction of face area, there is the so-called "combination" sash. That is, horizontal panels in a vertical rising sash frame system. Best of both worlds as long as you can instill and maintain the discipline required to keep the sash down when actively using the hood. The sash is only fully raised when there is a need to change or install equipment in the hood. A word of caution: Most fume hoods are bypass hoods (see the glossary) and this bypass area is open when the sash is down so the hood must be modified in order that the bypass is either severely restricted or closed. In this way, you are demanding that most of the exhaust air enter the hood through the face opening.

Now one step more—add auxiliary air. This is going to open a lot of options, improve safety, and save a bundle of money. Before you set your mind and say "no way," keep an open attitude and think along with me.

Two objections to auxiliary air hoods.

1. The Auxiliary Air Gives Me a Stiff Neck, 'Cause It's Always Blowing Down on Me.

This should not be the case if the hood is properly designed and constructed. The velocity of the air leaving the auxiliary air chamber should be in the area of 1.5 times the hood face velocity and of a uniform discharge velocity profile roughly ±20%. You should just barely feel this breeze on you. If it is a real strong breeze—it is too high.

2. The Auxiliary Air Is Cold on My Head.

Auxiliary air has to be very close to, if not a little bit above, room temperature in the winter or it can be uncomfortable. In the summer time it can be as much as 15°F above room temperature before it is noticeable by the hood user.

Table 16.1 shows potential savings in both initial investment and operating dollars. The data are from computer printouts for an installed and operating supply and exhaust system in a medium sized laboratory building, but

Table 16.1. Hood Exhaust Systems; Cost Analysis for Three Different Hood Face Velocities

Laboratory Set Up		Hood cfm	Auxiliary Air cfm	Total Tons/lab	HVAC Costs/lab	Auxiliary Air Costs/lab	Total Initial Costs/lab	Yearly Operating Costs/lab	Initial + 5 yr Operating Costs/lab
60 ft/min Face Velocity									
Vertical sash—full open	0% Auxiliary air	1620	0	9.31	$12,442	$0	$12,442	$571	$15,295
Horizontal sash—50% open	0% Auxiliary air	810	0	5.32	$ 7,768	$0	$ 7,768	$316	$ 9,350
Horizontal sash—50% open	50% Auxiliary air	810	405	3.74	$ 4,960	$ 2,182	$ 7,142	$272	$ 8,504
Vertical sash—full open	50% Auxiliary air	1620	810	6.68	$ 8,104	$ 3,864	$11,968	$498	$14,457
Vertical sash—full open	70% Auxiliary air	1620	1134	5.84	$ 5,952	$ 5,209	$11,161	$475	$13,535
Vertical sash—full open	95% Auxiliary air	1620	1539	4.79	$ 3,263	$ 6,891	$10,154	$446	$12,383
80-ft/min Face Velocity									
Vertical sash—full open	0% Auxiliary air	2160	0	11.97	$15,557	$0	$15,557	$740	$19,258
Horizontal sash—50% open	0% Auxiliary air	1080	0	6.65	$ 9,326	$0	$ 9,326	$401	$11,331
Horizontal sash—50% open	50% Auxiliary air	1080	540	4.72	$ 6,008	$ 2,742	$ 8,751	$347	$10,488
Vertical sash—full open	50% Auxiliary air	2160	1080	8.64	$10,200	$ 4,985	$15,185	$648	$18,425
Vertical sash—full open	70% Auxiliary air	2160	1512	7.52	$ 7,331	$ 6,779	$14,110	$617	$17,197
Vertical sash—full open	95% Auxiliary air	2160	2052	6.13	$ 3,745	$ 9,021	$12,766	$579	$15,661
100 ft/min Face Velocity									
Vertical sash—full open	0% Auxiliary air	2700	0	14.63	$18,673	$0	$18,673	$910	$23,222
Horizontal sash—50% open	0% Auxiliary air	1350	0	7.98	$10,884	$0	$10,884	$486	$13,313
Horizontal sash—50% open	50% Auxiliary air	1350	675	5.70	$ 7,056	$ 3,303	$10,359	$423	$12,472
Vertical sash—full open	50% Auxiliary air	2700	1350	10.60	$12,296	$ 6,106	$18,402	$798	$22,394
Vertical sash—full open	70% Auxiliary air	2700	1890	9.20	$ 8,710	$ 8,349	$17,058	$760	$20,858
Vertical sash—full open	95% Auxiliary air	2700	2565	7.46	$ 4,227	$11,152	$15,378	$712	$18,938

charged to only one laboratory. Standard ASHRAE degree days per full load cooling calculations are utilized and all building and environmental (geographical) conditions are included. The lab under investigation has two 6-ft fume hoods, operating 10 h/day and 220 days/yr, with face velocity of the hood(s) varied at 60, 80, and 100 fpm. The location for this comparison is the Chicago area of Illinois; costs are shown for six different types of hood systems.

Comment can be brief. Note that we can lay to rest the often repeated remark that you cannot save any money when you have two fan systems. Table 16.1 shows to the contrary that savings arise by exhausting more lower quality air than higher quality room air. The auxiliary air must have some quality; it is room temperature in the winter and no more than 15°F warmer than room air in the summer, with moisture content reduced to 80 grains/m³ (to keep the laboratory at an acceptable relative humidity). The hood diversity (see below) is postulated to be 75% and the first pass capture efficiency of the auxiliary air by the hood is estimated at 90% with added room heating/cooling capacity to condition this 10% sluff-off air.

To understand first pass capture efficiency, note that the auxiliary air is presented to the hood at a point approximately 1 ft above and in front of the sash opening. As this air is going past this face opening, 90% is captured into the hood on its first trip past the opening and 10% diffuses into the room.

The crowning statement that can be sustained is that properly designed and operated auxiliary air hoods are safer than nonauxiliary air hoods, and normally at lower face velocities.

To this end I present Fig. 16.9, which is a condensed chart of an ASHRAE 110-1985 test, showing a walk-in hood at a 60-fpm face velocity with and without auxiliary air. Monsanto Chemical has been doing a massive hood evaluation program and Dr. Abrahamson (Monsanto) (14) flatly states that "there is increasing evidence indicating that auxiliary air hoods provide better performing characteristics than nonauxiliary air hoods."

16.7.1 Face Velocity Monitor and Control Systems

There are systems now being marketed that will maintain a preset face velocity for a fume hood regardless of sash position. They fall into three general categories: (a) pressure differential, (b) hot wire, and (c) thermistor type systems. Each of the systems relates as to what velocity value it is monitoring and exercises control over air movement to regulate to this hood face velocity. My brief exposure to all three systems is insufficient to form a real set of values. I have some reservations about pressure differential control since it operates on a very small pressure differential, 0.006 in. of water. Hot wire control systems require periodic cleaning if they are to be reliable, and should be recalibrated occasionally. Thermistors are a new entry and could prove out quite well; time will tell.

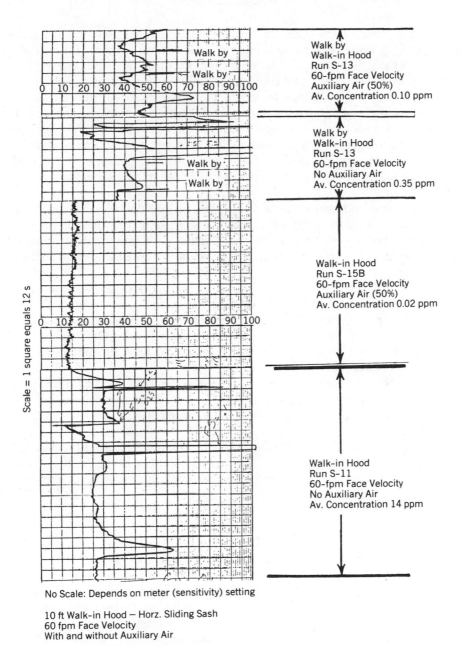

Scale = 1 square equals 12 s

0 10 20 30 40 50 60 70 80 90 100

Walk by

Walk by

Walk by

Walk by

0 10 20 30 40 50 60 70 80 90 100

Walk by
Walk-in Hood
Run S-13
60-fpm Face Velocity
Auxiliary Air (50%)
Av. Concentration 0.10 ppm

Walk by
Walk-in Hood
Run S-13
60-fpm Face Velocity
No Auxiliary Air
Av. Concentration 0.35 ppm

Walk-in Hood
Run S-15B
60-fpm Face Velocity
Auxiliary Air (50%)
Av. Concentration 0.02 ppm

Walk-in Hood
Run S-11
60-fpm Face Velocity
No Auxiliary Air
Av. Concentration 14 ppm

No Scale: Depends on meter (sensitivity) setting

10 ft Walk-in Hood — Horz. Sliding Sash
60 fpm Face Velocity
With and without Auxiliary Air

Figure 16.9. ASHRAE Standard 110-1985 test results on a composite chart.

307

Further, none of the three approaches are as simple as you may want to believe. They must be integrated into the total system, regulating the supply as well as the exhaust air. Yet no doubt the inventive and exploring ventilation engineer could very well combine one of these systems with horizontal sash and auxiliary air hoods to substantially reduce air usage. The main study would be a system cost analysis and this would vary by installation but certainly provides food for thought. Then, if you combine these systems into a heat–cool (energy) recovery system, the end result could be impressive.

One last item to consider is fume hood diversity, that is, the design criteria in regard to the maximum number of hoods to be in operation at any one time.

Many supply and exhaust systems are sized at 75% diversity (maximum of 75% of the hoods in operation simultaneously); others at 100%. This is a critical number in order to size the heating and air conditioning equipment for the building and this represents initial capital expense plus yearly operating expense.

A major industrial research complex in the upper midwest that was completed in 1981, was sized for 75% diversity. However, a 2-yr profile showed 41%. The supply system was of a variable volume design and the resulting savings were significant. They had more initial costs with the 75% parameter and it does allow for adding more hoods at a later date but the design did accept the savings. Good design.

There was an interesting paper presented to the 1983 ASHRAE meeting by Moyer (15) in which he plotted hood diversity at a large eastern university research facility (with 190 hoods). He found that diversity decreased markedly with staff lectures on the subject; in some cases as much as 30%.

Diversity is a most important energy consumption factor and should be carefully considered and measured against the mission of the facility. Over design can be as costly as under design—particularly if it is never utilized.

16.8   BLOWER SYSTEMS

There are two approaches to exhaust air systems: a single exhaust blower per hood or a manifolded system with several hoods interconnected to one exhaust blower. There are advantages and disadvantages to each approach.

A quick safety overview. If a one hood, one fan type system has a blower failure, then only one hood goes down. It is unusual to have a backup blower for this type of a system, it can be, but is rarely done. So if the blower fails, so does the hood. On a manifolded system you can (and should) have a second or backup blower. It may have full operating capacity or it may provide only a percentage of the full exhaust volume. Either way, the hood is not totally disabled should the primary blower fail.

There is a vocal controversy regarding maintenance. One side contends it is easier to monitor small fans versus large fans and replacement of operating parts is a one man job, not a major project. It has been my observation that maintenance is much better on central systems.

From a control standpoint the engineering fraternity would much prefer the larger central system to the one hood, one fan system. They contend, and with some justification, that more and better instrumentation systems are readily available.

The systems, either single fan or central blower, become somewhat more involved when auxiliary air is utilized. Needless to say that when you select a system, provide the best of controls.

Energy recovery is much more easily accomplished with central systems. The Austin Co., in designing the exhaust and makeup systems for the Upjohn Research Laboratories in Kalamazoo, MI, combined one hood, one fan, central manifolding, and heat recovery. The Abbott Laboratories in north Chicago accomplished somewhat the same results with a control system but with dampers on the exhaust ducting of the individual hoods. Both of these systems, and many others I am sure, achieve reliability and efficiency due to the expertise of well-qualified mechanical engineering people. I have worked with a number of engineers with whom I am most comfortable and would suggest that in new construction projects that a great deal of weight be given to the competing architect or engineering firms on the basis of past projects that do work well. Be sure and visit these projects and talk to a lot of people, get facts, and look out for snow jobs. Corporate people in purely administrative positions are somewhat reluctant to admit to anything less than perfection. However, the facilities maintenance and engineering personnel are not adverse to telling it as it is, since they normally did not design the system, but they must keep it working.

One last comment. Perchloric acid cannot be used in hoods connected to a central system. This hood must have its own duct work and blower and both must be equipped with a water wash down system.

16.9 MAINTENANCE OF TOTAL SYSTEMS

Here is a simplistic comparison of maintenance procedures on a product we can all relate to quickly and easily:

You and your neighbor buy new cars, same make, same model. You have the same initial problems of adjustments and after 3 or 4 months of consultation with the new car dealer, your machines are operating at an acceptable level for both families.

You have read the operators manual and have the recommended oil changes, lube jobs, tune-ups, tire rotation, and so on; also, should you feel your car is not up to standard, you return it to your dealer for help. Your neighbor did not read the book, changes the oil on a guess plan, foregoes tune-ups, tire rotation, and so on.

Two years pass and your car is doing well, acts and looks new. Your neighbor's car is a mess and he tells everyone what a lemon he purchased, how poor the manufacturer is, and that the dealer is a crook. You spent a few dollars and have received reliability as a result. Total hood systems are a direct comparison to these automobiles.

After the hood has been installed be sure it is operating to your satisfaction: the service valves do not leak, the exhaust volume (face velocity) is correct, the noise level is acceptable, the sash operates properly, the interior light intensity level is adequate, there is sufficient makeup air, and any special details peculiar to the hood are in place and functional.

Now establish your maintenance manual with yourself and your facilities people. You can use SAMA standard LF 10-1980 (1) as a basis for the practices and schedules and then monitor the system for compliance.

- Keep the hood clean. A couple of times each year take some time and wipe down the interior surface, clean the glass in the sash, and the light unit.
- Check the face velocity. Have a qualified person with a calibrated instrument check the face velocity at least once per year—I would really suggest twice per year if you do not have an indicating device on the hood that has some relationship for indicating face velocity.
- Check the fan(s). The maintenance people should check the fan drive, belts, oil the bearings, notice if there is vibration, and if possible, visually check the condition of the fan impeller wheel on a minimum of once per year.
- If you have a 24 hr/365 day system, be sure these inspections are made more frequently—say 3 times each year.
- Check the duct work proper. This may not be easily accomplished but once every year—or possibly every 2 yr—inspect the ducting. Joints can separate and holes can appear due to corrosion.
- Exmaine the hood proper. How does the sash operate—it may need adjustment or some oil on the pulleys. Valves that are leaking should be fixed before the problem becomes acute and Murphy's Law presents a major problem at a critical time. If you have indicating or alarm systems, be sure to check their operation on a weekly basis. If there is corrosion appearing in the hood in what you consider as an accelerated schedule, try to determine the reason before it is a major repair or replacement problem. A side note on corrosion: select the lining material with care to accommodate your procedures: Hood manufacturers can provide you with test data on various materials. Stainless steel is not necessarily stainless in the presence of certain acids and combinations of other reagents. This can be corrected in some areas (except where perchloric acid is used) by cleaning the metal and getting a small spray can of a clear vinyl

in a paint store or hobby shop and coating the stainless—it does not last forever since it will eventually peal off—but it can be stripped and reapplied. This vinyl spray approach applies equally to all metal parts inside of a hood. Do not use this vinyl spray in a perchloric acid hood. If the work surface or lining material has cracked, it may not be correctable over the long haul, and depending on the location of the crack it might not present any significant problem.

- If the crack allows liquids to leak down and out of the hood, have it either caulked or replaced. Hood work surfaces do crack because of physical or heat loading or both and until some of this type of stress can be minimized or eliminated, you might not stop these cracks from occurring.

There are two rather common types of hoods that require additional inspection and maintenance: radiochemical hoods and perchloric acid hoods.

Radiochemical hoods usually have high efficiency particle arrestor (HEPA) filters in the exhaust system. There should be an indicating manometer across the filter proper to indicate the pressure drop across the filter. There should also be a damper that allows you to maintain some semblance of an adequate face velocity as the filter gets dirty. This filter should be changed when the pressure drop is approximately double the clean filter reading (at the same face velocity) if you cannot adjust the volume to maintain an adequate face velocity, then the indicating manometers do not change much, just the volume decreases and you can be in real trouble.

Perchloric acid hoods must have a water wash down system that includes the area behind the hood baffles, the exhaust ducting, the exhaust blower, and the discharge stack from the blower proper. This wash down system must be used on a schedule relating to perchloric acid use frequency and the quantity of the acid used.

Each laboratory safety group or individual scientist makes his or her own schedules of wash down frequency. These run from daily to weekly and I was in one lab that activated the wash down twice each day. Besides the ducting you should wash (or wipe) the interior hood (stainless steel or PVC) liner (including the sash and light) on a frequent basis. Include lab apparatus since hot plates, and so on, can also become contaminated with perchlorates. *Remember,* perchlorates are very unstable, are explosive, and are water soluble. Do not put any organic materials in the hood or exhaust system. One time I stopped an architect's drawing showing a HEPA filter for a perchloric acid hood. Scrubber systems are fine and are recommended—but never HEPA filters.

Good maintenance and good housekeeping presents you with a good and continued safety device, your fume hood.

16.10   NOISE

Exhaust systems that include laboratory fume hoods do generate some noises. This noise can be acceptable, marginal, or annoying.

So many times you hear the comment that a specific manufacturer's hoods make a lot of noise. I shall defend all manufacturers to a point. Hoods themselves only make noise if the exhaust duct collar is undersized and the exhaust duct collar velocities are in excess of 1800 fpm. Hoods with exhaust fans or blowers that are part of or directly on top of the hood can have some blower noise that a properly designed system does not contain.

Noise that you hear emanating from a fume hood fals into two broad categories, duct velocity whistle and/or blower moan and rumble. Blower rumble can result from one or all of several problems: Imbalanced impeller wheels, improper impeller choices, too high a tip speed for the blower impeller, failure to isolate the blower from the duct and the duct from the hood or some part of the system that vibrates against the duct work and sounds like a drummer doing a constant rat-a-tat-tat.

How do you beat the annoying noises? Make your architect or mechanical engineer do his or her homework. Exhaust duct collars and in general the duct work should be sized to keep the velocities down in the 1500-fpm range. Pick a blower that is not running at top speed to move sufficient air— select a larger blower than needed and slow it down; all blower manufacturers can and will size blowers by capacity and by lack of noises. Select the proper blade—forward or backward inclined—but if you resort to a paddle wheel design you will have appreciable noise. Efficient fans make less noise than inefficient ones. Proper sized connections to the blower inlet and outlet can minimize noise. Mount the blower outside of the building and as far away from the hood as practical. Put the blower on a platform with isolation supports as part of the design. Have a rubber, fiberglass, Teflon, or a chemically acceptable flexible isolator to connect the exhaust duct to the blower. Make sure the blower impeller imbalance does not shake the blower when it is running. Use smooth V belts on the blower, not segmented V belts. Support the duct work independently of the other ducts. If you have any pneumatic lines to activate closure dampers on the blower exhaust, be sure to isolate them from the duct completely; otherwise you have hired a very noisy drummer.

Some hood manufacturers and HVAC system designers call for and insist on bell mouthed (radiused) duct collars on the hood. If the duct collar is properly sized or a bit oversized, this added expense of radiused collars is not necessary. It is one of those items that become "fancy" and not "fact" and are a salesman's dream.

Many laboratory people judge the face velocity of a hood by its noise level—the more noise the more air movement. Do not be misled. I visited one installation where seven exhaust ducts (all stainless steel) were ganged together support wise and were all noisy. The biggest fraction of the noise level came from but two of the blowers. Each hood had more than its share of noise even when its own individual blower was turned off and no air was moving through the particular hood.

If you have a noise condition that you feel needs correcting, start on an item by item inspection: duct size, blower size, face velocity, imbalanced impeller, system design, and so on. Each item may only contribute a little noise but in the collective end it sounds like a poorly trained brass band.

16.11 REMODELING

Needless to say, you must outline the laboratory's mission and requirements to even consider any type of renovation. Document this area since it is most pertinent to securing the needed monies and equipment. Once you have the necessary administrative support it is time to involve the engineering and purchasing people. Both groups must fully understand your needs and the reasons behind the specific items.

Let us synthesize a situation. You have a small 20-yr old lab, the new equipment you have acquired and will add to in the next 2 or 3 yr requires more space. Your hood is too small and the design is not of the airfoil shaped entrance; you also need a new walk-in hood.

The engineer must be told of the new exhaust air and makeup air requirements. Hood manufacturer's catalogs will supply this data. You should make a drawing of the area and locate the hoods as well as the other laboratory furniture and apparatus so the engineer can figure out the mechanical problems of servicing the new lab. There may have to be some compromise due to physical obstructions presented by beams, supporting walls, and so on.

Review available hood systems carefully and be absolutely certain that you go for safety in design and operation. This is why you must involve the person who places the order and the person who signs the check.

Visit other facilities that have similar (or dissimilar) missions and get overview opinions on types of hoods made by various manufacturers. Inspect hood lining materials, arrangement of electrical and service fittings, general construction, and product appearance. Call in the salespeople for the two or three manufacturers you consider as acceptable and listen to their sales presentation—do they know what they are talking about? Involve engineering and purchasing at these meetings and be leery of the salesperson who always reverts to "ours is better than his," but gives no supportive reasons.

Set up a constructive schedule for completion—haste can make waste and result in substandard equipment and construction. It is sad but true that many projects languish for months in a corporate environment and once the approval is finally given, then the purchasing and construction is predicated on delivery, not quality, and you get stuck with second best. Sure you want to get done and on with business but once the remodeling work is done, you are the one who works there and are saddled with the mistakes.

If you are primarily or solely concerned with the hood, you can either replace or upgrade, depending on the hood itself and its physical condition (16). To do this you have to consider previous parts of this chapter: exhaust and supply systems, lighting, entrance shape, location, mechanical fixtures. If the hood is in pretty bad shape, replace it. To try and reline an existing hood is dirty work and expensive. If the existing hood liner is asbestos bearing cement board you will have trouble finding a qualified person to really refurbish it and asbestos scrap is expensive to dispose of.

If you are replacing an existing perchloric acid hood brace yourself for some very expensive work. To remove it you should engage a firm knowledgeable in this area. It will be wet and messy work but by going through this you might not blow up the building. You will have a hard time locating a qualified contractor for the removal and I would suggest that you ask several hood manufacturers for the names of companies that are in this business; you will receive a lot of blank looks. If a salesperson says "no problem," be careful and secure a lot of details.

16.12   NEW CONSTRUCTION

Participating in the planning of a new laboratory is exciting, confusing, satisfying, depressing, frustrating, but it all culminates in pride. It is a lot of physical and mental work spaced over a considerable length of time. It is fun.

From the fume hood point of view, the proper selection of the architect-engineer is absolutely vital. Before the top executive level selects an architect/engineer because of some monument he or she has seen, go to the director of R&D, the top level scientific staff, and everyone who will listen and plead for technical competance. *Example:* There is a major drug research facility that was completed ~ 1980 that is now having a multimillion dollar renovation just to make the hoods work. The hoods are fine—the system totally inadequate. The building is very attractive from the exterior and the labs are most functional—the mechanical HVAC system was not what it should have been.

Once the architect or engineer is selected, make it a point to meet the engineer who will be in charge of the ventilation system. Then try and discern his or her pedigree and level of excellence. It has been my experience that, as good as some engineering people may be, they are not all that knowledgeable in the area of fume hood systems. Many architect/engineer firms rely on the "to be" occupants of the new lab to tell them (the architect/engineer) what is needed. What is really required is cross pollenization. Educate the engineer as to what a hood is, and does, and needs, and does not need. Let the engineer explain various options on controls, systems, and so on. Have some fume hood manufacturers make product presentations to the scientific and the engineering staff. Weed out those you feel are unqualified and have a say in manufacturer selection. A lot of architects get fixed after a period of time and think that their selection of the hood manufacturer supersedes everyone else's opinion. Products change and so do personnel—judge the manufacturer on his or her current capabilities and not just because he or she has been in business forever and a day.

Now you have impressed the architectural staff that you are interested and knowledgeable and the architect/engineer has discovered you are not an unbending egghead; the architect/engineer has proven the same qualities to you. Now you can get down to work.

Review with the appropriate people all that we have covered in this chapter. Take part in the decision making process of hood size, location, type, and so on—if you feel over a difference of opinion that you are right, stick to your guns and say no or yes or whatever. I cannot reiterate often enough that you, the user, live in a building long after the architect has gone, so be firm in what you want. Learn to say "NO" from a position of knowledge and mean it.

As the drawings progress get immersed in details: duct size, diffuser location relative to the hoods, have the engineer make a noise evaluation study of the ducting and blowers, evaluate lining materials, study valve locations, and so on. Then before bid time (on the hoods and furniture) go over the specifications so you are satisfied that you will get the quality level you want.

Once the bids have been received, get involved in the evaluation process. This is going to be tough but essential to getting what you want. Consider the situation: All the dollars from the general contractor are made known to you or to the construction management firm. The subcontract pricing and companies may or may not be revealed to you. Insist on knowing the proposed suppliers and the position (dollarwise) of alternate suppliers if you are not satisfied with the low price bidder. Buying by price alone can be a real mistake in both the short and the long haul. In fume hoods, you are buying something that is there to help spare you from injury. Safety is not decided on dollars but on expertise.

16.13 PUTTING A NEW OR RENOVATED FACILITY INTO OPERATION

The laboratory or laboratory building is now complete and the contractor(s) want to get their monies, give you the facilities, and get onto their next project. Needless to say, you want to move in and enjoy the fruit of your labors.

We shall make a few assumptions here, all the while remembering that making assumptions can be a dangerous practice.

The lights, water, air, electric outlets, fans, thermostat, and room air controls are all functional. Now comes a company that can make or break the laboratory or building: The air balancing engineers. Theirs is not an easy job because they have to set fan speeds, damper settings, control functions, pressure drops, face velocities (volumes), and so on. These parameters are directly related one to another, so setting one and then changing the setting will upset or at least alter some other function of the system. It is slow and tedious work and by necessity is quite exacting.

As in every profession, there are some good and some bad balancing companies. I have worked with some that were outstanding and others were disasters on wheels. Do not let the architect and the balancing group off the hook until you (in-house) or some one you have confidence in has set the back baffles and has checked the face velocity of each and every hood and found that they are in the specified range (60, 80, 100 fpm) you have chosen.

If they are not, have them properly balanced before signing off. Remember, when you readjust one hood on a system, you can alter another hood or two or three or more.

Believe it or not, a major problem in new buildings is duct leakage. Insist that the architect/engineer require an appropriate test (or visual inspection) to minimize the number of leaks and correct all major leaks.

How is the noise level—high or low? If it is objectionable, get to the architect and have it fixed—this is his or her area of expertise and if he or she did not size the system properly, it might just be his/her responsibility.

Any noticeable cross drafts in the room that would affect the hoods? If you think there might be, go back to the smoke bomb in the hood that we discussed in the early part of this chapter.

Now to the hood as a safety device—still assuming all the various control systems are functioning as designed. You should select a representative number of the hood population—at least 10%—and have them tested against the ASHRAE 110-1985 standard.

We discussed hood testing by the ASHRAE standard prior to purchase and now this hood is installed and should again pass the test, but perhaps at a slightly higher control level, say 4AM.05 compared to 4AU.10. But what if the AU test is well above an acceptable level?

You cannot blame the manufacturer since he or she has certified that the hood is capable of passing the procedure. Obviously the problem is inherent in the laboratory space. Cross drafts will have to be your problem and it is up to the architecct and/or HVAC subcontractor to find the problem and correct it. It could be as simple as relocating a diffuser that was improperly installed or cutting back on some makeup air. Whatever the cause, the owner and the users should not accept the project as being complete until the hood or hoods perform up to safety standards.

The break-in period following building acceptance will tell whether the control assumptions were valid or not. Does the temperature remain acceptably constant? Do the hoods have a consistent, not pulsing, face velocity? If you have an auxiliary air hood system, is the auxiliary air too cold, too fast, pulsating, noisy? If you notice control problems, go through your corporate or building chain of commands and have them corrected as soon as possible. There is less chance of attention by a building subcontractor as time passes.

16.14   EDUCATING THE TECHNICAL STAFF

There is no substitute for user knowledge when it comes to using fume hoods. Each laboratory, large or small, should have some form of a safety program for hood usage and testing.

Here are some suggestions:

1. 3M Inc., St. Paul, MN has an excellent 20-min AV film covering hood usage. It was made as part of their in-house training program.
2. Show the operator how to ascertain whether the exhaust blower is functioning or not, and at what level.
3. As per Fuller and Etchells (5) do no work in the hood within 6 in. of the face of the chamber.
4. If the hood exhaust volume is only adequate when the sash is partially closed, be absolutely sure that the operator knows this and acts accordingly.
5. If you have a fire in the hood, what does the operator do? It depends on type and circumstances and this should be part and parcel of a safety program.
6. Who does the operator alert and how rapidly when his or her hood is drawing strange odors into the laboratory? (It does happen and should receive instant attention.)
7. Is there a periodic testing of the face velocity? It should be done at least once per year.
8. Forbid the use of the hood(s) as a storage area. Enforce by periodic inspections.
9. Have a periodic discussion with lab personnel on turning hoods off when not in use.
10. Keep the back baffle slots in proper adjustment.

The final challenge that I should add is directed primarily to the mechanical engineers who are required to design facilities that house fume hoods. Please establish a good and knowledgeable foundation relative to the design, location, face velocities, testing, and so on, of fume hoods. Give to the laboratory staff a safe and adequate facility. Do not provide less than adequate laboratory (mechanical) systems.

16.15 GLOSSARY

Air foil: curved or angular member(s) at fume hood opening (face).

Air volume: quantity of air, normally expressed in cubic feet per minute (cfm).

Auxiliary air: supply or secondary air delivered external to the chamber of a fume hood to reduce room air consumption.

Baffle: a panel system located across the back interior portion of a fume hood to control entrance air patterns and distribution.

Bypass hood: a hood with a compensating opening that maintains a relative constant exhaust volume through a fume hood regardless of sash position.

Cross drafts: a flow of air that blows into or across the hood face.

Face: front or access opening of the fume hood.

Face velocity: the speed of the air moving into the hood through the front or access opening; normally expressed in feet per minute (fpm).

Hood diversity: the percentage of a total hood population that operate simultaneously.

Liner: interior lining material area for the sides, top, and baffles of a fume hood chamber.

Makeup (supply) air: air needed to replace the air taken from a room by the fume hood(s).

Manometer: a device used to measure air pressure differential, usually calibrated in inches of water.

Pitot tube: a device used for measuring the velocity pressure of air in a duct work system.

Sash: the movable transparent panel set in the fume hood entrance (face).

Static pressure: air pressure in the fume hood chamber or duct; should be negative but is expressed as a positive number in inches of water.

Superstructure: that portion of a fume hood supported by the base cabinets, the work surface, or by the floor.

Velometer: an instrument used to measure the velocity of air.

REFERENCES

1. SAMA Standard LF 10-1980, "Laboratory Fume Hoods," Scientific Apparatus Makers Association, Washington, DC.

2. Industrial Ventilation Manual, 19th ed., American Conference of Governmental Industrial Hygenists, P.O. Box 1937, Cincinnati, OH.

3. OSHA regulations; 29 CFR 1910.1003 through 1910.1016.

4. College of American Pathologists, "Laboratory Accreditation Program", Skokie, IL, 1972.

5. F.H. Fuller and A.W. Etchells, "The Rating of Laboratory Hood Performance," *ASHRAE J.*, pp. 49–53 (1979).

6. K.J. Caplan and G.W. Knutson, "Laboratory Fume Hoods, A Performance Test," *RP 70 ASHRAE Trans.*, **84**, (I) (1978).

7. E. Vernon Hill Co., Corte Madera, CA.

8. ASHRAE Standard 110-1985, *Method of Testing Laboratory Fume Hoods*, American Society of Heating, Refrigeration, and Air Conditioning Engineers, Atlanta, GA.

9. STEL and TLV-C Short Term Exposure Limit and Threshold Limit Ceiling Value as Defined by the American Conference of Governmental Industrial Hygienists, Cincinnati, OH.

10. Laboratory Fume Hood Standards as recommended for the U.S. EPA dated January 15, 1978, Contract No. 68-01-4661.

11. National Safety Council, Research and Development Section, "Survey Report on Laboratory Hoods," Chicago, 1984.

12. G.W. Knutson, "Effect of Slot Position on Laboratory Fume Hoods Performance," in *Heating, Piping, and Air Conditioning*, 93 (February 1984).

13. G.T. Saunders, "A No-Cost Method of Improving Fume Hood Performance," *American Laboratory*, 102 (June 1984).

14. S. Abrahamson, "Monsanto Dispells the Fume Hood Myth," *Facility Planning News* (February, 1986).

15. R.C. Moyer, "Fume Hood Diversity for Reduced Energy Consumption," *ASHRAE Trans.*, **89**, 2A and 2B (1983).

16. G.T. Saunders, "Updating Older Fume Hoods," *J. Chem. Educ.*, **62**, A178 (1985).

APPENDIX 4
RESOURCE MATERIALS

CHAPTER 17

Using Audiovisual Materials in Safety Training

Patricia A. Redden

St. Peter's College, Jersey City, New Jersey

17.1 THE ROLE OF AUDIOVISUAL MATERIALS

Instruction in chemical or laboratory safety may be aimed at a number of audiences and may use many pedagogical approaches. The use of audiovisual materials in such instruction serves both a functional and an economic role.

First consider the possible audiences for safety instruction. In an academic laboratory, the emphasis is on instructing students in safe procedures for handling, storing, and disposing of chemicals. The sophistication and chemical knowledge of the students range from those of a high school student or college freshman being exposed to chemistry for the first time, through an upper division science major or graduate student. In an industrial or academic setting, one must offer right-to-know training for employees at all levels, from office staff to maintenance and plant employees. Such training must deal with general information on topics such as the contents of an MSDS as well as with specific information on expected hazards, either of a chemical or physical nature. Finally, an instructor may be in the position of informing the general public on the nature of chemical hazards and safety.

A safety instructor obviously has a choice of several pedagogical approaches when providing instruction to any of these groups. The approaches run the spectrum from a pure lecture or written format through a presentation given entirely by way of audiovisual materials. Why, then, use audiovisual materials, which after all will entail some financial outlay for preparation, rental, or purchase?

The first, and probably most obvious, reason for using audiovisual materials in a safety training program is the visual impact they may provide. A film or slide showing the results of a peroxide caused explosion is far more effective than the simple statement that such explosions may occur. Another advantage to the use of audiovisual materials is that they provide for consistency of approach and content, which is particularly important when the training program must be replicated for groups of students or employees. In fact, replacing or supplementing the straight lecture may prove economical in such a case, since it minimizes the need for extended, repetitive use of trained safety instructors or consultants. Finally, audiovisual materials allow for easy review by individuals or small groups before a particular hazardous situation is dealt with. Similarly, new employees may be given training immediately upon hiring without the employer having to either supply individual training or wait for a large enough group to make training economically feasible.

17.2 AVAILABLE FORMATS FOR TRAINING MEDIA

The most common formats available are audiotapes, 16-mm films, videotapes, and 35-mm slide/tape programs. Each has particular advantages and disadvantages to be considered when making an initial purchase or setting up a safety program.

Audiotapes have two significant advantages, portability and moderate cost. The tapes may be used in any tape player and so can be used in a classroom situation or by an individual in the office, home, or car. As has been demonstrated quite effectively with taped tour materials in museums and national parks, audiotapes may even be used on a walk around of a facility, to point out special hazards or safety features. They are relatively inexpensive to purchase and can be prepared in-house quite easily to meet special needs. The disadvantage is that they use either no visual aids or, at best, printed materials.

Films are ideally suited for use on a large screen by moderately sized or large groups. This large screen format commands attention, a significant advantage when dealing with student groups in particular, but makes films awkward to use on an individual basis. Sound and movement, even slow motion, may be effectively used in this format. Films may often be purchased in videotape formats or, with the consent of the distributor, may be transferred to videotape.

Videotapes are becoming more popular than films, particularly given the relatively low cost of setting up a VCR/TV system. They are generally available in VHS, $\frac{3}{4}$ in., and Beta formats. Videotapes have all of the advantages of films except the large screen format, but they are much easier to use on an individual or small group basis. They can be stopped periodically to allow a lecturer to elucidate a point or ask relevant questions, and they can easily be backed up for a second look at a section of the tape. It is not really possible, however, to update material or correct some perceived deficiency in the presentation without intervention by the instructor.

Slide programs are generally accompanied by an audiotape that may either automatically advance to the next slide or provide an audiotone as a signal to the projectionist that the program should be advanced. A judicious choice of projecting equipment allows slides to be used on an individual or large group basis. Slide presentations allow a great deal of flexibility, since individual slides may be replaced to show facilities or personnel at the site itself. Similarly, it is possible to correct perceived deficiencies or update information by changing the relevant slides. If desired, personalized shows can be created by using selected slides from the program. (Please note that copyright laws may require written approval of the distributor to make such changes in the program.) This medium, however, does not allow motion to be effectively shown but must depend instead on showing a sequence of slides, so that the presentation may seem quite static. This format is also fairly expensive.

17.3 CONSIDERATIONS IN THE SELECTION OF MATERIALS

In selecting the particular audiovisual materials to be used in a presentation, one must first of all consider the audience, the topic to be presented, and the type of media to be used. A presentation that focuses on an industrial setting may be meaningless to a student or a member of the clerical staff in right-to-know training programs, even if the presentation itself is informative and well done. Other initial considerations are the method and frequency of the presentation and the amount of money available. Will a fairly expensive program be shown often enough to justify the initial purchase price? If not, it may be possible and more feasible to rent a program for use only when it is needed. Rental costs vary from $40 to $175, depending on the supplier and the duration of the rental, compared to purchase costs of about $400 per slide or videotape program. Teachers on the high school and college levels in particular may find that many available programs either are inappropriate in setting or content or too expensive for purchase. In particular situations, the preferred language of instruction may not be English. Audiovisual companies currently provide Spanish versions of many of their programs, with French also available on occasion.

The author strongly urges preview and review of a program before purchase. Companies will generally allow previews, occasionally at no cost but more usually for between $20 and $75; this preview fee can usually be applied to purchase of the program. In the author's experience, the quality of many of the audiovisual materials available is quite variable. Comments solicited from viewers at several presentations of safety materials, sponsored by the Division of Chemical Health and Safety at national meetings of the American Chemical Society, confirm this opinion. Many excellent programs are on the market, but careful examination of the contents of others will bring out deficiencies in either content or presentation. For example, the use of safety glasses or goggles, particularly in older programs, is not neces-

sarily universal. One older program aimed at the academic laboratory referred to the use of benzene as a solvent; another aimed at the industrial
setting referred to the carcinogenicity of cyclamate. In a presentation on
safety in working with acids and corrosives, one program focused on a slide
showing a droplet of liquid on an undamaged hand while talking about the
hazards of hydrofluoric acid, leading to a possible misconception about the
serious hazards of this compound. A safety instructor may not necessarily
reject a program because of such difficulties, particularly if they are only a
very minor part of an otherwise ideal program; if present, it will be useful in
the presentation to point out and correct any misconceptions that arise as a
result.

In areas other than content, careful review is equally necessary. Is the
presentation professional in both audio and visual components, or is that
necessary in an individual situation? Some presentations have been perceived as "talking down" to the audience, admittedly a subjective response
but one that must be considered. Sex and ethnicity of the characters in a
program are often important considerations, particularly at sites with high
numbers of minority and/or female employees. Videotapes based on older
slide shows may remain essentially slide shows, ignoring the animation possible in the new format. The use of cartoons to illustrate a point may be
attractive to some viewers and purchasers but a negative point for others.

17.4 SOURCES OF AUDIOVISUAL MATERIALS

Rather than recommend specific audiovisual materials for use and/or purchase, the remainder of this chapter will review some major sources of
audiovisual materials. The list is not meant to be, and in fact could not be,
exhaustive. New materials are being produced constantly in light of increasing emphasis on laboratory safety, handling of hazardous waste, and right-
to-know legislation. Comments on the quality of the materials is based on
personal review by the author and by other individuals involved in safety
instruction; for reasons stated previously, those opinions are often somewhat subjective.

The price of many of the audiovisual packages may be prohibitively high
for a school or small industrial firm. It would be wise for potential users with
limited budgets to rent materials or to contact local professional groups such
as the American Chemical Society local sections to determine whether materials can be obtained on loan. One specific source for academic users is the
Laboratory Safety Workshop (Curry College, Milton, MA 02186), which
will loan items from its audiovisual library to schools and colleges. Currently, there is no charge for this service, thanks to contributions from the
Cabot Corporation Foundation, the northeastern secton of the American
Chemical Society, and the Union Carbide Company. Other sources for

loans might be chemical companies or laboratory safety suppliers in the user's geographical area.

17.4.1 Commercial Distributors

Distributors are the most obvious sources of audiovisual materials. In the fields of safety in working with hazardous materials and of industsrial hygiene, four are most prominent. These are ITS, Tel-a-Train, BNA Communications, and 3M. At present, most offerings from these companies are in videotape or 35-mm slide formats.

ITS, the acronym for Industrial Training Systems Corporation, (20 West Stow Road, Marlton, NJ 08053) offers a large selection of videotape and slide programs in the areas of safety, industrial hygiene, RCRA and environmental protection, hazard communication, RF energy, and proper use and maintenance of equipment. Handouts and reference materials are available to users of their programs, and ITS publishes a free newsletter on safety-related issues. Some programs are in French or Spanish as well as English. Their programs are generally excellent, although by price (over $400 each) and content they are aimed at an industrial rather than an academic setting.

Tel-a-Train, Inc. (P.O. Box 4752, Chattanooga, TN 37405) offers a three part series on chemical safety, stressing proper handling of chemicals, health hazards, and fires and explosions. They also have programs on general industrial safety, problems in confined spaces, and the proper use of various mechanical systems. Some of the programs are accompanied by manuals for the instructor and by tests that can be used in right-to-know training or in a class on safety. The generally excellent chemical safety series, although still slanted toward industry, could be used in an academic setting, but the cost of each program in the series is again over $400. Some programs are available in Spanish.

BNA Communications, Inc. (9439 Key West Avenue, Rockville, MD 20850) distributes their own audiovisual materials as well as materials prepared by other companies such as Allied Corporation, Media Arts, and Eastman Kodak. Topics include chemical safety, handling hazardous waste, material handling, and employee motivation. Perhaps because of the fact that the company offers materials from several sources, the quality is variable and a preview is strongly recommended. Again, programs cost more than $400 each, but the company offers a library subscription plan for the benefit of purchasers with small budgets.

3M, Inc. (220-7W 3M Center, Saint Paul, MN 55144) has a relatively small catalog, concentrating on particular chemical or physical hazards as well as on general laboratory safety. Their prices are significantly lower than those of the other distributors, in the range of $225 to $350, and the content is generally very good. The videotape presentations, however, are often basically animated slide shows, without live action.

17.4.2 Other Sources of Commercial Materials

Companies and organizations whose main business is not the preparation and distribution of audiovisual materials often sell programs at a more moderate cost. This is true of all of the following sources except the National Safety Council.

Prentice-Hall Media, Inc. (Box 1050, Mount Kisco, NY 10549) sells a slide/tape series on "Safety in Organic Chemistry Laboratories," Series 9155. The material is aimed at students but is not limited to organic courses. The first program in the series in particular, "General Laboratory Safety Procedures," can be used with any chemistry class. The age of the programs does mean that there are some references to using materials such as asbestos gloves or benzene. However, the content is otherwise excellent for this often neglected student audience.

The American Chemical Society (1155 Sixteenth Street, NW, Washington, DC 20036) has two programs available for the chemistry laboratory. "Laboratory Techniques in Organic Chemistry" is a multipart slide set, while "Laboratory Safety and Health" is a new (i.e., fall 1986) audiotape series. The older slide set is excellent, and as in the previous case its use is not limited to organic chemistry laboratories. The audiotape series has not been reviewed by the author of this article but has been well recommended by colleagues involved in laboratory safety.

The National Fire Protection Association (Batterymarch Park, Quincy, MA 02269) concentrates, as might be expected, on audiovisual materials related to fires and fire fighting. Within this framework, their programs are recommended.

The National Safety Council (444 North Michigan Avenue, Chicago, IL 60611) offers audiovisual materials at two levels. For $200 each one can purchase programs that deal with safety in a very general way. Programs on chemical safety and hazardous materials are significantly higher priced, costing about $350 to $500 each for members of the National Safety Council, with a 20% or higher surcharge for nonmembers. The reaction of viewers to the latter programs has been very mixed: the consensus is that these materials should be individually reviewed before purchase.

17.4.3 Other Sources

Fisher Scientific Company, Inc. (52 Fadem Road, Springfield, NJ 07081) offers two 16-mm films, available on free loan or by purchase. The older film is "Twenty-Eight Grams of Prevention," the newer and better one is "Safety, Isn't It Worth It." Both are motivational as well as content oriented and can be used for any audience. They may be borrowed through arrangement with a local Fisher sales representative.

NIOSH (Cincinnati, OH 45226) offers material for the academic user primarily. One program, "Safety in the School Science Laboratory," is a videotape version of a safety workshop for teachers at various levels. Also available is a slide set for secondary school teachers and administrators, "The Buck Stops Here." The author has not reviewed these materials herself but they have been recommended by others. Both may be borrowed from the Laboratory Safety Workshop at Curry College.

Another training program for chemistry instructors is the videotape, "Safety in Academic Labs," which is accompanied by an instructor's guide and workbooks. Prepared by Project Teach (Chemistry Department, University of Nebraska—Lincoln, Lincoln, NE 68588), this is a good combination of content and motivational tool, although the production is not slickly professional.

ACKNOWLEDGMENTS

The author wishes to thank the Division of Chemical Health and Safety of the American Chemical Society for sponsoring viewings of many of these materials at national meetings in New York City and Anaheim; participants in these viewings who contributed their comments; Dr. Richard Uriarte of Saint Peter's College for organizing the viewings at Anaheim; and Dr. James Kaufman of Curry College for supplying information about the Laboratory Safety Workshop's Library.

CHAPTER 18

Laboratory Safety Library Holdings

Jay A. Young

Chemical Safety and Health Consultant
Silver Spring, Maryland

18.1 INTRODUCTION

The encouragingly increasing variety and utility of reference works in chemical health and safety during the past few years has led to the pleasant problem: Which of these should I select for my own chemical safety and health library? Obviously, one's own choices are subjective. In the list that follows, my subjective choices were made for the reasons summarized in the comments. Absence of mention signifies only that; and certainly does not suggest that the omitted work might be unsuitable for others. Also, some of the references identified here have been cited in preceding chapters for the purpose of those chapters; their repetition here signifies that they have a broader application as well.

The following annotated bibliographic listings are divided into four groups; each group is further subdivided into three classes.

To each of the four groups two other sources of information should be added, precautionary labels on the containers and material safety data sheets. Naive reliance on either of these sources is unwise. Instead, by comparing a few labels and MSDSs from various suppliers with the information supplied in the cited bibliographic references for the same chemicals, determine which suppliers provide reliable information.

The three classes subdividing each of the groups are Class I, references considered to be essential, in the same category, that is, as labels and MSDSs; Class II, references that should be available, and used; and Class III, other references that would be found to be useful if they were available.

18.2 REFERENCES FOR THE ELEMENTARY SCHOOL GENERAL SCIENCE LABORATORY OR CLASSROOM

Class I

Gerlovich, J.A. et al., *School Science Safety,* Flinn Scientific, Batavia, IL, 1984. In two volumes, one for elementary science, the other for secondary. Written to and for teachers. A practical, sensible, useful guide; one of the two best sources of information currently available to teachers.

Mackison, F.W. et al., Eds., *NIOSH/OSHA Pocket Guide to Chemical Hazards,* 5th printing, NIOSH Division of Technical Services, Publication No. 78-210, Cincinnati, OH, 1985.
A handy, and accurate, summary of the hazardous properties of chemicals for which OSHA has assigned permissible exposure limits. (It actually will, almost, fit in your pocket.)

School Science Laboratories; A Guide to Some Hazardous Substances, Consumer Product Safety Commission, Washington, DC, 1984.
The other best reference for teachers. Loaded with practical suggestions. *Example:* A complete set of sensible, practical laboratory safety rules on pp. 17 and 18.

Standard First Aid and Personal Safety, American Red Cross, Stock No. 321116, 1981. Available from local Red Cross chapters.
A well-known standard reference. Local Red Cross chapters also usually have other programs dealing with first aid that teachers will find useful.

Class II

Council Committee on Chemical Safety, *Safety in Academic Chemistry Laboratories,* 4th ed., American Chemical Society, Washington, DC, 1985.
A recognized classic authority; concise, useful.

Practice for Occupational and Educational Eye and Face Protection, Z 87.1, American National Standards Institute, New York, NY, current edition.
The standard for eye and face protection. Do not rely on any eye or face protection devices unless they can meet this standard; all that do meet the current requirements are marked "Z 87" on the device.

Reese, K. M., *Health and Safety Guidelines for Chemistry Teachers,* American Chemical Society, Washington, DC, 1980.
Decidedly informative; some of the recommendations are challenging.

Class III

The Merck Index, 10th ed., Merck, Rahway, NJ, 1983.
Accurate descriptions of many of the hazardous properties of many chemicals, unfortunately not always describing all of the hazardous properties that concern laboratory workers. All in all, probably better than the so-called *Chemical Dictionaries,* lists of "Dangerous Properties," and so on, which may describe hazardous properties that a chemical does not possess.

18.3 REFERENCES FOR SECONDARY SCHOOL CHEMISTRY LABORATORIES

Class I   All of the previous, and:

Fire Protection for Laboratories Using Chemicals, also known as "NFPA Code No. 45," National Fire Protection Association, Batterymarch Park, Quincy, MA, current edition.
The national safety code as it applies to laboratory fire prevention and protection.

Class II

Bretherick, L., *Handbook of Reactive Chemical Hazards,* 3rd ed., Butterworths, Stoneham, MA, 1985.
The list of all reported in the literature incompatable chemical combinations. Absolutely essential information.

Fire Protection Guide on Hazardous Materials, 8th ed., National Fire Protection Association, Batterymarch Park, MA, 1984.
A compilation of data gathered from other sources. Useful.

Kaufmann, J. A., *Laboratory Safety Guidelines,* Curry College, Milton, MA 02186.
Free for the asking; a handy-dandy practical list of corrective steps to be taken, some of them at very low cost.

Prudent Practices for the Disposal of Hazardous Chemicals in Laboratories, National Academy of Sciences, Washington, DC, 1983.
A useful book; how to properly dispose of hazardous waste chemicals. Check local regulations before adopting the recommendations in this reference.

Class III

Catalog Handbook of Fine Chemicals, revised annually, Aldrich Chemical Co., Milwaukee, WI.
Contains useful general data on physical properties, very brief descriptions of toxic and reactive potentials, and a reference to disposal procedures for each of the 10,000 entries, and it is free! However, do not rely on chemical disposal procedures described in this or other supplier's catalogs; many of the disposal procedures described in these sources violate current environmental protection laws and regulations.

Gosselin, et al., R. E., *Clinical Toxicology of Commercial Products,* 5th ed., Williams & Wilkins, Baltimore, MD, 1984.
A marvelous source of information, generally considered to be as reliable as Patty (Clayton and Clayton), see below.

Prudent Practices for Handling Hazardous Chemicals in Research Laboratories, National Academy of Sciences, Washington, DC, 1981.
Highly regarded by most users; intended to be the replacement for the 1972

edition of the Manufacturing Chemists Association classic *Guide for Safety in the Chemical Laboratory,* see below.

18.4 REFERENCES FOR COLLEGE OR UNIVERSITY UNDERGRADUATE CHEMISTRY LABORATORIES

Class I   All of the preceding, except Gerlovich. And also:

Steere, N.V., Ed., *Safety in the Chemical Laboratory,* Vols. 1–3; M. M. Renfrew, Ed., Vol. 4, Division of Chemical Education, ACS, Easton, PA (1967, 1971, 1974, 1981), and continuing columns by the same title in the *Journal of Chemical Education,* M. M. Renfrew, Ed.

The four volumes are reprints of columns from monthly issues of *J. Chem. Educ.* Usually, more than a nugget of utility in each column, although the level is variable. All in all, a veritable gold mine.

Class II

Clayton, G. D., and Clayton, F. E., Eds., "Toxicology", in *Patty's Industrial Hygiene and Toxicology,* Vols. 2A, 2B, and 2C, 3rd ed., Wiley-Interscience, New York, 1981/1982.

Authoritative, indispensable; unquestionably the best available general reference on the toxicology of specific elements and compounds.

Dutch Association of Safety Experts, *Handling Chemicals Safely,* 2nd ed. (in English), Dutch Safety Institute, Amsterdam, 1980.

The only independently prepared and reliable compilation of material safety data sheets available today. This is the single necessary book for that isolated laboratory on the desert island.

Guide to the Safe Handling of Compressed Gases, Matheson Gas Products, Seacaucus, NJ, 1983; and *Handbook of Compressed Gases,* 2nd ed. Van Nostrand Reinhold, New York, 1980.

Considered essential for persons working with compressed gases.

Smith et al., S. L., *A Management System for Occupational Safety and Health Programs for Academic Research Laboratories,* NIOSH Division of Technical Services, DHHS Publication No. 79-121, Cincinnati, OH, 1979.

Boring to read; essential for academic research managers.

Class III

Best Safety Directory, Vols. 1 and 2, A. M. Best Co., Oldwick, NJ, current edition. A buyers guide for sources of safety and personal protective equipment, with extensive text discussions. Better than a Sears Roebuck catalog if you want to decide which and what to buy for chemical health and safety purposes.

Collings, A. J., and Luxon, S. G., Eds., *Safe Use of Solvents,* Academic, New York, 1982.
The proceedings of the IUPAC Symposium at Brighton, UK in September, 1982 covering many aspects of use (and misuse) of this large and important group of industrial and laboratory materials.

Fawcett, H. H., and Wood, W. S., Eds., *Safety and Accident Prevention in Chemical Operations,* 2nd ed., Wiley, New York, 1982.
The best available book that addresses chemical health and safety as it applies to chemical industrial operations.

"Fourth Annual Report on Carcinogens," U.S. Department of Health and Human Services, Washington, DC, 1985.
An "official" list of human carcinogens and suspected carcinogens, as of late 1985, with a discussion of each; up-dated more or less annually.

Parmeggiani, L., Technical Ed., *Encyclopaedia of Occupational Health and Safety,* 3rd ed., International Labour Office, Geneva, Switzerland, 1983.
The scope includes more than chemical health and safety; the articles on chemical health and safety topics are uniformly informative.

Pitt, M. J., and Pitt, E., *Handbook of Laboratory Waste Disposal,* Wiley-Halstead, New York, 1985.
A comprehensive treatment of attitudes, organization, and practical details necessary for development of an effective waste disposal system. Not a cookbook for disposal, but will equip laboratory workers to solve their particular problems for themselves.

Tatken, R. L., and Lewis, R. J., Eds., *Registry of Toxic Effects of Chemical Substances (RTECS),* Vols. 1–3, and Supplement Vols. 1 and 2, U.S. Government Printing Office, Washington, DC, 1981–1982; supplement: 1984.
The last of the Mohicans; from now on subsequent editions will be available in microfiche or via computer/modem, not in printed book format. A remarkable compendium of almost all published toxic information, true or not. Very useful, but read the prefatory material before using the information.

18.5 REFERENCES FOR OTHER CHEMISTRY LABORATORIES

Class I   All of the preceding except Gerlovich, the Consumer Product Safety Commission Guide, Reese, and perhaps a few others. And also:

Walters, D. B., *Safe Handling of Chemical Carcinogens, Mutagens, Teratogens and Highly Toxic Substances,* Vols. 1 and 2, Ann Arbor Press, Ann Arbor, MI, 1981.
Considered by many to be the best available on this topic.

Class II

Bretherick, L., Ed., *Hazards in the Chemical Laboratory,* 4th ed., The Chemical
    Society, London, 1986.
Considered as the standard general reference for chemical laboratory health
and safety in Great Britain—and worthy of such respect; pretty darn useful
on this side of the Atlantic, too.

Class III

Manufacturing Chemists Association (now: Chemical Manufacturers Association),
    Guide for Safety in the Chemical Laboratory, 2nd ed. Van Nostrand Reinhold,
    New York, 1972.
A classic and still useful, even though quite old.

*Staying Out of Trouble: What You Should Know About the New Hazardous Waste
    Law,* and *1986 Supplement,* National Association of Manufacturers, Washington,
    DC, 1984 and 1986.
What to do and what to avoid in order to comply with federal waste disposal
regulations.

I wish to acknowledge the helpful suggestions of Leslie Bretherick in the
preparation of this annotated bibliography.

Index